Engineering Mechanics 1: Statics

Christian Mittelstedt

Engineering Mechanics 1: Statics

Christian Mittelstedt
Fachbereich Maschinenbau
Technical University of Darmstadt
Darmstadt, Hessen, Germany

ISBN 978-3-662-71851-3 ISBN 978-3-662-71852-0 (eBook)
https://doi.org/10.1007/978-3-662-71852-0

Translation from the German language edition: "Technische Mechanik 1: Statik" by Christian Mittelstedt,

This Springer imprint is published by the registered company Springer-Verlag GmbH, DE, part of Springer Nature.
The registered company address is: Heidelberger Platz 3, 14197 Berlin, Germany

Preface

This book is the first volume in a series of books on engineering mechanics and deals with the fundamentals of statics. The other volumes are dedicated to elastostatics and strength of materials as well as dynamics and are also published by Springer.

I hope you enjoy working through the contents of this book. Of course, I am always happy to receive feedback of any kind.

Darmstadt, Germany
Spring 2025

Christian Mittelstedt

Competing Interests The author has no competing interests to declare that are relevant to the content of this manuscript.

Contents

1 Fundamentals

In this introductory chapter, we introduce some basic concepts that are necessary for the further understanding of this book. These are the concepts of force, rigid body, reaction forces and degrees of freedom of bodies. In addition, we briefly discuss free-body diagrams and the principle of action and reaction, two aspects that are of outstanding importance in applied mechanics and which we will use in many ways in this book. The chapter concludes with a presentation of some basic relations of vector calculus as they are relevant to the contents of this book.

1.1 Force

The concept of force is common to every person. Forces occur in many places in our everyday lives and can be experienced by every person, although forces cannot be seen and are not directly observable, but their effects can be. An important example is the force of weight, which can be experienced by every person—the force of weight is overcome when we lift an object in the Earth's gravitational field using our muscle power or drop it under the influence of gravity. We overcome the weight of our own body when we get up from a sitting position to a standing position or when we walk. If we press our hands against a rigid wall, we can feel the counter-pressure that the wall exerts on our hands (known as the principle of action and reaction, see Sect. 1.5), and the muscle tension in our body is a measure of how much pressure we exert on the wall (Fig. 1.1). Forces also occur when we move at a certain speed, e.g. in a vehicle, and accelerate or decelerate. Of course, other types of forces also occur, e.g. electric forces or magnetic forces which we will not deal with in this book. Forces can also be time-dependent, but in the following we will focus exclusively on forces that do not

C. Mittelstedt, *Engineering Mechanics 1: Statics*,
https://doi.org/10.1007/978-3-662-71852-0_1

Fig. 1.1 Compressive force F applied to a wall

show any time dependence. In this book, we will focus on statics, i.e. bodies and structures that are at rest and do not move. The investigation of bodies in motion concerns dynamics, which we discuss in detail in Volume 3.

In order to introduce a general definition for the purposes of this book, we make the following definition. A force is an effect on a body that changes the state of this body, whereby this can mean a resting state or the state of a uniform movement, whereby forces can have both an accelerating and a decelerating effect. In this book, however, we will limit ourselves to bodies at rest, i.e. to statics. What both cases, i.e. statics and dynamics, have in common is that the occurrence of forces and their magnitude can be inferred by looking at the change in the state of motion of a body.

Forces are specified in a suitable unit. The unit Newton[1] [N] has become established for this purpose. The Newton is the SI unit for force and is expressed in the units kilogram [kg], metre [m] and second [s] as follows:

$$1\text{N} = \frac{\text{kg} \cdot \text{m}}{\text{s}^2}. \tag{1.1}$$

A force of 1N is therefore the force that causes an acceleration of $1\frac{\text{m}}{\text{s}^2}$ with a mass of 1 kg. However, forces depend not only on their magnitude but also on their direction and the point at which they act (the so-called force application point). This can easily be seen from the situation in Fig. 1.2, in which a cuboid on a smooth surface is subjected to a force. Since the force depends on its magnitude and direction, it is represented as an arrow with a line of action. A common symbol for a force is F. If the force F is applied as shown in Fig. 1.2, top left, then the cuboid will move in the direction of the $x-$axis. If, on the other hand, the force F is applied eccentrically as shown in Fig. 1.2, top right, then the cuboid will not only move in the $x-$direction, but will also rotate around the $z-$axis. This clearly shows that the point of application of a force plays a very important role when analysing the effects of forces. In the event that the force F acts as shown in Fig. 1.2, bottom left, the cuboid will move against the $y-$direction. Obviously, the direction of the force is very important for its effect. If, on the other hand, the force points against the $z-$axis (Fig. 1.2, bottom right), then there will be no movement at all, but the cuboid will be pressed against the surface.

Based on the above conclusions, it is obvious that a force F is a vector that depends on its magnitude (i.e. the amount of the force) and its direction (i.e. the direction along the line of

[1] Sir Isaac Newton, 1643–1727, English polymath.

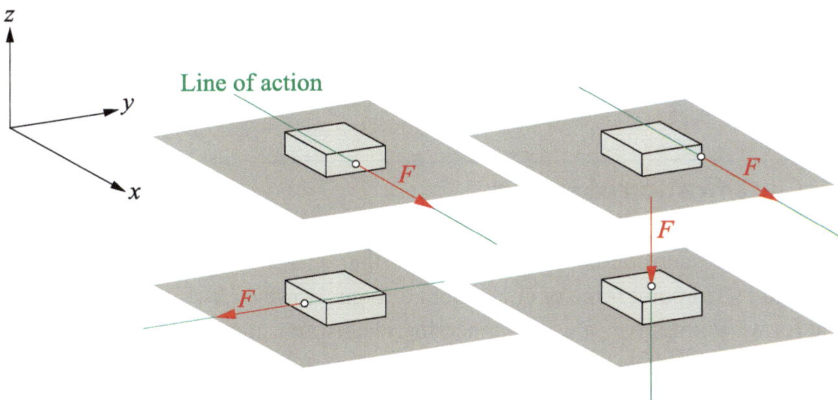

Fig. 1.2 Cuboid on a smooth surface under the force F

action, see Fig. 1.2). Since the point of application of the force is also of central importance, a force is also referred to as a point-bound vector. Some basic principles of vector calculus are summarised in Sect. 1.6. In the following, we use a symbol with an underscore for vectors, i.e. $\underline{F}$, the magnitude of the force vector is $\|\underline{F}\| = F$. In many cases, only the magnitude F is specified in a graphical representation of a force, the vector character results from the representation as a directional arrow. We will use both forms of representation in this book.

A force $\underline{F}$ can be described in Cartesian coordinates x, y, z in the spatial case (Fig. 1.3) by the three vectors $\underline{F}_x$, $\underline{F}_y$, $\underline{F}_z$ (the so-called components of the vector $\underline{F}$) with the magnitudes F_x, F_y, F_z (the so-called coordinates of the vector $\underline{F}$), which are parallel to the unit vectors $\underline{e}_x$, $\underline{e}_y$, $\underline{e}_z$ and are therefore multiples of these. Thus:

$$\underline{F} = \underline{F}_x + \underline{F}_y + \underline{F}_z = F_x\underline{e}_x + F_y\underline{e}_y + F_z\underline{e}_z. \tag{1.2}$$

Fig. 1.3 Force vector $\underline{F}$

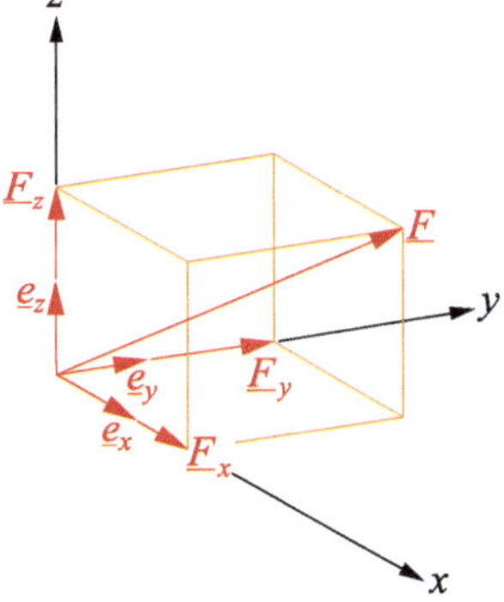

The magnitude F of the force vector $\underline{F}$ follows as:

$$|\underline{F}| = F = \sqrt{F_x^2 + F_y^2 + F_z^2}. \tag{1.3}$$

1.2 Classification of Forces

The form of force considered so far is a force that acts at one point. We therefore also speak of a so-called point force or concentrated force. However, forces can occur in several different forms. Point forces are an engineering idealisation and do not actually exist in this form in reality. Nevertheless, point forces are in many cases a suitable engineering model to approximate reality and are frequently used in many technical fields of application.

One class of forces are the so-called volume forces, which are distributed in a body over the entire volume or only a part of it (Fig. 1.4). An example of a volume force is the weight of a body. In Fig. 1.4, the volume force $\underline{f}$ points in the direction of the acceleration due to gravity g and is therefore representative of the weight of the body. Volume forces are specified in the unit of a force per unit volume, e.g. in $\left[\frac{\text{N}}{\text{m}^3}\right]$. Other examples of volume forces are magnetic forces and electric forces.

Furthermore, in reality, so-called surface forces or area forces occur. Figure 1.5 shows a plate, i.e. a very thin flat structure, which is loaded by a surface load p, where p can be a function of the two coordinates x and y: $p = p(x, y)$. We will not use a vector notation at this point. Surface forces are specified in a corresponding unit, i.e. as force per unit area, e.g. in $\left[\frac{\text{N}}{\text{m}^2}\right]$. Examples of surface loads are snow loads on a roof structure or surface pressures between two bodies.

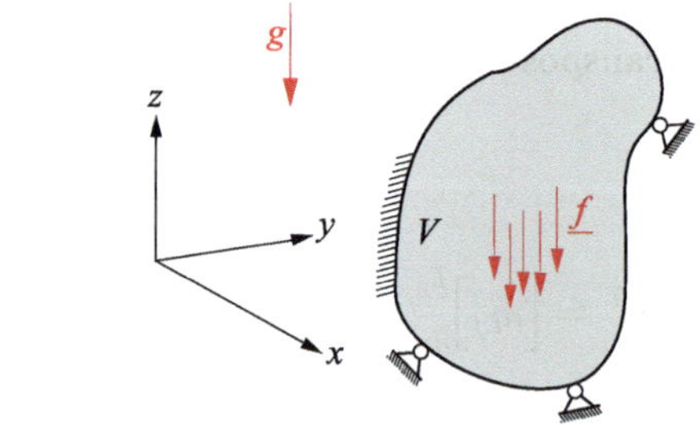

Fig. 1.4 Body under volume force $\underline{f}$

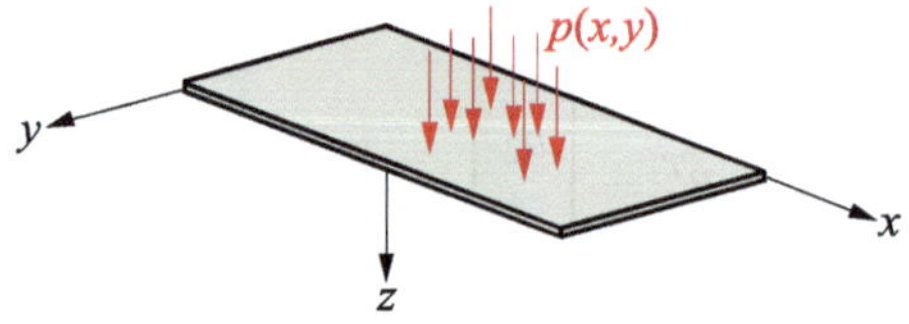

Fig. 1.5 Plate under surface load p

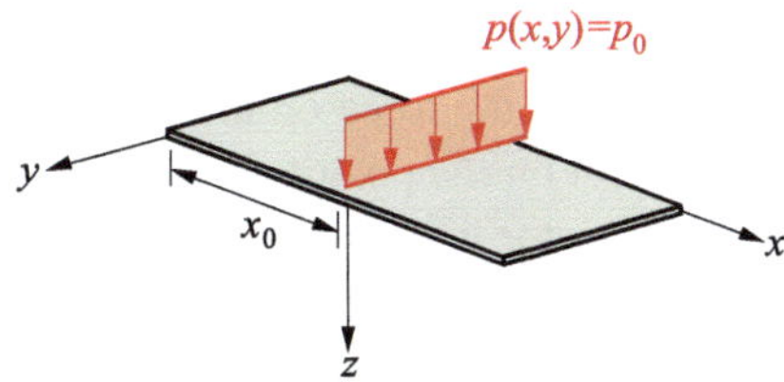

Fig. 1.6 Plate under line load p_0

If we think of a surface load p as being concentrated on a line, this results in the idealisation of the so-called line load, which can be specified in the unit of a force per unit length, e.g. in $\left[\frac{\mathrm{N}}{\mathrm{m}}\right]$. An example is the plate in Fig. 1.6, which is loaded by a line load p_0.

Forces are also differentiated according to so-called active forces and so-called reaction forces. Active forces are forces that have a physical cause. Examples include gravitational forces or effects on structures or buildings such as weight, wind forces, snow forces, and the like. Reaction forces, on the other hand, arise when a body is restricted in its ability to move. In statics, these are referred to as bearing or support reactions or simply reaction forces, which we will discuss in detail in Chap. 4.

A further distinction is made between external and internal forces. External forces are forces that act on a given structure and reaction forces. Internal forces, on the other hand, are forces that act inside the body under consideration and can be visualised by cutting them free. We will discuss this in detail in Chaps. 5 and 6.

If several forces occur simultaneously, this is referred to as a force system. These are divided into planar and spatial force systems as well as coplanar/concurrent or central force systems and general force systems and are discussed in detail in Chap. 2. A concurrent or central force system is a system of forces that have a common point of intersection of their lines of action, whereas a general force system is a force system in which the lines of action of the forces involved do not have a common point of intersection.

1.3 The Rigid Body

Every real body or structure undergoes deformation under load. A rubber band that we pull apart by applying a force will stretch. A bridge that is crossed by vehicles will bend. We will neglect the deformations of bodies in this book; these will be discussed in detail in Volume 2. In this book, we always want to assume so-called rigid bodies, i.e. bodies that do not deform under load and where the distances between different body points remain the same even under load. This is an engineering idealisation, but in many cases it is a good approximation of the real conditions: A bridge, for example, is always designed in such a way that its deflections remain very small. The rubber band mentioned above, on the other hand, is a case in which very small deformations can generally no longer be assumed. In all the following explanations, however, we want to assume that we can neglect deformations of a body under consideration, and such a body is referred to as a rigid body.

With a rigid body, a single force can be displaced along its line of action without changing the effect of the force on the body. The effect of the force on a rigid body is therefore not dependent on the position of the point of application of the force on the line of action, i.e. a force vector is bound by its line of action but not its point of application on a rigid body. This conclusion can be visualised using the example of Fig. 1.2 if the cuboid is assumed to be rigid. For example, if we consider the situation of Fig. 1.2, top left, the cuboid will move in the $x-$direction. If we were to move the force along its line of action and let it act on the opposite cuboid surface, for example, the same state would occur due to the rigidity of the cuboid. It is easy to realise that this would not be the case with a deformable cuboid, as it would still move in the $x-$direction depending on the point of force application, but would deform differently. However, as we have already established, parallel displacements of forces on rigid bodies are not permissible without further ado, as can be seen by comparing the situations in Fig. 1.2, top left and right: The centrically acting force (Fig. 1.2, top left) will cause a movement in the $x-$direction, while the eccentrically acting force (Fig. 1.2, top right) will also cause a rotation around the $z-$axis.

1.4 Degrees of Freedom and Reaction Forces

A rigid body has three degrees of freedom in the plane. The cuboid in Fig. 1.2 can move both in the $x-$direction and in the $y-$direction. Such a rigid body displacement is called a translation. A rigid body in the plane can also rotate about the axis that is orientated normal to the plane. This is the $z-$axis in this case, and such a motion is referred to as rotation. A rigid body in the plane therefore has three degrees of freedom, namely two translations and one rotation. Any movement in the plane can then be represented by a superposition of these two translations and the rotation. By analogy, a rigid body has six degrees of freedom in space, namely three translations in the three spatial directions x, y, z and three rotations around the $x-$, $y-$ and $z-$axes. If one or more degrees of freedom are obstructed, then the corresponding degrees of freedom are restricted and as a result the so-called reaction forces occur. This can be illustrated using the example of an object of mass m, which will have the weight force $G = mg$ in the Earth's gravitational field. If this object is lifted from the ground and left to its own devices at a certain height, it will start to move and fall towards the centre of the earth. If we want to prevent the object from doing this, we will hinder this degree of freedom of downward motion, e.g. by holding it with our own muscle power or by using a constructive device. So if we hold this object in our hand, the weight force G exerts pressure on our hand. At the same time, our hand exerts a pressure of exactly the same magnitude in the opposite direction on the object, and this pressure force is a reaction force. This is shown in Fig. 1.7, in which a person balances an object of mass m in the Earth's gravitational field (gravitational acceleration g) with one hand overhead. The weight force $G = mg$ will then exert a compressive force V (the letter V indicates the vertical direction of the force) on the person's balancing hand, which, however, also acts as a counterforce on

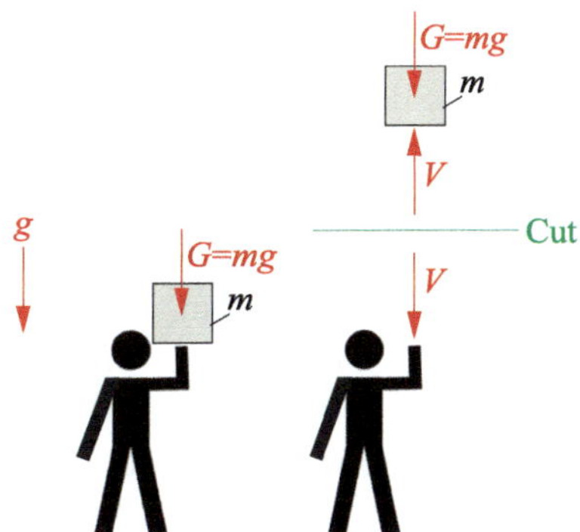

Fig. 1.7 Reaction force

the object itself. This compressive force can be visualised by making an imaginary cut by which a so-called free-body diagram is created. We can visualise that V acts downwards on the hand, whereas V acts upwards as a counterforce on the object under consideration. This is referred to as the so-called interaction principle or the law of action and reaction, which we will discuss in the next section. It is also immediately clear without any calculation that V corresponds to the weight $G = mg$ of the object.

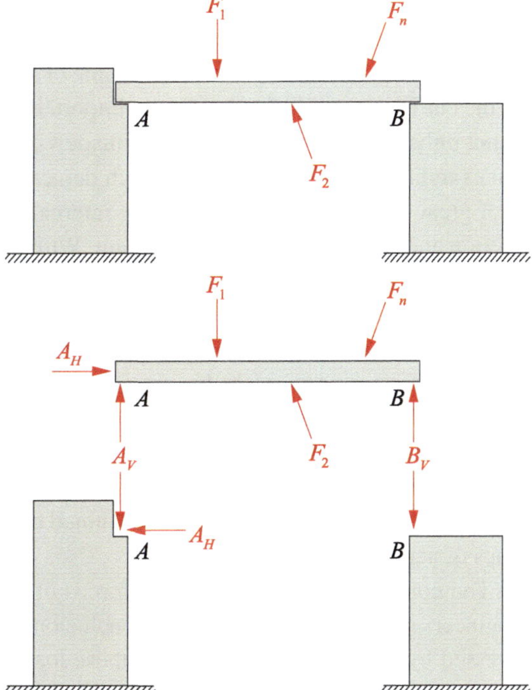

Fig. 1.8 Beam on two supports

Another example is shown in Fig. 1.8. Let a beam be given which is supported at the points A and B on two supports. It is clear that this beam can neither move nor rotate in its plane. All three degrees of freedom are obstructed here. As a result, reaction forces must occur at the two points A and B, which are referred to as bearings or supports. These reaction or support forces can be visualised by releasing the beam at the support points (Fig. 1.8, bottom) and applying the resulting support forces. This is also referred to as cutting free or isolating of the beam, and the resulting sketch is referred to as the so-called free-body diagram. While only the vertical degree of freedom is restricted at support point B, the horizontal translation is blocked at support point A in addition to vertical translation. As a result, two support forces occur in support A, which are usually labelled in the same way as the support point and given the indices H and V (for 'horizontal' and 'vertical'), i.e. A_H and A_V. Similarly, only the vertical support force B_V occurs at the support point B. In practice, this type of construction would have to be secured against lifting by suitable means if, for example, the force F_2 becomes very large. The types of supports and how to calculate these support reactions are discussed in Chap. 4.

1.5 Law of Action and Reaction

We have already seen that forces can be made visible by cutting a body and creating a free-body diagram. This is a very basic principle of statics that we will use frequently. It is known as the cutting principle and is of great importance for the statics of rigid bodies. However, it is not only reaction forces that can be made visible by cutting free, but also internal forces. An example is shown in Fig. 1.9, which depicts a beam under load. For such a structure, it is of great importance to determine the internal forces, which can be visualised by cutting a section and creating a free-body diagram. Without going into detail here, it should be noted that not only internal forces in the form of the so-called normal force N and the so-called shear force Q occur in a beam in its plane, but also a so-called bending moment M. The calculation of these quantities is discussed in detail in Chaps. 5 and 6. It should be noted here that two segments of the beam are created by making the cut as shown in Fig. 1.9, both of which must be in equilibrium. This is only made possible by applying the internal forces and moments. It should also be noted that the principle that forces can be shifted along their lines of action is not valid when considering internal forces, in contrast to external forces. The position of the inner forces is determined by the cut; they may not be shifted along their lines of action.

The consideration of the internal forces of a structure is of paramount importance for engineers in practice, as they allow conclusions to be drawn about how much a structure is stressed by an applied load, whether these loads can be withstood, or whether failure will occur. Detailed considerations on this topic will follow in Volume 2.

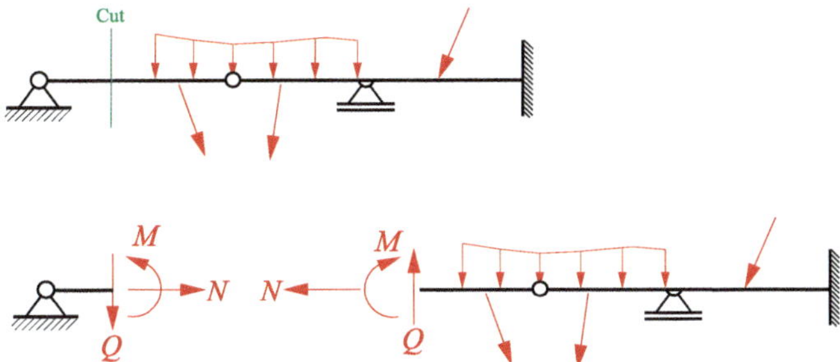

Fig. 1.9 Beam under load (top), free body diagram (bottom)

In the free-body diagram in Fig. 1.9, bottom, the internal forces and moments released in this way are to be applied in opposite pairs. The principle underlying this idea is the so-called interaction principle. This is based on the experience that a force acting on a body always causes a counterforce of the same magnitude but in the opposite direction. One example is given in Fig. 1.1 where a person pushes against a rigid wall. The applied muscle force is translated into a force F that acts on the wall. At the same time, a force F acts as a counterforce on the pushing hand, which is perceived as pressure in the palm of the hand and which is exactly opposite to the pressure on the wall. This principle is also known as Newton's third fundamental law and can be verbalised as follows: The forces that two bodies exert on each other are always of the same magnitude and opposed to each other. This principle is also known as 'Actio est Reactio'. For a given force, there is therefore always a counterforce that acts in exactly the opposite direction and has the same magnitude. We have also applied the principle of interaction in the free-body diagram in Fig. 1.9 by applying the released forces N and Q and the bending moment M to the two beam segments at the same magnitude but in opposite directions. The reaction forces A_H, A_V and B_V of Fig. 1.8 act both as support forces on the beam and as opposing forces on the columns. The interaction principle is a so-called axiom. An axiom is a law of nature that cannot be proven, but which has been confirmed by multiple observations of nature or by experiments.

By cutting a body free at its bearing points or dividing it into two segments through a cut, a displacement is made possible that the stationary body does not actually exhibit. We always assume that the body or the body segments are 'solidified' or 'frozen' in the given configuration. This principle is also referred to as the so-called solidification principle, and only in this way is it possible to calculate forces with the means of statics.

We will see later in this book that we will work with statements on equilibrium to determine support reactions and internal forces. A body at rest is always in equilibrium if it remains at rest even when a group of forces is applied. Such a group of forces is referred to as an equilibrium group. Accordingly, for example, for the beam at rest in Fig. 1.9 even if it is cut into two segments, both beam segments must remain at rest. This requirement makes it possible to calculate these internal forces and moments that we have released by cutting.

1.6 Fundamentals of Vector Calculus

In this section, we provide some basics of vector calculus as they are generally taught in the first semester at engineering faculties and as they are relevant to the contents of this book. In doing so, we expressly limit ourselves to those topics that we will need in the further course of this book. Readers who are already familiar with these fundamentals can skip this section.

1.6.1 Fundamentals

Physical quantities that are only described by their magnitude are referred to as scalars in physics. Scalar quantities are, for example, the time or mass of a body, which are fully described by specifying a numerical value. On the other hand, there are also quantities that are characterised not only by their magnitude but also by their direction. Such quantities are called vectors, and we will encounter vectors several times in this book. Examples of vectors are forces that are characterised by their magnitude, but also by their direction. Vectors are labelled with an underscore in this book. One example is the vector $\underline{F}$ of a force. In the literature, a symbol is often written in bold to signify a vector: $\underline{F} = \mathbf{F}$. Vectors are represented graphically as arrows, as in Fig. 1.3 using the example of the force vector $\underline{F}$. The magnitude $F = |\underline{F}|$ of the vector corresponds to its length. A vector with a length of 1 is referred to as a unit vector.

As already shown in Fig. 1.3 for the force vector $\underline{F}$, it can be represented as a multiple of the so-called base vectors $\underline{e}_x, \underline{e}_y, \underline{e}_z$ and correspondingly has the three spatial components $\underline{F}_x, \underline{F}_y, \underline{F}_z$ with the coordinates F_x, F_y, F_z, whereby F_x, F_y, F_z are also often labelled as components of the vector $\underline{F}$:

$$\underline{F} = \underline{F}_x + \underline{F}_y + \underline{F}_z = F_x\underline{e}_x + F_y\underline{e}_y + F_z\underline{e}_z. \tag{1.4}$$

The base vectors in the order $\underline{e}_x, \underline{e}_y, \underline{e}_z$ form a so-called right-handed system, and the so-called right-hand rule applies, according to which the thumb, index finger and middle finger of the right hand can be brought into line with $\underline{e}_x, \underline{e}_y, \underline{e}_z$. It is possible to use the following notation for vectors, in which the coordinates F_x, F_y, F_z of the vector are arranged in round brackets (so-called column vector):

$$\underline{F} = \begin{pmatrix} F_x \\ F_y \\ F_z \end{pmatrix}. \tag{1.5}$$

The magnitude F of the vector $\underline{F}$ results from the spatial theorem of Pythagoras[2] (see Fig. 1.3) and reads:

$$\left|\underline{F}\right| = F = \sqrt{F_x^2 + F_y^2 + F_z^2}. \tag{1.6}$$

1.6.2 Elements of Vector Algebra

Some important elements of vector algebra are presented below, as they are relevant for the contents of this book. We limit ourselves to what is absolutely necessary; for more detailed information, the reader is referred to the wide range of specialist literature available.

Multiplication with a Scalar

Let a vector $\underline{a}$ be given as follows:

$$\underline{a} = \begin{pmatrix} a_x \\ a_y \\ a_z \end{pmatrix}. \tag{1.7}$$

If this vector is multiplied by a scalar λ, a new vector $\underline{b} = b_x\underline{e}_x + b_y\underline{e}_y + b_z\underline{e}_z = \lambda\underline{a} = \lambda a_x\underline{e}_x + \lambda a_y\underline{e}_y + \lambda a_z\underline{e}_z$ is created, which has the magnitude $\left|\underline{b}\right| = |\lambda| \left|\underline{a}\right|$ (Fig. 1.10):

$$\underline{b} = \begin{pmatrix} b_x \\ b_y \\ b_z \end{pmatrix} = \begin{pmatrix} \lambda a_x \\ \lambda a_y \\ \lambda a_z \end{pmatrix} = \lambda \begin{pmatrix} a_x \\ a_y \\ a_z \end{pmatrix}. \tag{1.8}$$

If $\lambda < 0$, the direction of the vector is reversed. If $\lambda = 0$, the result is the so-called zero vector $\underline{0}$, i.e. a column vector with zero entries.

Vector Addition and Vector Subtraction

Let two vectors $\underline{a}$ and $\underline{b}$ be given. If these two vectors are added together, a new vector $\underline{c}$ is created as follows:

$$\underline{c} = \begin{pmatrix} c_x \\ c_y \\ c_z \end{pmatrix} = \begin{pmatrix} b_x \\ b_y \\ b_z \end{pmatrix} + \begin{pmatrix} a_x \\ a_y \\ a_z \end{pmatrix} = \begin{pmatrix} a_x + b_x \\ a_y + b_y \\ a_z + b_z \end{pmatrix}. \tag{1.9}$$

Vector addition corresponds to the process of adding the vector $\underline{a}$ to the vector $\underline{b}$ at its end point. The newly created vector $\underline{c}$ then points from the starting point of $\underline{a}$ to the end point of $\underline{b}$ (so-called vector parallelogram, Fig. 1.11). It should be noted that both representations

[2] Pythagoras of Samos, ca. 570 BC—after 510 BC, Greek polymath.

Fig. 1.10 Multiplication of a vector by a scalar

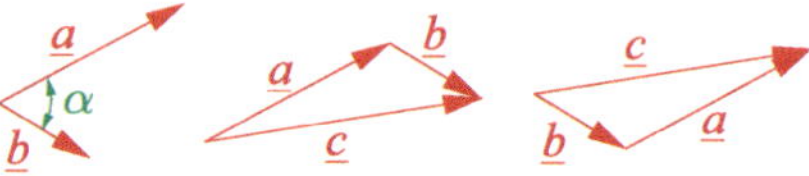

Fig. 1.11 Vector addition

of the vector parallelogram are equivalent; the commutative law applies:

$$\underline{a} + \underline{b} = \underline{b} + \underline{a}. \tag{1.10}$$

The vectors $\underline{a}$ and $\underline{b}$ are referred to as components of the vector $\underline{c}$.

The vector subtraction $\underline{c} = \underline{a} - \underline{b}$ can be understood as the addition of a vector -$\underline{b}$:

$$\underline{c} = \underline{a} - \underline{b} = \underline{a} + \left(-\underline{b}\right), \tag{1.11}$$

i.e.:

$$\begin{pmatrix} c_x \\ c_y \\ c_z \end{pmatrix} = \begin{pmatrix} a_x \\ a_y \\ a_z \end{pmatrix} - \begin{pmatrix} b_x \\ b_y \\ b_z \end{pmatrix} = \begin{pmatrix} a_x - b_x \\ a_y - b_y \\ a_z - b_z \end{pmatrix}. \tag{1.12}$$

Scalar Product

Given two vectors $\underline{a}$ and $\underline{b}$, which include the angle α (Fig. 1.11). The so-called scalar product (also called inner product or dot product) is a measure for the included angle α and is defined as:

$$\underline{a} \cdot \underline{b} = \left|\underline{a}\right| \left|\underline{b}\right| \cos\alpha = ab\cos\alpha. \tag{1.13}$$

Here, the expression $\left|\underline{b}\right| \cos\alpha$ is the projection of the vector $\underline{b}$ onto the vector $\underline{a}$ (Fig. 1.12). The magnitude of a vector $\underline{a}$ results from the scalar product as follows:

$$\left|\underline{a}\right| = \sqrt{\underline{a} \cdot \underline{a}}. \tag{1.14}$$

The commutative law also applies to the scalar product, i.e.:

$$\underline{a} \cdot \underline{b} = \underline{b} \cdot \underline{a}. \tag{1.15}$$

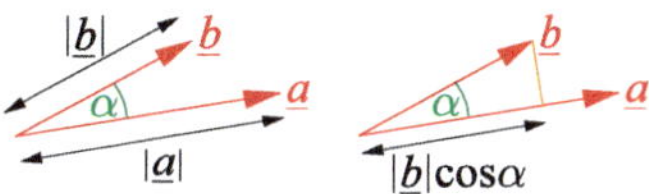

Fig. 1.12 Projection of the vector $\underline{b}$ onto the vector $\underline{a}$

At this point, we once again consider the orthonormal basis vectors $\underline{e}_x, \underline{e}_y, \underline{e}_z$, which have the magnitudes $\left|\underline{e}_x\right| = \left|\underline{e}_y\right| = \left|\underline{e}_z\right| = 1$. The following applies:

$$\begin{aligned} \underline{e}_x \cdot \underline{e}_x &= 1, \quad \underline{e}_x \cdot \underline{e}_y = 0, \quad \underline{e}_x \cdot \underline{e}_z = 0, \\ \underline{e}_y \cdot \underline{e}_x &= 0, \quad \underline{e}_y \cdot \underline{e}_y = 1, \quad \underline{e}_y \cdot \underline{e}_z = 0, \\ \underline{e}_z \cdot \underline{e}_x &= 0, \quad \underline{e}_z \cdot \underline{e}_y = 0, \quad \underline{e}_z \cdot \underline{e}_z = 1. \end{aligned} \tag{1.16}$$

The scalar product $\underline{a} \cdot \underline{b}$ of two vectors $\underline{a}$ and $\underline{b}$ can then be written as:

$$\begin{aligned} \underline{a} \cdot \underline{b} &= \begin{pmatrix} a_x \\ a_y \\ a_z \end{pmatrix} \cdot \begin{pmatrix} b_x \\ b_y \\ b_z \end{pmatrix} \\ &= \left(a_x \underline{e}_x + a_y \underline{e}_y + a_z \underline{e}_z\right) \cdot \left(b_x \underline{e}_x + b_y \underline{e}_y + b_z \underline{e}_z\right) \\ &= a_x b_x + a_y b_y + a_z b_z. \end{aligned} \tag{1.17}$$

The length of a vector $\underline{a}$ can thus be specified as follows:

$$\left|\underline{a}\right| = \sqrt{a_x^2 + a_y^2 + a_z^2}. \tag{1.18}$$

Vector Product

The vector product (often also referred to as cross product) results in a vector $\underline{c}$ through the arithmetic operation $\underline{a} \times \underline{b}$:

$$\underline{a} \times \underline{b} = \underline{c}. \tag{1.19}$$

This vector $\underline{c}$ is perpendicular to $\underline{a}$ and $\underline{b}$, and the three vectors $\underline{a}$, $\underline{b}$ and $\underline{c}$ form a right-handed system. The magnitude $\left|\underline{c}\right|$ of the vector $\underline{c}$ indicates the area spanned by the two vectors $\underline{a}$ and $\underline{b}$. The following applies:

$$\left|\underline{c}\right| = \left|\underline{a} \times \underline{b}\right| = \left|\underline{a}\right| \left|\underline{b}\right| \sin\alpha. \tag{1.20}$$

The commutative law does not apply here, and instead we have $\underline{a} \times \underline{b} = -\underline{b} \times \underline{a}$. If the two vectors $\underline{a}$ and $\underline{b}$ are parallel to each other, the vector product results in the zero vector.

For the vector products of the orthonormal basis vectors $\underline{e}_x, \underline{e}_y, \underline{e}_z$ follows:

$$\underline{e}_x \times \underline{e}_x = \underline{0}, \quad \underline{e}_x \times \underline{e}_y = \underline{e}_z, \quad \underline{e}_x \times \underline{e}_z = -\underline{e}_y,$$
$$\underline{e}_y \times \underline{e}_x = -\underline{e}_z, \quad \underline{e}_y \times \underline{e}_y = \underline{0}, \quad \underline{e}_y \times \underline{e}_z = \underline{e}_x,$$
$$\underline{e}_z \times \underline{e}_x = \underline{e}_y, \quad \underline{e}_z \times \underline{e}_y = -\underline{e}_x, \quad \underline{e}_z \times \underline{e}_z = \underline{0}. \tag{1.21}$$

Then for the vector product $\underline{a} \times \underline{b}$ the following applies:

$$\begin{aligned}
\underline{c} &= \underline{a} \times \underline{b} \\
&= \left(a_x \underline{e}_x + a_y \underline{e}_y + a_z \underline{e}_z\right) \times \left(b_x \underline{e}_x + b_y \underline{e}_y + b_z \underline{e}_z\right) \\
&= \left(a_y b_z - a_z b_y\right) \underline{e}_x + \left(a_z b_x - a_x b_z\right) \underline{e}_y + \left(a_x b_y - a_y b_x\right) \underline{e}_z \\
&= \begin{pmatrix} a_y b_z - a_z b_y \\ a_z b_x - a_x b_z \\ a_x b_y - a_y b_x \end{pmatrix}.
\end{aligned} \tag{1.22}$$

The vector product can also be specified as a determinant as follows:

$$\begin{aligned}
\underline{c} &= \underline{a} \times \underline{b} \\
&= \begin{vmatrix} \underline{e}_x & \underline{e}_y & \underline{e}_z \\ a_x & a_y & a_z \\ b_x & b_y & b_z \end{vmatrix} \\
&= \left(a_y b_z - a_z b_y\right) \underline{e}_x + \left(a_z b_x - a_x b_z\right) \underline{e}_y + \left(a_x b_y - a_y b_x\right) \underline{e}_z \\
&= \begin{pmatrix} a_y b_z - a_z b_y \\ a_z b_x - a_x b_z \\ a_x b_y - a_y b_x \end{pmatrix}.
\end{aligned} \tag{1.23}$$

2 Force Systems and Equilibrium

This chapter is dedicated to the treatment of force systems and deals first of all with plane central force systems, i.e. force systems in which the lines of action of the forces all intersect at one point. We deal with how forces can be summarised and also decomposed and also introduce the very basic concept of equilibrium. The considerations are then extended to spatial central force systems before we turn to the treatment of general plane force systems. The concepts of couple and moment are introduced, and we also consider moment systems. In addition, we address the question of how forces can be displaced in parallel, introduce the moment of a force and expand the concept of equilibrium. The chapter concludes with the treatment of general spatial force systems.

2.1 Plane Central Force Systems

A central force system is a system of forces whose lines of action all intersect at one point. In this section, we will first consider such central force systems in the plane; spatial force systems will be discussed later.

2.1.1 Force Parallelogram

We consider two forces F_1 and F_2 that act on a rigid body (Fig. 2.1) and include the angle α. We want to consider how these two forces F_1 and F_2 can be transformed into a single force, the so-called resultant R. We already know that we can treat forces on a rigid body

C. Mittelstedt, *Engineering Mechanics 1: Statics*,
https://doi.org/10.1007/978-3-662-71852-0_2

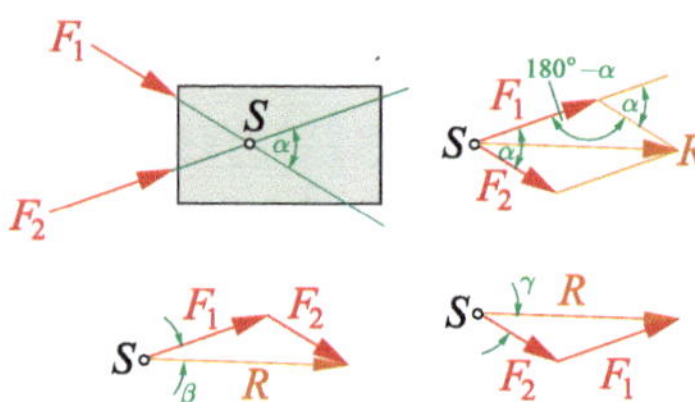

Fig. 2.1 Rigid body under two forces F_1 and F_2 (top left), force parallelogram (top right), force plots (bottom)

as line-following, so that we can shift the two forces until they both act at the intersection point S of the two lines of action. The two forces F_1 and F_2 thus form a so-called central force system.

We now utilise the fact that the static effect of the two forces F_1 and F_2 is equivalent to the effect of a single force, the resultant R, which also acts in S. The magnitude and direction of R result from the diagonal of the parallelogram spanned by F_1 and F_2. This statement is also known as the axiom of the force parallelogram.

In vectorial representation, it becomes clear that the determination of the resultant $\underline{R}$ is the addition of the two force vectors $\underline{F}_1$ and $\underline{F}_2$, i.e. the following applies:

$$\underline{R} = \underline{F}_1 + \underline{F}_2. \tag{2.1}$$

There are basically two ways to determine the resultant R, namely graphical determination on the one hand and mathematical determination on the other. For the graphical determination, the two forces F_1 and F_2 are plotted on a scale to be chosen appropriately and form the graph shown in Fig. 2.1. Alternatively, we can use a so-called force plan (Fig. 2.1, bottom), in which the two forces F_1 and F_2 are placed next to each other in a force polygon. In the resulting force polygon of Fig. 2.1, bottom left, the vector pointing from S to the end of the force F_2 is the resultant R. It is irrelevant whether the force diagram in Fig. 2.1, bottom right or left, is used, both representations are correct and equivalent, since the order of addition of the force vectors is irrelevant for determining the resultant according to (2.1), i.e. the commutative law $\underline{F}_1 + \underline{F}_2 = \underline{F}_2 + \underline{F}_1$ applies.

The resultant can be calculated with the help of elementary trigonometric considerations. To do this, let us take another look at the layout plan in Fig. 2.1, top right. The following then follows from the cosine theorem:

$$R = \sqrt{F_1^2 + F_2^2 - 2F_1F_2\cos(180° - \alpha)}. \tag{2.2}$$

With $\cos(180° - \alpha) = -\cos\alpha$ this results in:

$$R = \sqrt{F_1^2 + F_2^2 + 2F_1F_2\cos\alpha}. \tag{2.3}$$

We can determine the direction of the resultant R from the angles β or γ in the force diagrams in Fig. 2.1, bottom, by applying the sine theorem. It follows on the force plan of

Fig. 2.1, bottom left, with $\sin(180° - \alpha) = \sin\alpha$:

$$\frac{F_2}{\sin\beta} = \frac{R}{\sin(180° - \alpha)}, \tag{2.4}$$

or:

$$\sin\beta = \frac{F_2 \sin\alpha}{R}, \tag{2.5}$$

with R according to (2.3). The same applies to the angle γ:

$$\sin\gamma = \frac{F_1 \sin\alpha}{R}. \tag{2.6}$$

Example 2.1

Let the forces $F_1 = 2F_0$ and $F_2 = F_0$ be given, which are orientated at the angle $\alpha = 60°$ to each other and act at the point S (Fig. 2.2). Determine the magnitude of the resultant R and its direction.
Solution:

We first determine the resultant R and its direction graphically. We form the force parallelogram from the two forces F_1 and F_2 (Fig. 2.2, top right) and can read off the magnitude of the resultant R as $R = 2.65F_0$. The direction of the resultant can be described by the two angles β and γ, which we can read off from 2.2, bottom, as $\beta = 19°$ and $\gamma = 41°$. The sum must be $\alpha = \beta + \gamma$, which is fulfilled here.

We also determine the magnitude of the resultant mathematically and obtain from (2.3):

$$\begin{aligned} R &= \sqrt{F_1^2 + F_2^2 + 2F_1F_2\cos\alpha} \\ &= \sqrt{F_0^2 + (2F_0)^2 + 2 \cdot F_0 \cdot 2F_0 \cos 60°} = \sqrt{7}F_0 = 2.65F_0. \end{aligned} \tag{2.7}$$

The angles β and γ can be determined from (2.5) and (2.6), whereby it should be noted here that it is sufficient to determine only one of these two angles to determine the direction of the resultant. The following applies:

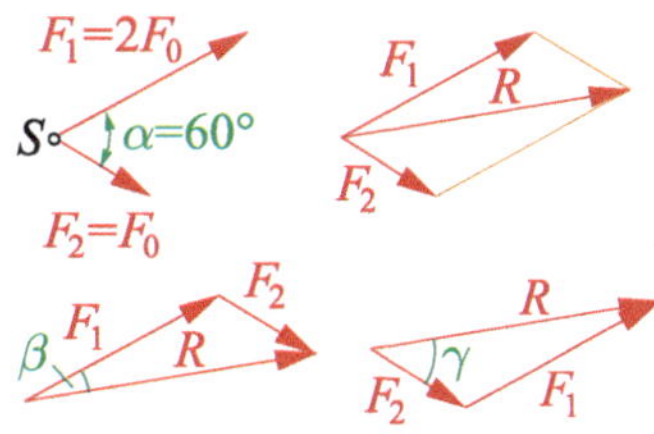

Fig. 2.2 Forces F_1 and F_2 with common intersection point S (top left), force parallelogram (top right), force plans (bottom)

$$\sin\beta = \frac{F_2 \sin\alpha}{R} = \frac{F_0 \sin 60°}{2.65 F_0} = 0.33 \quad \rightarrow \quad \beta = 19.1°,$$
$$\sin\gamma = \frac{F_1 \sin\alpha}{R} = \frac{2 F_0 \sin 60°}{2.65 F_0} = 0.65 \quad \rightarrow \quad \gamma = 40.8°. \tag{2.8}$$

These results agree with the previously graphically determined results with sufficient accuracy. Deviations result from the underlying drawing accuracy.

◀

2.1.2 Resultant of a Plane Central Force System

We now consider the case that we are dealing with a plane central force system composed of n forces F_1, F_2, ..., F_n (Fig. 2.3). We are looking for the magnitude of the resultant R and its direction.

We first discuss the graphical determination and create a force diagram with a suitable scale and determine the resultant force R by determining partial resultants and thus working our way forward step by step. Firstly, we consider the two forces F_1 and F_2 and determine the resultant from this, which we want to designate as R_1 (Fig. 2.3, top right). We determine its magnitude and direction by measuring it. In the next step, we consider the triangle of forces from R_1 and F_3 and determine the partial resultant R_2, which is also shown in Fig. 2.3, on the right. We successively form all other partial resultants one after the other until we arrive at the partial resultant R_{n-1}, which represents the resultant force R we are looking for. We determine its magnitude and direction by reading it off the force diagram. The force polygon with the partial resultants and the required resultant force R is shown in Fig. 2.4. The resultant R is therefore the force that runs from the point S to the end point of the force F_n. Its direction is opposite to the direction of rotation of the forces in the force polygon. As this is a central force system, the resultant also runs through the point S. It is

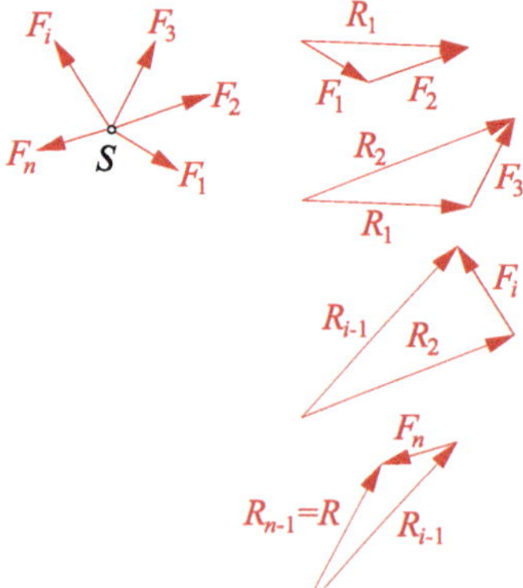

Fig. 2.3 Plane central force system consisting of the forces F_1, F_2, ..., F_n

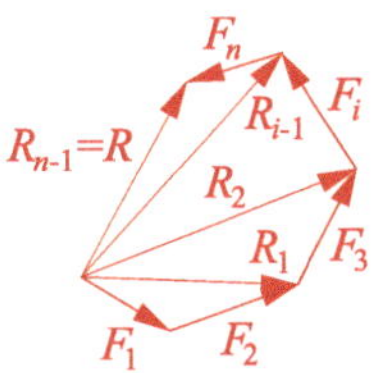

Fig. 2.4 Force polygon with partial resultants and resultant force R

statically equivalent to the n forces F_1, F_2, ..., F_n. Since we have summarised, i.e. reduced, the n forces F_1, F_2, ..., F_n to a single resultant force R, this process is also referred to as reduction.

The above procedure and the illustration in Fig. 2.4 show that the resultant $\underline{R}$ results from the addition of the individual force vectors $\underline{F}_1$, $\underline{F}_2$, $\underline{F}_3$, ..., $\underline{F}_i$, ..., $\underline{F}_n$:

$$\underline{R} = \underline{F}_1 + \underline{F}_2 + \underline{F}_3 + ... + \underline{F}_i + ... + \underline{F}_n. \tag{2.9}$$

Due to the commutative law, the order of addition is arbitrary and has no influence on the result. This can be seen in Fig. 2.5, in which we have carried out the order of formation of the partial resultants differently. Of course, this results in the same resultant force R as in Fig. 2.4.

Example 2.2

Consider a central plane force system consisting of the forces $F_1 = F_0$, $F_2 = 2F_0$, $F_3 = 3F_0$, $F_4 = 2F_0$, which are oriented under the angles $\alpha_1 = -50°$, $\alpha_2 = 45°$, $\alpha_3 = 160°$, $\alpha_4 = -135°$ to the horizontal and act at the common point S (Fig. 2.6). Determine the magnitude of the resultant R and its direction, described by the angle α_R to the horizontal, graphically.

Solution:

For the given problem, we draw the forces in the order F_1, F_2, F_3, F_4, taking into account their directions α_1, α_2, α_3, α_4, and draw the force polygon as shown in Fig. 2.6, bottom. It should be noted again that the order in which the forces are arranged in the force polygon is arbitrary. We can read off both the magnitude and the direction of the resultant from this,

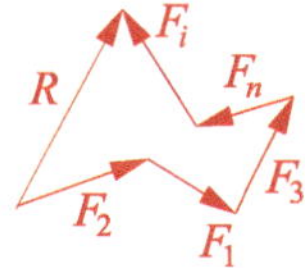

Fig. 2.5 Force polygon with alternative order of forces

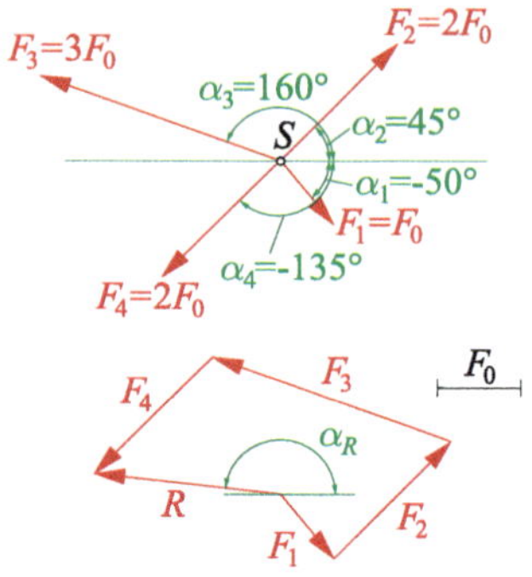

Fig. 2.6 Plane central force system, consisting of the forces F_1, F_2, F_3, F_4 (top), force polygon and determination of the resultant (bottom)

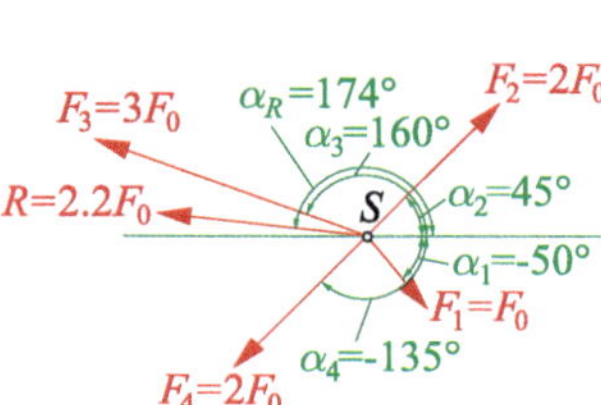

Fig. 2.7 Force plan with resulting force

although it should be noted that these results are only valid within the limits of the drawing accuracy. We obtain:

$$R = 2.2F_0, \quad \alpha_R = 174^\circ. \tag{2.10}$$

The resultant is shown in Fig. 2.7.

◀

2.1.3 Force Decomposition in the Plane

We now want to address the question of how a given force $\underline{F}$ can be decomposed. For this purpose, we use a Cartesian coordinate system consisting of $x-$ and $y-$axes, and we want to decompose a force $\underline{F}$ acting in the $xy-$plane into its components $\underline{F}_x$ and $\underline{F}_y$ in the direction

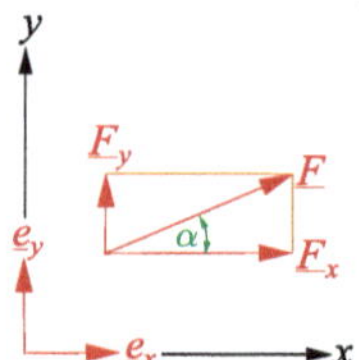

Fig. 2.8 Decomposition of a force $\underline{F}$

of the $x-$axis and the $y-$axis (Fig. 2.8), which are characterised by the unit vectors $\underline{e}_x$ and $\underline{e}_y$. Such a decomposition into two components in the plane is always possible, and the components $\underline{F}_x$ and $\underline{F}_y$ have the magnitudes F_x and F_y:

$$\underline{F}_x = F_x \underline{e}_x, \quad \underline{F}_y = F_y \underline{e}_y. \tag{2.11}$$

The force vector $\underline{F}$ can thus be expressed as follows by adding the components $\underline{F}_x$ and $\underline{F}_y$:

$$\underline{F} = \underline{F}_x + \underline{F}_y = F_x \underline{e}_x + F_y \underline{e}_y. \tag{2.12}$$

Let F be the magnitude of the force vector $\underline{F}$. Then the following applies to the magnitudes of the components $\underline{F}_x$ and $\underline{F}_y$:

$$F_x = F \cos\alpha, \quad F_y = F \sin\alpha, \tag{2.13}$$

and the magnitude F of the vector $\underline{F}$ is then:

$$|\underline{F}| = F = \sqrt{F_x^2 + F_y^2}. \tag{2.14}$$

The angle α, which indicates the direction of the force $\underline{F}$, can be determined as follows:

$$\tan\alpha = \frac{F_y}{F_x} \quad \rightarrow \quad \alpha = \arctan\left(\frac{F_y}{F_x}\right). \tag{2.15}$$

We can also write the vector $\underline{F}$ according to (2.12) as a column vector as follows:

$$\underline{F} = F_x \underline{e}_x + F_y \underline{e}_y = \begin{pmatrix} F_x \\ F_y \end{pmatrix}. \tag{2.16}$$

We now consider a plane central force group consisting of the forces $\underline{F}_1, \underline{F}_2, \ldots, \underline{F}_n$ and having the resultant $\underline{R}$. We can proceed in a similar way when decomposing the resultant and add up the components of the forces involved. This is shown in Fig. 2.9 for the example of a force group consisting of two forces $\underline{F}_1, \underline{F}_2$. The forces $\underline{F}_1, \underline{F}_2$ have the components $\underline{F}_{1x} = F_{1x}\underline{e}_x, \underline{F}_{2x} = F_{2x}\underline{e}_x$ with respect to the $x-$direction and $\underline{F}_{1y} = F_{1y}\underline{e}_y, \underline{F}_{2y} = F_{2y}\underline{e}_y$ with respect to the $y-$direction. The resultant

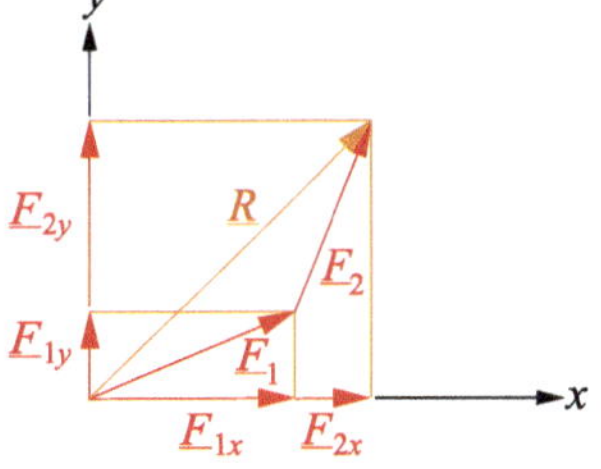

Fig. 2.9 Decomposition of the resultant $\underline{R}$ of a plane central force group

$$\underline{R} = R_x \underline{e}_x + R_y \underline{e}_y = \begin{pmatrix} R_x \\ R_y \end{pmatrix} \tag{2.17}$$

can then be determined from the vector addition of $\underline{F}_1$ and $\underline{F}_2$ as:

$$\begin{aligned} \underline{R} &= \underline{F}_1 + \underline{F}_2 = \underline{F}_{1x} + \underline{F}_{2x} + \underline{F}_{1y} + \underline{F}_{2y} \\ &= F_{1x} \underline{e}_x + F_{2x} \underline{e}_x + F_{1y} \underline{e}_y + F_{2y} \underline{e}_y \\ &= (F_{1x} + F_{2x}) \underline{e}_x + \left(F_{1y} + F_{2y}\right) \underline{e}_y. \end{aligned} \tag{2.18}$$

The components R_x and R_y of the resultant therefore follow as:

$$R_x = F_{1x} + F_{2x}, \quad R_y = F_{1y} + F_{2y}. \tag{2.19}$$

If there is the case of a resultant $\underline{R}$ of n forces, then the following applies:

$$\begin{aligned} \underline{R} &= R_x \underline{e}_x + R_y \underline{e}_y = \begin{pmatrix} R_x \\ R_y \end{pmatrix} \\ &= \sum_{i=1}^{n} \underline{F}_i = \sum_{i=1}^{n} (F_{ix}) \underline{e}_x + \sum_{i=1}^{n} \left(F_{iy}\right) \underline{e}_y. \end{aligned} \tag{2.20}$$

The components R_x, R_y of the resultant are therefore as follows:

$$R_x = \sum_{i=1}^{n} F_{ix}, \quad R_y = \sum_{i=1}^{n} F_{iy}, \tag{2.21}$$

and magnitude and direction can be specified as:

$$R = \left|\underline{R}\right| = \sqrt{R_x^2 + R_y^2}, \quad \tan \alpha_R = \frac{R_y}{R_x}. \tag{2.22}$$

Example 2.3

We repeat Example 2.2 and use the component representation of the resultant introduced above.

Solution:

The given planar central force system is represented by a reference system x, y as shown in Fig. 2.10. The component R_x can be determined as follows:

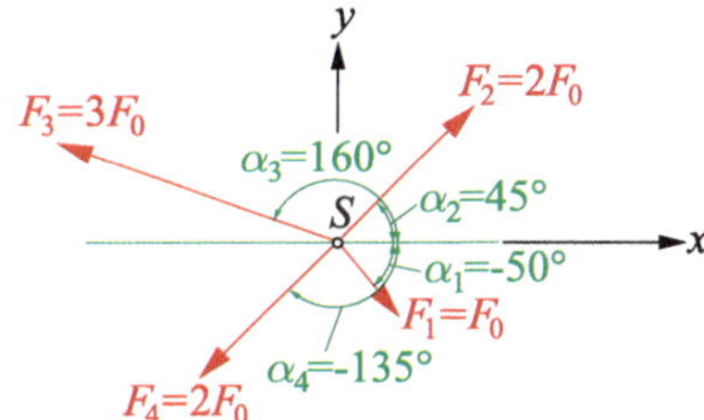

Fig. 2.10 Given plane central force system with coordinate system

$$\begin{aligned} R_x &= F_{1x} + F_{2x} + F_{3x} + F_{4x} \\ &= F_1 \cos\alpha_1 + F_2 \cos\alpha_2 + F_3 \cos\alpha_3 + F_4 \cos\alpha_4 \\ &= F_0 \cos(-50°) + 2F_0 \cos 45° + 3F_0 \cos 160° + 2F_0 \cos(-135°) \\ &= -2.18 F_0. \end{aligned} \tag{2.23}$$

The same applies to the component R_y:

$$\begin{aligned} R_y &= F_{1y} + F_{2y} + F_{3y} + F_{4y} \\ &= F_1 \sin\alpha_1 + F_2 \sin\alpha_2 + F_3 \sin\alpha_3 + F_4 \sin\alpha_4 \\ &= F_0 \sin(-50°) + 2F_0 \sin 45° + 3F_0 \sin 160° + 2F_0 \sin(-135°) \\ &= 0.26 F_0. \end{aligned} \tag{2.24}$$

The magnitude R of the resultant can thus be specified as:

$$R = \sqrt{R_x^2 + R_y^2} = 2.2 F_0. \tag{2.25}$$

It can be seen that this result agrees with (2.10). For the direction α_R of the resultant follows:

$$\tan\alpha_R = \frac{R_y}{R_x} = -\frac{0,26}{2,18} \quad \rightarrow \quad \alpha_R = -6.80°. \tag{2.26}$$

This result agrees with $180° - 6.80° = 173.20°$ with the angle of $\alpha_R = 174°$ determined in Example 2.2.

◄

2.1.4 Equilibrium

Consider a rigid body at rest subject to two opposing forces $\underline{F}_1$ and $\underline{F}_2$ on the same line of action (Fig. 2.11). It is clear that the rigid body on which these two forces act remains at rest exactly when the magnitudes F_1 and F_2 of the forces $\underline{F}_1$ and $\underline{F}_2$ are identical, i.e. when the following applies:

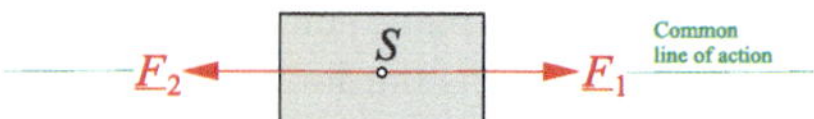

Fig. 2.11 Rigid body under two forces $\underline{F}_1$ and $\underline{F}_2$

$$\left|\underline{F}_1\right| = F_1 = F_2 = \left|\underline{F}_2\right|, \tag{2.27}$$

so that:

$$\underline{F}_2 = -\underline{F}_1. \tag{2.28}$$

This results in the following definition of equilibrium: Two forces are in equilibrium with each other if they lie on a common line of action and are oppositely directed with the same magnitude. Accordingly, the resultant of these two forces defined in this way disappears:

$$\underline{R} = \underline{F}_1 + \underline{F}_2 = \underline{0}. \tag{2.29}$$

For a central force system consisting of n forces, it applies analogously that such a force system is in equilibrium if the resultant of all forces involved disappears, i.e. if the vector sum over all force vectors results in the zero vector:

$$\underline{R} = \begin{pmatrix} R_x \\ R_y \end{pmatrix} = \underline{F}_1 + \underline{F}_2 + \ldots + \underline{F}_n = \sum_{i=1}^{n} \underline{F}_i = \begin{pmatrix} \sum_{i=1}^{n} F_{ix} \\ \sum_{i=1}^{n} F_{iy} \end{pmatrix} = \underline{0}. \tag{2.30}$$

If we again consider a force polygon consisting of n forces (Fig. 2.12), this means that this force polygon must be closed for the state of equilibrium, i.e. the resultant is exactly zero. A central force system that fulfils this condition is referred to as an equilibrium group. Accordingly, the requirement for equilibrium means that the components $R_x = \sum_{i=1}^{n} F_{ix}$ and $R_y = \sum_{i=1}^{n} F_{iy}$ become zero. The following so-called equilibrium conditions therefore apply in the $xy-$plane:

$$\sum_{i=1}^{n} F_{ix} = 0, \quad \sum_{i=1}^{n} F_{iy} = 0. \tag{2.31}$$

As a conclusion, it can be stated that a plane central force system is in equilibrium exactly when the two sums of the force components with respect to the $x-$direction and the $y-$direction disappear. The summation limits are often omitted, so that one also writes

$$\overset{\rightarrow}{\sum} F_{ix} = 0, \quad \overset{\uparrow}{\sum} F_{iy} = 0. \tag{2.32}$$

The arrows above the sums indicate the positive counting direction.

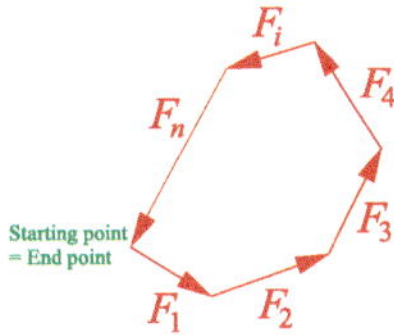

Fig. 2.12 Equilibrium at the closed force polygon

Example 2.4

Consider a central plane force system consisting of the forces $F_1 = F_0$, $F_2 = 2F_0$, $F_3 = 3F_0$, which are orientated at the angles $\alpha_1 = -50°$, $\alpha_2 = 45°$, $\alpha_3 = 160°$ to the horizontal and are applied at the common point S (Fig. 2.13). How large must an additional force F_4 be and at what angle to the horizontal must it be orientated so that the forces F_1, F_2, F_3, F_4 form an equilibrium group?

Solution:

We first solve the problem graphically and create a force polygon that is closed by the force F_4 (Fig. 2.13, bottom). The magnitude of the force F_4 and its direction can be read from this as:

$$F_4 = 1.8F_0, \quad \alpha_4 = -65°. \tag{2.33}$$

The force F_4 can be calculated from the equilibrium conditions (2.32). We obtain the sum of all forces in the $x-$ direction:

$$\sum F_{ix} = 0: \quad F_{1x} + F_{2x} + F_{3x} + F_{4x} = 0. \tag{2.34}$$

This expression can be solved for the force component F_{4x}:

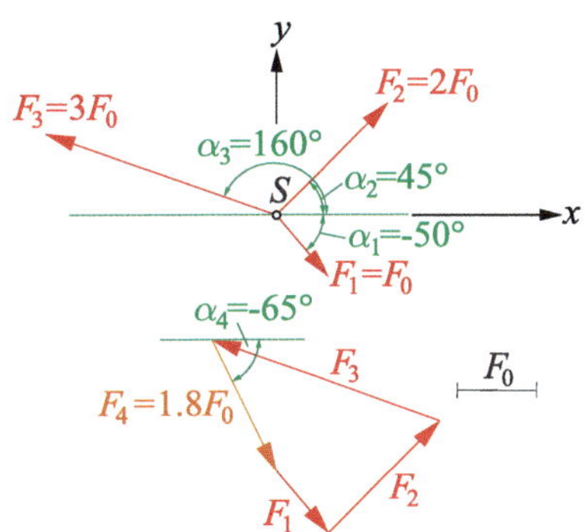

Fig. 2.13 Plane central force system consisting of the forces F_1, F_2, F_3 (top), closed force polygon (bottom)

$$\begin{aligned} F_{4x} &= -F_{1x} - F_{2x} - F_{3x} \\ &= -F_1 \cos\alpha_1 - F_2 \cos\alpha_2 - F_3 \cos\alpha_3 \\ &= -F_0 \cos(-50^\circ) - 2F_0 \cos 45^\circ - 3F_0 \cos 160^\circ \\ &= -0.64F_0 - 1.41F_0 + 2.81F_0 = 0.77F_0. \end{aligned} \tag{2.35}$$

The same applies to the equilibrium condition with regard to the y−direction:

$$\sum F_{iy} = 0: \quad F_{1y} + F_{2y} + F_{3y} + F_{4y} = 0. \tag{2.36}$$

We can solve this expression for F_{4y} and obtain

$$\begin{aligned} F_{4y} &= -F_{1y} - F_{2y} - F_{3y} \\ &= -F_1 \sin\alpha_1 - F_2 \sin\alpha_2 - F_3 \sin\alpha_3 \\ &= -F_0 \sin(-50^\circ) - 2F_0 \sin 45^\circ - 3F_0 \sin 160^\circ \\ &= 0.77F_0 - 1.41F_0 - 1.02F_0 = -1.67F_0. \end{aligned} \tag{2.37}$$

The magnitude of the force F_4 then follows as:

$$F_4 = \sqrt{F_{4x}^2 + F_{4y}^2} = \sqrt{(0.77F_0)^2 + (-1.67F_0)^2} = 1.84F_0. \tag{2.38}$$

The orientation to the horizontal is:

$$\tan\alpha_4 = \frac{F_{4y}}{F_{4x}} \quad \rightarrow \quad \alpha_4 = -65.25^\circ. \tag{2.39}$$

These results agree with the values already determined graphically.

◀

Example 2.5

The bar system shown in Fig. 2.14 is considered, which is loaded by the two forces $F_1 = F_0$ and $F_2 = F_0$. We want to determine the bar forces.

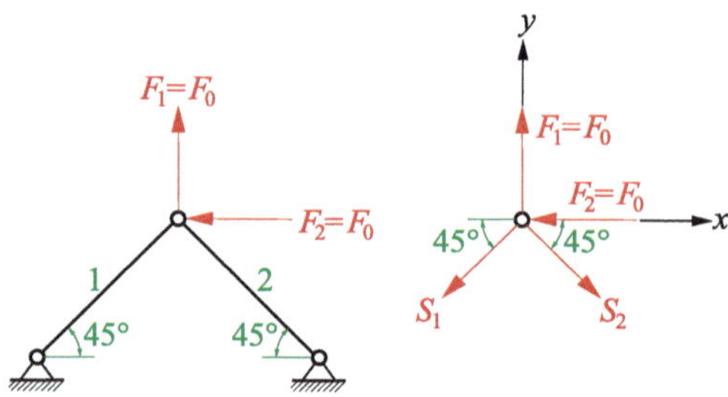

Fig. 2.14 Truss (left), free-body diagram (right)

Solution:

The given structure, which consists of two bars simply supported at both ends, is a so-called truss. Such structures are discussed in detail in Chap. 5. At this point, it should suffice to note that only forces that point in the axial direction of both bars 1 and 2 (the so-called bar forces) occur in the bars of such a truss. In the free-body diagram of the hinge point of Fig. 2.14, right, it is assumed that both bar forces S_1 and S_2 point away from the loaded hinge point, i.e. cause tension in the respective bar. The sign of the calculation results will later show whether this assumption was justified.

We form the sums of forces on the free-body diagram of Fig. 2.14, right, with respect to the x−direction and the y−direction, whereby we provide the sum signs with directional arrows that indicate the respective positive counting direction. We obtain:

$$\overset{\rightarrow}{\sum} F_{ix} = 0: \quad -S_1 \cos 45° - F_2 + S_2 \cos 45° = 0,$$
$$\overset{\uparrow}{\sum} F_{iy} = 0: \quad -S_1 \sin 45° + F_1 - S_2 \sin 45° = 0. \tag{2.40}$$

With $\cos 45° = \sin 45° = \frac{1}{\sqrt{2}}$ and $F_1 = F_0$, $F_2 = F_0$ we have:

$$-S_1 \frac{1}{\sqrt{2}} - F_0 + S_2 \frac{1}{\sqrt{2}} = 0,$$
$$-S_1 \frac{1}{\sqrt{2}} + F_0 - S_2 \frac{1}{\sqrt{2}} = 0. \tag{2.41}$$

This is a linear system of equations consisting of two equations for the two unknown rod forces S_1 and S_2. A static problem in which there are as many equilibrium conditions as there are unknown forces is called statically determinate. If the number of unknowns exceeds the number of equilibrium conditions, the problem is said to be statically indeterminate. The latter problem class will be discussed in detail in Volume 2.

We can solve the system of Eq. (2.41) for the bar forces S_1 and S_2 and obtain

$$S_1 = 0, \quad S_2 = \sqrt{2} F_0. \tag{2.42}$$

The remarkable result is that bar 1 remains completely unloaded, while bar 2 has to bear a bar force $S_2 = \sqrt{2} F_0$. As this bar force has a positive sign, it is a tensile force.

◀

Example 2.6

We consider a plane central force system consisting of a vertically acting force F_1 and two forces F_2 and F_3 acting at the angles α_2 and α_3 (Fig. 2.15). Determine the two angles α_2 and α_3 so that equilibrium prevails for given values for F_1, F_2 and F_3.

Solution:

We introduce a reference system x, y as shown in Fig. 2.15 and form the force sums with respect to the x−direction and the y−direction:

$$\overset{\rightarrow}{\sum} F_{ix} = 0: \quad F_2 \cos\alpha_2 - F_3 \cos\alpha_3 = 0,$$
$$\overset{\uparrow}{\sum} F_{iy} = 0: \quad -F_1 + F_2 \sin\alpha_2 + F_3 \sin\alpha_3 = 0. \tag{2.43}$$

This is a system of equations from which the two unknown angles α_2 and α_3 can be determined for given forces F_1, F_2 and F_3. We reformulate the two equations above as follows:

$$F_3 \cos\alpha_3 = F_2 \cos\alpha_2,$$
$$-F_1 + F_3 \sin\alpha_3 = -F_2 \sin\alpha_2. \tag{2.44}$$

If we square these equations and add them together, we obtain the following result for the angle α_3:

$$\sin\alpha_3 = \frac{F_1^2 - F_2^2 + F_3^2}{2F_1F_3} \quad \rightarrow \quad \alpha_3 = \arcsin\left(\frac{F_1^2 - F_2^2 + F_3^2}{2F_1F_3}\right). \tag{2.45}$$

We also obtain:

$$\sin\alpha_2 = \frac{F_1^2 + F_2^2 - F_3^2}{2F_1F_2} \quad \rightarrow \quad \alpha_2 = \arcsin\left(\frac{F_1^2 + F_2^2 - F_3^2}{2F_1F_2}\right). \tag{2.46}$$

◀

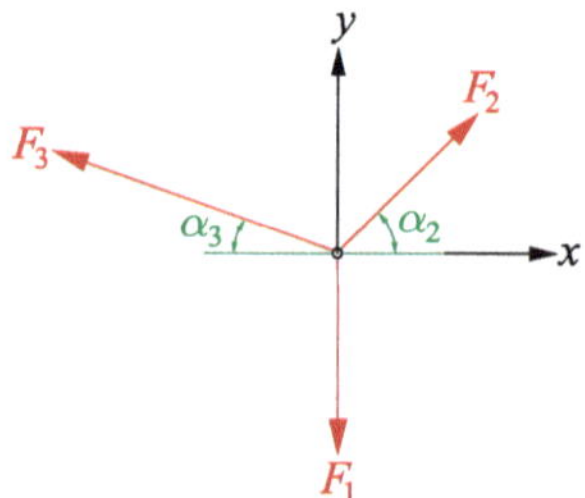

Fig. 2.15 Given planar central force system

2.2 Spatial Central Force Systems

2.2.1 Resultant of a Spatial Central Force System

If we consider a system of forces in which not all forces lie in the same plane, this is referred to as a spatial force system. If all forces have a common intersection point S of their lines of action, then, analogous to the previous considerations, we speak of a spatial central force system.

In the following, we consider the representation of a vector $\underline{F}$ in Cartesian coordinates x, y, z, which are characterised by the unit vectors $\underline{e}_x, \underline{e}_y, \underline{e}_z$:

$$\underline{e}_x = \begin{pmatrix} 1 \\ 0 \\ 0 \end{pmatrix}, \quad \underline{e}_y = \begin{pmatrix} 0 \\ 1 \\ 0 \end{pmatrix}, \quad \underline{e}_z = \begin{pmatrix} 0 \\ 0 \\ 1 \end{pmatrix}. \tag{2.47}$$

A vector $\underline{F}$ in space can then be represented by its three spatial coordinates F_x, F_y, F_z as (Fig. 2.16):

$$\begin{aligned} \underline{F} &= F_x\underline{e}_x + F_y\underline{e}_y + F_z\underline{e}_z \\ &= \underline{F}_x + \underline{F}_y + \underline{F}_z \\ &= \begin{pmatrix} F_x \\ 0 \\ 0 \end{pmatrix} + \begin{pmatrix} 0 \\ F_y \\ 0 \end{pmatrix} + \begin{pmatrix} 0 \\ 0 \\ F_z \end{pmatrix} \\ &= \begin{pmatrix} F_x \\ F_y \\ F_z \end{pmatrix}. \end{aligned} \tag{2.48}$$

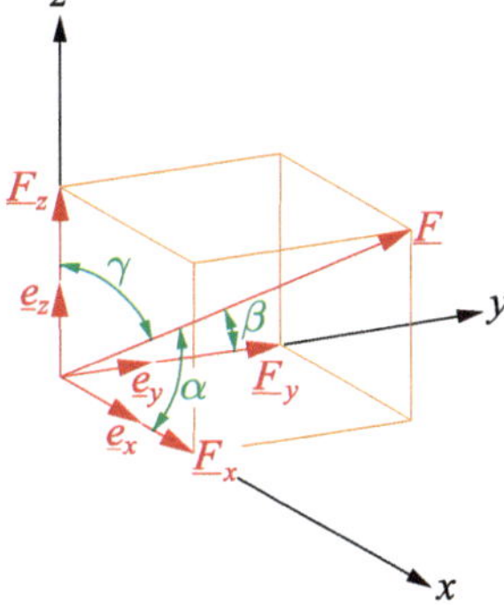

Fig. 2.16 Spatial force

The vector $\underline{F}$ includes the direction angles α, β and γ with the coordinate axes x, y, z as shown in Fig. 2.16, which is used to define the direction of $\underline{F}$ in space. Its magnitude F is:

$$F = |\underline{F}| = \sqrt{F_x^2 + F_y^2 + F_z^2}. \tag{2.49}$$

This allows the coefficients F_x, F_y, F_z to be written as:

$$F_x = F\cos\alpha, \quad F_y = F\cos\beta, \quad F_z = F\cos\gamma. \tag{2.50}$$

The angles α, β, γ are not independent of each other, as can be seen by squaring (2.49) and inserting (2.50). The following applies:

$$\cos^2\alpha + \cos^2\beta + \cos^2\gamma = 1. \tag{2.51}$$

Now let a spatial force system consisting of the forces $\underline{F}_1$, $\underline{F}_2$, ..., $\underline{F}_n$ be given, whose lines of action all intersect at a point S (Fig. 2.17). This is a central force system for which a resultant $\underline{R}$ can be determined. This resultant follows as in the plane case by vector addition, i.e. by successive application of the principle of the force parallelogram:

$$\begin{aligned} \underline{R} &= \underline{F}_1 + \underline{F}_2 + \ldots + \underline{F}_n = \sum_{i=1}^{n} \underline{F}_i \\ &= \sum_{i=1}^{n} F_{ix}\underline{e}_x + \sum_{i=1}^{n} F_{iy}\underline{e}_y + \sum_{i=1}^{n} F_{iz}\underline{e}_z && (2.52) \\ &= R_x\underline{e}_x + R_y\underline{e}_y + R_z\underline{e}_z && (2.53) \\ &= \underline{R}_x + \underline{R}_y + \underline{R}_z. && (2.54) \end{aligned}$$

The components of the resultant thus follow as:

$$R_x = \sum_{i=1}^{n} F_{ix}, \quad R_y = \sum_{i=1}^{n} F_{iy}, \quad R_z = \sum_{i=1}^{n} F_{iz}, \tag{2.55}$$

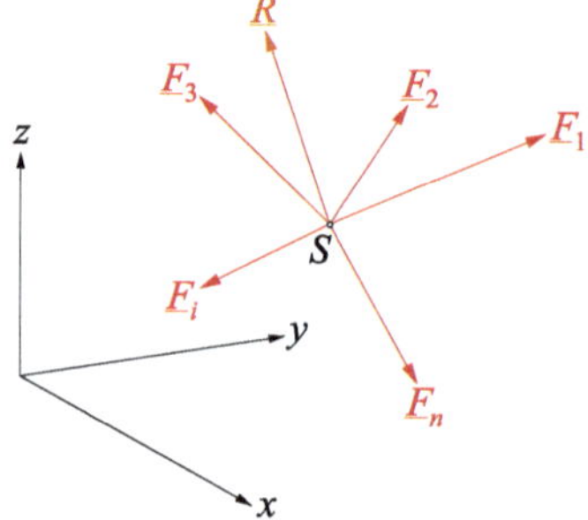

Fig. 2.17 Spatial central force system

wherein

$$F_{ix} = F_i \cos\alpha_i, \quad F_{iy} = F_i \cos\beta_i, \quad F_{iz} = F_i \cos\gamma_i. \tag{2.56}$$

The magnitude R of the resultant follows as:

$$R = |\underline{R}| = \sqrt{R_x^2 + R_y^2 + R_z^2}. \tag{2.57}$$

For the direction angles α_R, β_R, γ_R we obtain:

$$R_x = R\cos\alpha_R, \quad R_y = R\cos\beta_R, \quad R_z = R\cos\gamma_R. \tag{2.58}$$

2.2.2 Equilibrium

Analogously to our considerations on plane central force systems, it also applies to spatial force systems that equilibrium prevails exactly when the resultant $\underline{R}$ disappears. In vector form we thus have:

$$\underline{R} = \underline{F}_1 + \underline{F}_2 + \ldots + \underline{F}_n = \sum_{i=1}^{n} \underline{F}_i = \underline{0}. \tag{2.59}$$

This results in the following scalar equilibrium conditions:

$$\begin{aligned} R_x &= F_{1x} + F_{2x} + \ldots + F_{nx} = \sum_{i=1}^{n} F_{ix} = 0, \\ R_y &= F_{1y} + F_{2y} + \ldots + F_{ny} = \sum_{i=1}^{n} F_{iy} = 0, \\ R_z &= F_{1z} + F_{2z} + \ldots + F_{nz} = \sum_{i=1}^{n} F_{iz} = 0. \end{aligned} \tag{2.60}$$

Thus, equilibrium exists for a rigid body that is under a central spatial force system if the sum of all forces with respect to all three spatial directions x, y, z is zero.

Example 2.7

Consider a spatial truss consisting of three members connected to each other at a point S (Fig. 2.18). The two forces F_1 and F_2 are applied at point S. We are looking for the bar forces S_1, S_2 and S_3.

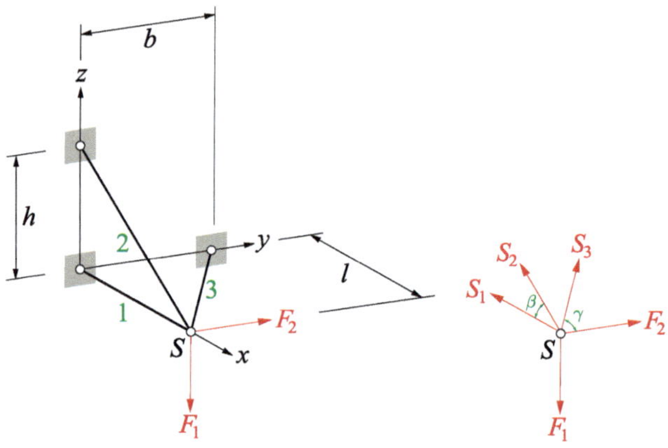

Fig. 2.18 Spatial truss

Solution:

To determine the bar forces S_1, S_2 and S_3, we cut the point S free and apply the bar forces. This type of structure is a spatial truss, and trusses are discussed in detail in Chap. 5. At this point, it is sufficient to note that only bar forces that are orientated parallel to the respective member axis occur in a truss. Positive barforces are assumed to be tensile forces, i.e. they point away from the point S.

We first determine the two angles β and γ that we need to formulate the equilibrium conditions. They follow as:

$$\tan\beta = \frac{h}{l} \quad \rightarrow \quad \beta = \arctan\left(\frac{h}{l}\right),$$
$$\tan\gamma = \frac{l}{b} \quad \rightarrow \quad \gamma = \arctan\left(\frac{l}{b}\right). \tag{2.61}$$

We formulate the spatial equilibrium conditions (2.60) for the given situation and obtain:

$$\sum F_{ix} = 0: \quad -S_1 - S_2\cos\beta - S_3\sin\gamma = 0,$$
$$\sum F_{iy} = 0: \quad S_3\cos\gamma + F_2 = 0,$$
$$\sum F_{iz} = 0: \quad S_2\sin\beta - F_1 = 0. \tag{2.62}$$

This provides three equations to determine the three unknown rod forces. We obtain:

$$S_1 = -\frac{F_1}{\tan\beta} + \frac{F_2}{\cot\gamma}, \quad S_2 = \frac{F_1}{\sin\beta}, \quad S_3 = -\frac{F_2}{\cos\gamma}. \tag{2.63}$$

We have solved this example without performing a vectorial representation of the forces. How to do this is shown below. We first write the forces as column vectors:

$$\underline{S}_1 = \begin{pmatrix} S_{1x} \\ S_{1y} \\ S_{1z} \end{pmatrix}, \quad \underline{S}_2 = \begin{pmatrix} S_{2x} \\ S_{2y} \\ S_{2z} \end{pmatrix}, \quad \underline{S}_3 = \begin{pmatrix} S_{3x} \\ S_{3y} \\ S_{3z} \end{pmatrix}. \tag{2.64}$$

The coordinates of these vectors are shown in Fig. 2.18. The force vector $\underline{S}_1$ only has one component in the negative $x-$direction, so the following applies:

$$\underline{S}_1 = S_1 \begin{pmatrix} -1 \\ 0 \\ 0 \end{pmatrix}. \tag{2.65}$$

The force vector $\underline{S}_2$ has a component S_{2x} in the negative $x-$direction and a component S_{2z} in the $z-$direction; a component S_{2y} in the $y-$direction does not exist. To determine the two components S_{2x} and S_{2z}, we decompose the force vector and obtain $S_{2x} = -S_2 \cos\beta$ and $S_{2z} = S_2 \sin\beta$. The force vector $\underline{S}_2$ can thus be written as:

$$\underline{S}_2 = S_2 \begin{pmatrix} -\cos\beta \\ 0 \\ \sin\beta \end{pmatrix}. \tag{2.66}$$

We proceed in the same way for $\underline{S}_3$ and obtain:

$$\underline{S}_3 = S_3 \begin{pmatrix} -\sin\gamma \\ \cos\gamma \\ 0 \end{pmatrix}. \tag{2.67}$$

The two force vectors $\underline{F}_1$ and $\underline{F}_2$ are as follows:

$$\underline{F}_1 = F_1 \begin{pmatrix} 0 \\ 0 \\ -1 \end{pmatrix}, \quad \underline{F}_2 = F_2 \begin{pmatrix} 0 \\ 1 \\ 0 \end{pmatrix}. \tag{2.68}$$

With the force vectors determined in this way, we can evaluate and obtain the equilibrium condition (2.59):

$$\underline{S}_1 + \underline{S}_2 + \underline{S}_3 + \underline{F}_1 + \underline{F}_2 = \underline{0}, \tag{2.69}$$

or

$$S_1 \begin{pmatrix} -1 \\ 0 \\ 0 \end{pmatrix} + S_2 \begin{pmatrix} -\cos\beta \\ 0 \\ \sin\beta \end{pmatrix} + S_3 \begin{pmatrix} -\sin\gamma \\ \cos\gamma \\ 0 \end{pmatrix} + F_1 \begin{pmatrix} 0 \\ 0 \\ -1 \end{pmatrix} + F_2 \begin{pmatrix} 0 \\ 1 \\ 0 \end{pmatrix} = \begin{pmatrix} 0 \\ 0 \\ 0 \end{pmatrix}. \tag{2.70}$$

This represents a system of three equations that we can write as follows:

$$
\begin{aligned}
-S_1 - S_2 \cos\beta - S_3 \sin\gamma &= 0, \\
S_3 \cos\gamma + F_2 &= 0, \\
S_2 \sin\beta - F_1 &= 0.
\end{aligned}
\tag{2.71}
$$

Apparently, this result agrees with (2.62).

◀

2.3 Plane General Force Systems

In this section we consider plane force systems in which the lines of action of the forces involved do not all intersect at a common point. Such a force system is called a general force system or also a non-central force system.

2.3.1 Resultant of Two Parallel Forces

We have seen in the previous sections that we can summarise a central force group into a resultant. We now want to address the question of how we can combine two parallel forces $\underline{F}_1$ and $\underline{F}_2$, which act on a rigid body and do not lie on a line of action (Fig. 2.19, left), into a resultant $\underline{R}$. The two forces have the distance a. At this point, we introduce an equilibrium group consisting of the two oppositely directed forces $\underline{H}$ and $-\underline{H}$. This is shown in Fig. 2.19, centre. We already know that such an equilibrium group has no influence on the state of a rigid body, these two forces are in equilibrium with each other. We now determine the partial resultants $\underline{R}_1 = \underline{F}_1 + \underline{H}$ and $\underline{R}_2 = \underline{F}_2 + (-\underline{H})$. The lines of action of these two partial resultants intersect at a point S, i.e. they form a central force system. The two forces

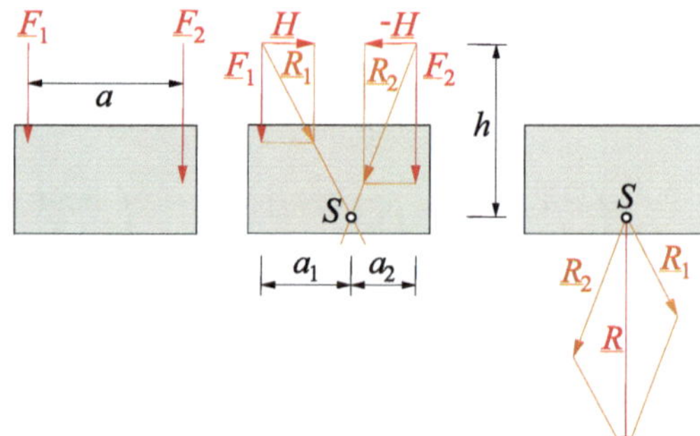

Fig. 2.19 Resultant of two parallel forces

$\underline{F}_1$ and $\underline{F}_2$ have the distances a_1 and a_2 from the point S, where $a = a_1 + a_2$. We now move the two partial resultants $\underline{R}_1$ and $\underline{R}_2$ to the point S (Fig. 2.19, right) and form the resultant $\underline{R} = \underline{R}_1 + \underline{R}_2$. Since the two auxiliary forces $\underline{H}$ and $-\underline{H}$ cancel each other out, the resultant $\underline{R}$ has the same direction as $\underline{F}_1$ and $\underline{F}_2$, and $\underline{R} = \underline{F}_1 + \underline{F}_2$ obviously applies. The resultant $\underline{R}$ is therefore statically equivalent to the sum of the two forces $\underline{F}_1$ and $\underline{F}_2$. The magnitude of $\underline{R}$ is correspondingly $R = F_1 + F_2$. From Fig. 2.19 we obtain:

$$\frac{h}{a_1} = \frac{F_1}{H}, \quad \frac{h}{a_2} = \frac{F_2}{H}. \tag{2.72}$$

With $a = a_1 + a_2$ this can be solved for a_1 and a_2, and the following result is obtained:

$$a_1 = \frac{F_2}{R}a, \quad a_2 = \frac{F_2}{R}a. \tag{2.73}$$

The ratio of a_2 and a_1 is as follows:

$$\frac{a_2}{a_1} = \frac{F_1}{F_2}, \tag{2.74}$$

or:

$$F_1 a_1 = F_2 a_2. \tag{2.75}$$

This expression is known as the lever law of Archimedes.[1]

If $F_1 = F_2$ is the case, then $R = 2F$ and $a_1 = a_2 = \frac{a}{2}$ results. Another special case is $F_1 = -F_2$. Then it follows that the resultant disappears: $R = 0$. Furthermore, this results in $a_1 \to \infty$ and $a_2 \to \infty$. In this case, we speak of a so-called force pair or a couple, which obviously cannot be reduced to a resultant. This case has considerable technical significance and is discussed in the following section.

2.3.2 Couple and Moment of a Couple

Consider a rigid body on a smooth surface (Fig. 2.20). We now attach a pair of forces or a so-called couple to this rigid body, i.e. two parallel forces F at a distance a, which are directed in opposite directions, have the same magnitude and do not lie on a common line of action. This is shown in Fig. 2.20, top left. It can be seen that this couple has a different effect on the rigid body than previously discussed. While the two forces F cancel each other out in their effect due to their equal magnitudes and thus no displacement of the body takes place, a pair of forces as shown has the tendency to rotate the body. If the pair of forces acts in the xy-plane as shown here, then this rotation will take place around the z-axis. Such a

[1] Archimedes of Syracuse, 287 BC to 212 BC, Greek polymath.

rotation is counted as positive if it takes place in the sense of a clockwise rotation around the z-axis.

It is clear that the same rotation occurs when the pair of forces is displaced in one direction as shown in Fig. 2.20, top right. The effect on the rigid body is also unchanged if the force pair is rotated in the xy−plane as shown in Fig. 2.20, bottom left. However, it can also be seen that the effect of the couple depends very much on the plane in which it acts. Figure 2.20, bottom right, shows the same couple, but now acting in the yz−plane. The result will be different here: The body will now be rotated around the x−axis and will tilt. Obviously, not only the plane of action of a couple is relevant, but also its direction of rotation: While the situation of Fig. 2.20, top left, leads to a positive rotation around the z−axis, a reversal of the two forces F would lead to a negative rotation. A couple is therefore characterised by the following quantities: Magnitude and direction of the forces F, distance a, plane of action.

At this point, a new force quantity is introduced, namely the so-called moment M of a couple. The moment M of a couple is defined as the product of force F and perpendicular distance a, i.e.

$$M = Fa. \tag{2.76}$$

A moment therefore has the unit of a force F multiplied by a length a, i.e. a product of a unit of force and a unit of length. The so-called Newton metre [Nm] is commonly used here. Since the moment is characterised by both its magnitude M and its direction of rotation, but also its plane of action, it is a vectorial quantity. It can be both positive and negative.

A couple can be shifted in its plane of action without changing its effect and thus its moment. This has already been shown in Fig. 2.20 and is considered in more detail below (Fig. 2.21). Consider a couple consisting of the two force vectors $\underline{F}$ and $-\underline{F}$, which have the distance a_1. They generate a moment with the magnitude $M = Fa_1$, where F is the magnitude of the force vector $\underline{F}$. We agree that the moment of a couple in the plane is

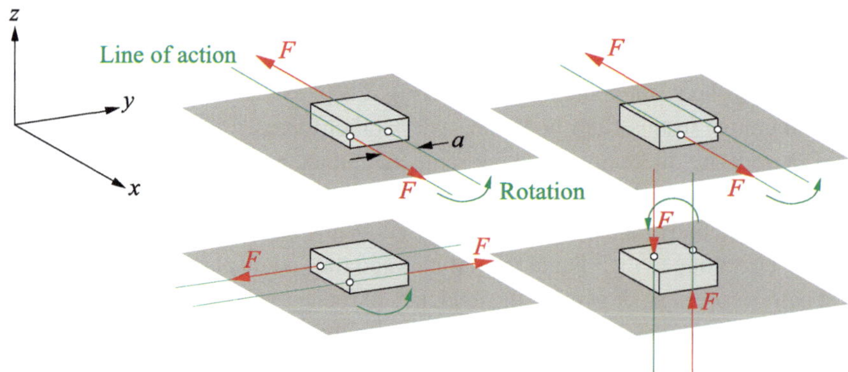

Fig. 2.20 Rigid body under a pair of forces

always positive when it rotates anti-clockwise. We now move the two force vectors along their lines of action and introduce an equilibrium group consisting of $\underline{H}$ and $-\underline{H}$ as shown in Fig. 2.21, bottom. Let the resulting forces be $\underline{R} = \underline{F} + \underline{H}$ or $-\underline{R} = -\underline{F} - \underline{H}$. This results in a new pair of forces consisting of $\underline{R}$ and $-\underline{R}$ with the distance a_2. We read from Fig. 2.21:

$$\cos\alpha = \frac{a_2}{a_1} = \frac{F}{R}, \tag{2.77}$$

so that:

$$a_2 = \frac{F}{R}a_1. \tag{2.78}$$

This results in the moment of the couple as follows:

$$M = Ra_2 = R \cdot \frac{F}{R}a_1 = Fa_1. \tag{2.79}$$

Obviously, neither the magnitude nor the direction of rotation of the moment M changes as a result of the transformation carried out. Since the introduced equilibrium group $\underline{H}$, $-\underline{H}$ is also arbitrary, a couple can be shifted arbitrarily in its plane of action without changing its effect. This is in clear contrast to force vectors on rigid bodies, which can be shifted along their lines of action but cannot be moved arbitrarily in the plane. Accordingly, two moments are statically equivalent to each other if they lie in the same plane of action, their magnitudes are identical and they have identical senses of rotation. The location of the representation of a pair of forces on a rigid body is arbitrary in the plane of action, as can be seen in Fig. 2.21.

A moment is characterised by a so-called moment arc as shown in Fig. 2.22 indicating the magnitude $M = Fa$. According to the above, it is irrelevant at which point of the rigid body the moment is drawn in its plane of action.

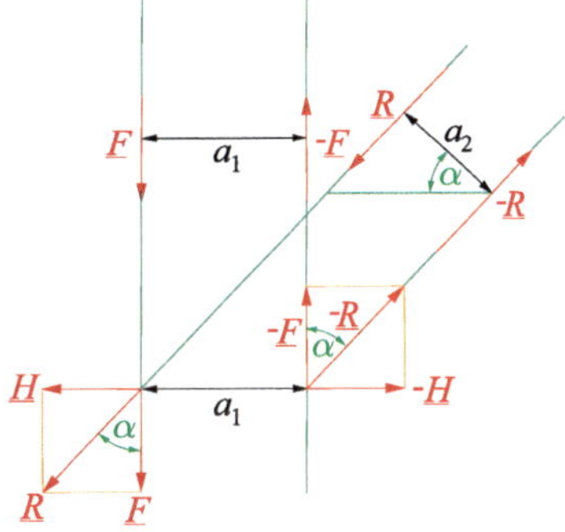

Fig. 2.21 Displacement of a pair of forces

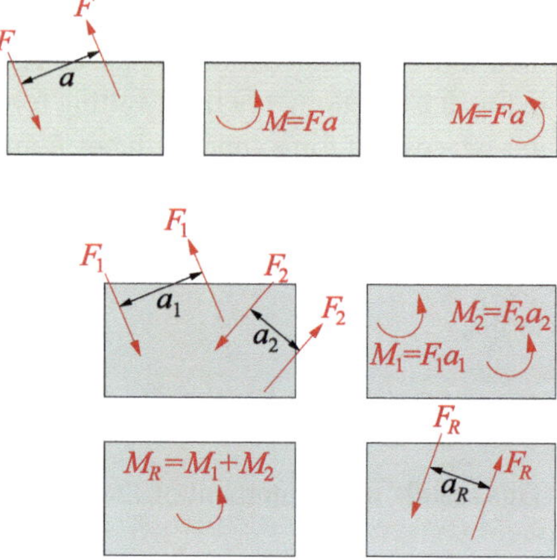

Fig. 2.22 Couple (left); statically equivalent moments (centre and right)

Fig. 2.23 Rigid body with two couples, reduction to one resulting moment and couple

2.3.3 Moment Systems

If n couples or moments with a common plane of action are applied to a rigid body, these can be converted into a resulting moment M_R, whereby the total moment M_R results from the sum of the n individual moments:

$$M_R = M_1 + M_2 + \ldots + M_n. \tag{2.80}$$

This is shown in Fig. 2.23, in which two couples $F_1 a_1$ and $F_2 a_2$ act on a rigid body. The two couples can be translated into the moments $M_1 = F_1 a_1$ and $M_2 = F_2 a_2$ (Fig. 2.23, top right), which in turn can be combined to form a resulting moment $M_R = M_1 + M_2$ (Fig. 2.23, bottom left). The position at which the two moments M_1 and M_2 and the resulting moment M_R are plotted on the rigid body is irrelevant. Ultimately, the resulting moment M_R can be understood as a statically equivalent pair of forces F_R with the distance a_R, whereby $F_R a_R = M_R$ must apply.

A system of n moments is in equilibrium if the resulting moment disappears, i.e. if the following applies:

$$M_R = M_1 + M_2 + \ldots + M_n = 0. \tag{2.81}$$

Example 2.8

Consider a rigid body on which the two couples $F_1 a_1$ and $F_2 a_2$ and two moments M_3 and M_4 act (Fig. 2.24). We want to determine the resulting moment M_R.

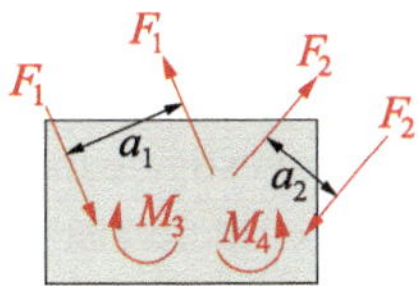

Fig. 2.24 Rigid body with two couples and two moments

Solution:

The two couples can be converted into moments $M_1 = F_1 a_1$ and $M_2 = -F_2 a_2$, whereby it should be noted that both moments M_1 and M_2 have opposite senses of rotation. We can now add up the moments M_1, M_2, M_3 and M_4 to the resulting moment M_R, taking into account the sense of rotation of the moments (positive in anti-clockwise direction):

$$M_R = F_1 a_1 - F_2 a_2 - M_3 + M_4. \tag{2.82}$$

As long as $F_1 a_1 + M_4 > F_2 a_2 + M_3$ applies, M_R is a positive moment, otherwise it is negative.

◀

Example 2.9

We consider the rigid body from Example 2.8 under the two couples $F_1 a_1$ and $F_2 a_2$ and the moments M_3 and M_4 (Fig. 2.25). How large must the additional couple $F_5 a_5$ be for equilibrium to prevail?

Solution:

For equilibrium, the resulting moment M_R must disappear:

$$M_R = F_1 a_1 - F_2 a_2 - M_3 + M_4 - F_5 a_5 = 0. \tag{2.83}$$

This expression can be solved immediately for the couple $F_5 a_5$:

$$F_5 a_5 = F_1 a_1 - F_2 a_2 - M_3 + M_4. \tag{2.84}$$

◀

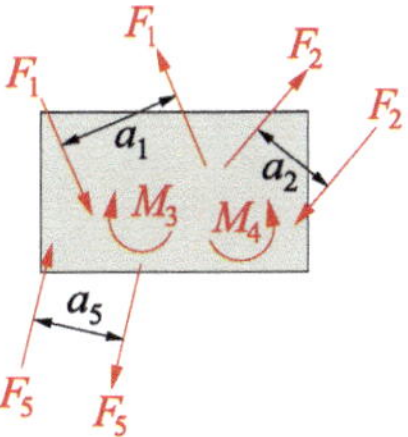

Fig. 2.25 Rigid body with two pairs of forces and two moments as well as an additional pair of forces $F_5 a_5$

2.3.4 Parallel Shift and Moment of a Force

We investigate the situation in Fig. 2.26, left, in which a rigid body is depicted under a force F. Let there also be a point A at an arbitrary position that is at a perpendicular distance a from the line of action of the force F. We now want to consider how F can be moved to the point A so that the static effect remains unchanged. We first apply two opposing forces with the magnitude F at point A as shown in Fig. 2.26, centre. This step is permissible, as these two forces cancel each other out. In addition, the original force F can be shifted arbitrarily along its line of action without changing its effect on the rigid body under consideration. We see that there is now a pair of forces Fa that can be converted into an equivalent moment $M = Fa$ (Fig. 2.26, right). This means that a force F can be shifted by the distance a to a new, parallel line of action at a point A, taking into account that a moment $M_A = Fa$ must also be applied for a statically equivalent effect. It is customary to label the corresponding moment with an index, in this case the index A, to clarify which reference point is involved. The moment depends on the distance a, its sign is determined by the direction of rotation of the force F. The distance a is also called the lever arm of the force F, and the moment M_A is referred to as the moment of the force F with respect to the point A. If the reference point A lies on the line of action of the force F, then the moment M_A is identical to zero.

It is often useful to break down a force in Cartesian coordinates into its components (Fig. 2.27) and use this to consider the statically equivalent effect. From Fig. 2.27 we obtain:

$$\sin \alpha = \frac{F_y}{F}, \quad \cos \alpha = \frac{F_x}{F}. \tag{2.85}$$

The lever arm a of the force F with respect to the point A is as follows:

$$a = x \sin \alpha - y \cos \alpha. \tag{2.86}$$

For the moment M_A then follows:

$$M_A = Fa = F\,(x \sin \alpha - y \cos \alpha) = F\left(x\frac{F_y}{F} - y\frac{F_x}{F}\right) = F_y x - F_x y. \tag{2.87}$$

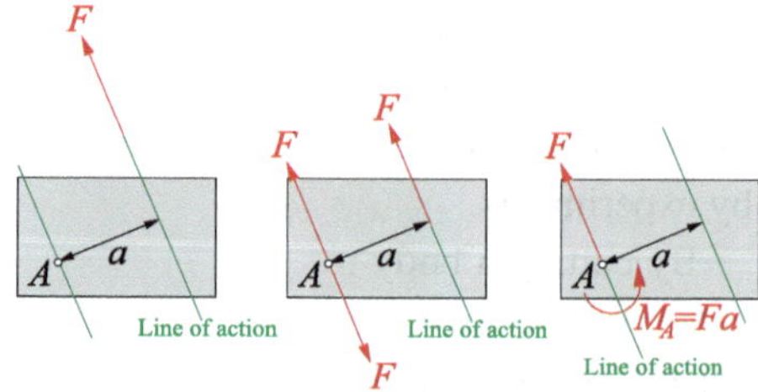

Fig. 2.26 Rigid body under force F

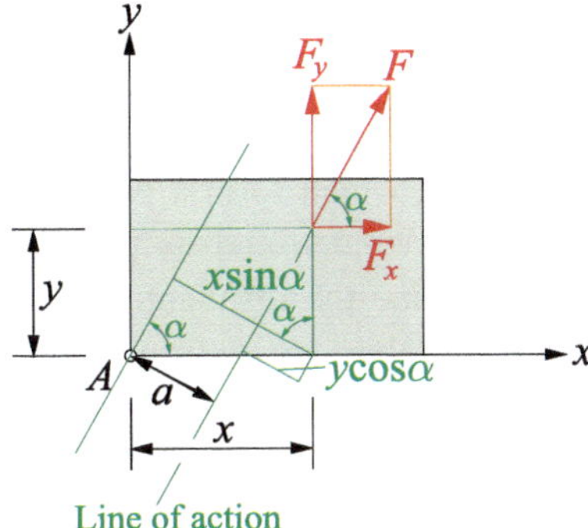

Fig. 2.27 Rigid body under force F

The moment of the force F with respect to the point A can therefore also be formed by the components F_x and F_y of F, whereby attention must be paid here to the correct consideration of the sense of rotation.

Two forces F_1 and F_2 are now considered in Fig. 2.28. These two forces can be broken down into their components F_{1x}, F_{2x}, F_{1y}, F_{2y}. The moments M_{1A} and M_{2A} of these two forces are as follows:

$$M_{1A} = F_{1y}x - F_{1x}y, \quad M_{2A} = F_{2y}x - F_{2x}y. \tag{2.88}$$

The resulting moment M_R results from the sum of M_{1A} and M_{2A}:

$$M_R = M_{1A} + M_{2A} = \left(F_{1y} + F_{2y}\right)x - \left(F_{1x} + F_{2x}\right)y = R_y x - R_x y. \tag{2.89}$$

It is therefore obviously irrelevant whether the two forces are added first and then the resulting moment is determined, or whether the sum of the individual moments is calculated. This also applies if there are n forces. Accordingly, the resulting moment M_R is identical to the moment of the resultant R.

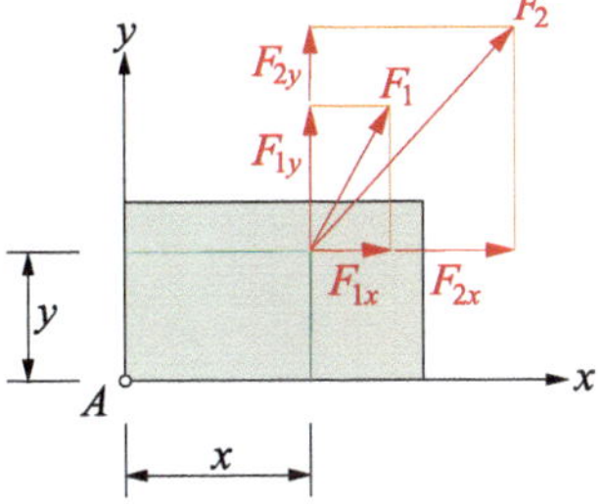

Fig. 2.28 Rigid body under two forces F_1 and F_2

2.3.5 Reduction to a Total Resultant

Consider the rigid body of Fig. 2.29, top left, which is loaded by the n forces $F_1, F_2, ..., F_i, ..., F_n$, which form a plane non-central force system. The force F_i has the angle α_i to the horizontal, and its point of application has the coordinates x_i and y_i. We want to determine the resultant R and the resulting moment M_{RA} with respect to the point A so that they are statically equivalent to the original non-central force system. To do this, we move all forces $F_1, F_2, ..., F_i, ..., F_n$ to the point A and also apply the moments $M_{1A}, M_{2A}, ..., M_{iA}, ..., M_{nA}$ (Fig. 2.29, top right). The forces $F_1, F_2, ..., F_i, ..., F_n$ are combined to form the resultant R (Fig. 2.29, bottom left):

$$\underline{R} = \underline{F}_1 + \underline{F}_2 + ... + \underline{F}_i + ... + \underline{F}_n = \sum_{i=1}^{n} \underline{F}_i. \tag{2.90}$$

The resulting moment M_{RA} is also formed with respect to the point A:

$$M_{RA} = M_{1A} + M_{2A} + ... + M_{iA} + ... + M_{nA} = \sum_{i=1}^{n} M_{iA}. \tag{2.91}$$

In general, the following applies to the components R_x and R_y of the resultant R of a non-central plane force group:

$$R_x = F_{1x} + F_{2x} + ... + F_{ix} + ... + F_{nx} = \sum_{i=1}^{n} F_{ix},$$

$$R_y = F_{1y} + F_{2y} + ... + F_{iy} + ... + F_{ny} = \sum_{i=1}^{n} F_{iy}. \tag{2.92}$$

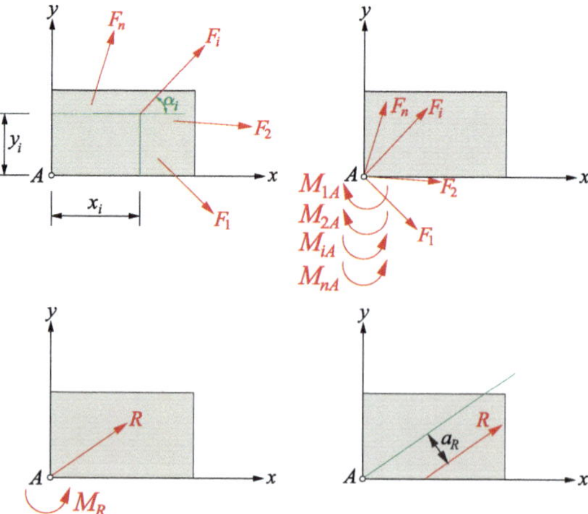

Fig. 2.29 Non-central plane force system, reduction to a total resultant

The magnitude and direction angle of R are given by

$$R = \sqrt{R_x^2 + R_y^2}, \quad \tan \alpha_R = \frac{R_y}{R_x}. \tag{2.93}$$

The resulting moment M_{RA} follows as:

$$M_{RA} = \sum_{i=1}^{n} M_{iA} = \sum_{i=1}^{n} \left(F_{iy} x_i - F_{ix} y_i \right). \tag{2.94}$$

Finally, the resultant can be shifted to a line of action outside A so that the moment M_{RA} becomes zero (Fig. 2.29, bottom right). The resulting moment with respect to the new line of action must become zero, so the following must apply:

$$M_{RA} - R a_R = 0. \tag{2.95}$$

This can be solved for a_R:

$$a_R = \frac{M_{RA}}{R}. \tag{2.96}$$

We call the resultant R determined in this way at a vertical distance a_R from the point A the total resultant.

Example 2.10

For the rigid body of Fig. 2.30 under the four forces $F_1 = F_0$, $F_2 = 4F_0$, $F_3 = 2F_0$, $F_4 = 3F_0$ with the distances $x_1 = a_0$, $x_2 = 2a_0$, $x_3 = 3a_0$, $x_4 = 4a_0$ to the point A, the total resultant is sought.

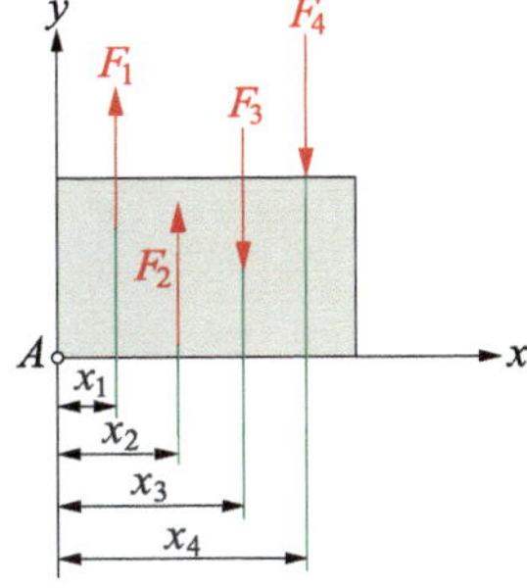

Fig. 2.30 Rigid body under four forces F_1, F_2, F_3, F_4

Solution:

The components of the resultant R are given by

$$R_x = F_{1x} + F_{2x} + F_{3x} + F_{4x} = 0,$$
$$R_y = F_{1y} + F_{2y} + F_{3y} + F_{4y} = F_0 + 4F_0 - 2F_0 - 3F_0 = 0. \quad (2.97)$$

Obviously, the resulting force R is identical to zero in this particular case. The resulting moment M_{RA} results in:

$$M_{RA} = F_1x_1 + F_2x_2 - F_3x_3 - F_4x_4 = F_0 \cdot a_0 + 4F_0 \cdot 2a_0 - 2F_0 \cdot 3a_0 - 3F_0 \cdot 4a_0$$
$$= F_0a_0 + 8F_0a_0 - 6F_0a_0 - 12F_0a_0 = -9F_0a_0. \quad (2.98)$$

◀

Example 2.11

A rigid body with the dimensions $4l$ and $4h$ is loaded by four forces $F_1 = F_2 = F_3 = F_4 = F_0$ (Fig. 2.31). We are looking for the components R_x, R_y of the resulting force $\underline{R}$ and the resulting moment M_{RA} with respect to the point A. In addition, the total resultant is sought.

Solution:

We first determine the components of the resultant R and obtain:

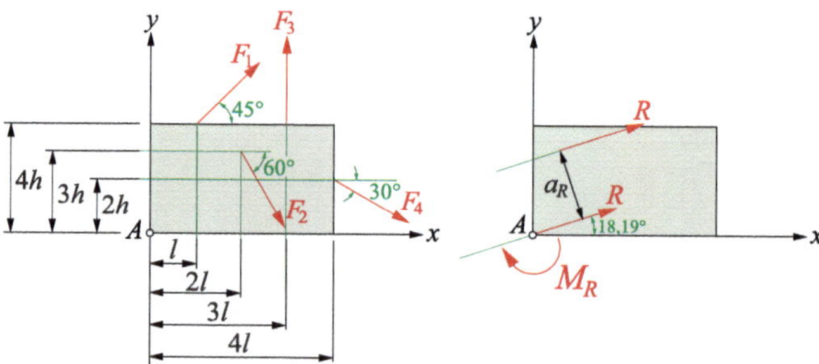

Fig. 2.31 Rigid body under four forces F_1, F_2, F_3, F_4 (left), resultant and resulting moment as well as position of the total resultant (right)

$$
\begin{aligned}
R_x &= F_1 \cos 45^\circ + F_2 \cos 60^\circ + F_4 \cos 30^\circ \\
&= \frac{\sqrt{2}F_0}{2} + \frac{F_0}{2} + \frac{\sqrt{3}F_0}{2} \\
&= \frac{F_0}{2}\left(1 + \sqrt{2} + \sqrt{3}\right) = 2.07F_0, \\
R_y &= F_1 \sin 45^\circ - F_2 \sin 60^\circ + F_3 - F_4 \sin 30^\circ \\
&= \frac{\sqrt{2}F_0}{2} - \frac{\sqrt{3}F_0}{2} + F_0 - \frac{F_0}{2} \\
&= \frac{F_0}{2}\left(1 + \sqrt{2} - \sqrt{3}\right) = 0.68F_0.
\end{aligned}
\tag{2.99}
$$

The magnitude of the resultant then follows as:

$$
R = \sqrt{R_x^2 + R_y^2} = \sqrt{(2.07F_0)^2 + (0.68F_0)^2} = 2.18F_0. \tag{2.100}
$$

It is orientated at the angle α_R as follows:

$$
\tan \alpha_R = \frac{F_y}{F_x} = \frac{0.68F_0}{2.07F_0} \quad \rightarrow \quad \alpha_R = 18.19^\circ. \tag{2.101}
$$

We also determine the resulting moment M_{RA} with respect to the point A and obtain (positive anti-clockwise direction of rotation):

$$
\begin{aligned}
M_{RA} &= F_1 \sin 45^\circ \cdot l - F_1 \cos 45^\circ \cdot 4h - F_2 \sin 60^\circ \cdot 2l - F_2 \cos 60^\circ \cdot 3h \\
&\quad + F_3 \cdot 3l - F_4 \sin 30^\circ \cdot 4l - F_4 \cos 30^\circ \cdot 2h \\
&= F_0 \left[\left(1 + \frac{\sqrt{2}}{2} - \sqrt{3}\right) l - \left(\frac{3}{2} + 2\sqrt{2} + \sqrt{3}\right) h\right] \\
&= -F_0\,(0.02l + 6.06h)\,.
\end{aligned}
\tag{2.102}
$$

The position of the line of action of the total resultant is as follows:

$$
a_R = \frac{M_{RA}}{R} = 0.92\,(0.01l + 3.03h)\,. \tag{2.103}
$$

If, for example, the special case $h = l$ is considered, then we obtain the following:

$$
a_R = 2.80h. \tag{2.104}
$$

The total resultant is shown in Fig. 2.31, right.

◀

2.3.6 Equilibrium

In this section, we turn to the equilibrium conditions for a non-central plane force system consisting of the forces F_1, F_2, ..., F_i, ..., F_n. We require that both the resultant $\underline{R}$ and the resulting moment M_{RA} vanish with respect to a point A:

$$R_x = 0, \quad R_y = 0, \quad M_{RA} = 0. \tag{2.105}$$

This can also be written in the force components and partial moments involved, and we obtain the following equilibrium conditions:

$$\sum_{i=1}^{n} F_{ix} = 0, \quad \sum_{i=1}^{n} F_{iy} = 0, \quad \sum_{i=1}^{n} M_{iA} = 0. \tag{2.106}$$

The summation limits are often omitted at the summation signs and the positive counting direction is indicated by arrows:

$$\overset{\rightarrow}{\sum} F_{ix} = 0, \quad \overset{\uparrow}{\sum} F_{iy} = 0, \quad \overset{\curvearrowleft}{\sum} M_{iA} = 0. \tag{2.107}$$

A rigid body is therefore in static equilibrium when both the sums of all forces in the $x-$direction or in the $y-$direction and the sum of moments with respect to any point A become zero. The choice of the reference point A is arbitrary, this point can lie both inside and outside the rigid body. There are therefore exactly as many equilibrium conditions available as the rigid body under consideration has degrees of freedom in the plane.

It can be shown that other equations can be used in addition to the above equilibrium conditions. For example, instead of the moment conditions relating to A, we can also use a moment sum relating to any other point B. The equilibrium conditions are then as follows

$$\overset{\rightarrow}{\sum} F_{ix} = 0, \quad \overset{\uparrow}{\sum} F_{iy} = 0, \quad \overset{\curvearrowleft}{\sum} M_{iB} = 0. \tag{2.108}$$

We can also dispense with the force sum in the $y-$direction and instead use two moment sums in relation to the points A and B, which must not be perpendicular to each other:

$$\overset{\rightarrow}{\sum} F_{ix} = 0, \quad \overset{\curvearrowleft}{\sum} M_{iA} = 0, \quad \overset{\curvearrowleft}{\sum} M_{iB} = 0. \tag{2.109}$$

In the same way, we can leave the force sum with regard to the $x-$direction out of consideration and use two moment sums for this, whereby in this case A and B must not be at the same height:

$$\overset{\uparrow}{\sum} F_{iy} = 0, \quad \overset{\curvearrowleft}{\sum} M_{iA} = 0, \quad \overset{\curvearrowleft}{\sum} M_{iB} = 0. \tag{2.110}$$

Finally, three moment equilibrium conditions can also be used with regard to the points A, B, C, whereby A, B and C must not be on a straight line:

$$\overset{\curvearrowleft}{\sum} M_{iA} = 0, \quad \overset{\curvearrowleft}{\sum} M_{iB} = 0, \quad \overset{\curvearrowleft}{\sum} M_{iC} = 0. \tag{2.111}$$

Finally, a note on notation should be made. In many application cases, when considering equilibrium conditions, one will not refer to a coordinate system, but rather distinguish between the horizontal and vertical direction, for example, and also consider the moment equilibrium with respect to one or more points. It is therefore possible to use the following notations for the equilibrium conditions, whereby positive counting directions are again indicated by arrows:

$$\overset{\rightarrow}{\sum} H = 0, \quad \overset{\uparrow}{\sum} V = 0, \quad \overset{\curvearrowleft}{\sum} M_A = 0. \tag{2.112}$$

We will make repeated use of this type of notation throughout the rest of this book.

Example 2.12

A beam of length l (Fig. 2.32) is loaded by two forces F. Let the beam be supported at the two points A and B on two rigid walls. We are looking for the reaction forces at the points A and B.

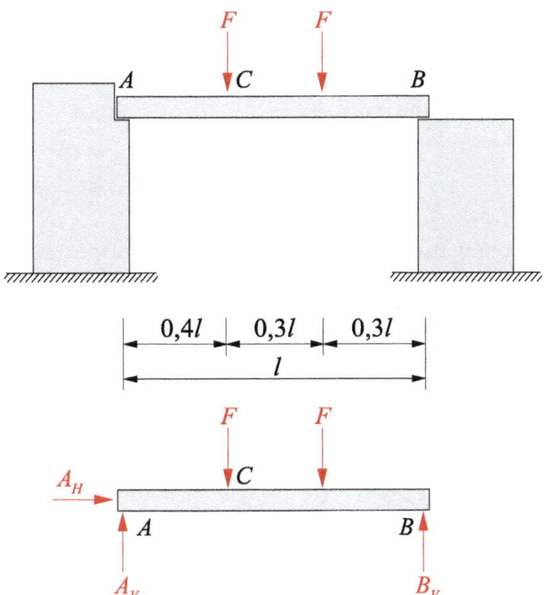

Fig. 2.32 Beam under two forces F

Solution:

We cut the beam free at points A and B (Fig. 2.32, bottom). This releases a horizontal and a vertical reaction force at point A, which we will call A_H and A_V. A vertical reaction force B_V occurs at point B. We now use the equilibrium conditions (2.107) to determine the support reactions:

$$\overset{\rightarrow}{\sum} H = 0: \quad A_H = 0,$$

$$\overset{\uparrow}{\sum} V = 0: \quad A_V + B_V - F - F = 0,$$

$$\overset{\curvearrowleft}{\sum} M_A = 0: \quad B_V \cdot l - F \cdot 0.4l - F \cdot 0.7l = 0. \tag{2.113}$$

Obviously, the first of the above equations immediately results in $A_H = 0$, which is a plausible result: There is no horizontal force acting on this beam, so the horizontal reaction force is identical to zero. In the second equation, both unknown reaction forces A_V and B_V occur, so that we cannot draw any conclusions from this for the time being. The reaction force B_V results from the third equation as:

$$B_V l = 0.4Fl + 0.7Fl \quad \rightarrow \quad B_V = 1.1F. \tag{2.114}$$

We can now use the second equilibrium condition in (2.113) to determine the still unknown reaction force A_V:

$$A_V + 1.1F - F - F = 0: \quad \rightarrow \quad A_V = 0.9F. \tag{2.115}$$

The same result for A_V would be obtained if one were to consider the sum of moments with respect to the point B:

$$\overset{\curvearrowright}{\sum} M_B = 0: \quad A_V \cdot l - F \cdot 0.6l - F \cdot 0.3l = 0 \quad \rightarrow \quad A_V = 0.9F. \tag{2.116}$$

We now use the sum of moments with respect to the arbitrarily chosen point C to check the results. It follows:

$$\overset{\curvearrowright}{\sum} M_C = 0: \quad A_V \cdot 0.4l + F \cdot 0.3l - B_V \cdot 0.6l = 0. \tag{2.117}$$

By inserting A_V and B_V we see that the left-hand side of this equation becomes zero, the equilibrium condition is fulfilled.

◀

Example 2.13

The lever in Fig. 2.33, which is loaded by a force F, is held by a cable between the points A and B. Determine the cable force and the reaction forces at point C.

Solution:

A cable is only capable of absorbing tensile forces so that a cable force occurs in the cable between points A and B that acts parallel to the direction of the cable. The bearing point C is mounted in such a way that both horizontal and vertical displacements of the lever are prevented at this point. As a result, both a horizontal and a vertical reaction force will occur at this bearing point. To solve the problem, we cut the lever free so that both the cable force S and the two reaction forces C_H and C_V are released (Fig. 2.33, bottom). We first calculate the sum of moments with respect to point C and obtain:

$$\overset{\curvearrowleft}{\sum} M_C = 0: \quad S \cdot l - F_0 \cdot 2l = 0 \quad \rightarrow \quad S = 2F_0. \tag{2.118}$$

We also evaluate the force equilibria with regard to the horizontal direction and the vertical direction:

$$\begin{aligned} \overset{\rightarrow}{\sum} H = 0: &\quad C_H - S = 0 \quad \rightarrow \quad C_H = S = 2F_0, \\ \overset{\uparrow}{\sum} V = 0: &\quad C_V - F_0 = 0 \quad \rightarrow \quad C_V = F_0. \end{aligned} \tag{2.119}$$

◀

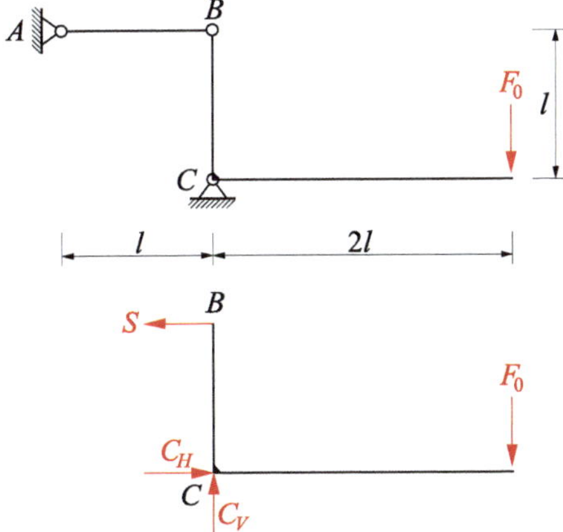

Fig. 2.33 Lever under force F (top), free-body diagram (bottom)

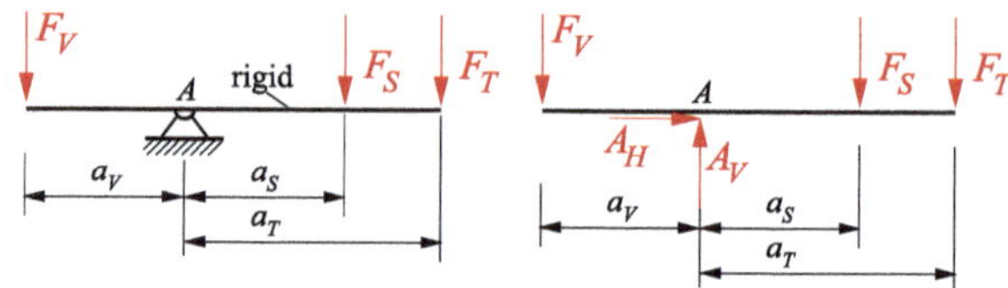

Fig. 2.34 Beam (left), free-body diagram (right)

Example 2.14

Three people with the weight forces F_V, F_S and F_V sit on a beam that is assumed to be rigid (Fig. 2.34). How large must the distance a_V be so that equilibrium prevails for given a_S and a_T? How large are the reaction forces at point A?

Solution:

We cut the beam free at the support point A and thus release the two bearing forces A_H and A_V. The equilibrium conditions then result in:

$$\overset{\rightarrow}{\sum} H = 0: \quad A_H = 0,$$

$$\overset{\uparrow}{\sum} V = 0: \quad A_V - F_V - F_S - F_T = 0 \quad \rightarrow \quad A_V = F_V + F_S + F_T,$$

$$\overset{\curvearrowleft}{\sum} M_A = 0: \quad F_V a_V - F_S a_S - F_T a_T = 0 \quad \rightarrow \quad a_V = \frac{F_S a_S + F_T a_T}{F_V}. \tag{2.120}$$

The horizontal support reaction A_H is zero, which is an obvious result because there are no forces acting in the horizontal direction. The support reaction A_V, on the other hand, simply corresponds to the sum of the weight forces of the persons on the beam. The distance a_V is largely controlled by the weight force F_V: A lighter person will have to position themselves further away from point A on the beam, whereas a heavier person will have to sit closer to A to ensure balance.

◀

2.4 Spatial General Force Systems

2.4.1 Moment Vector

We want to show below that a moment represents a vector, i.e. that it is a directional quantity with a magnitude. If we look again at the relationship (2.89), we see that a resulting moment

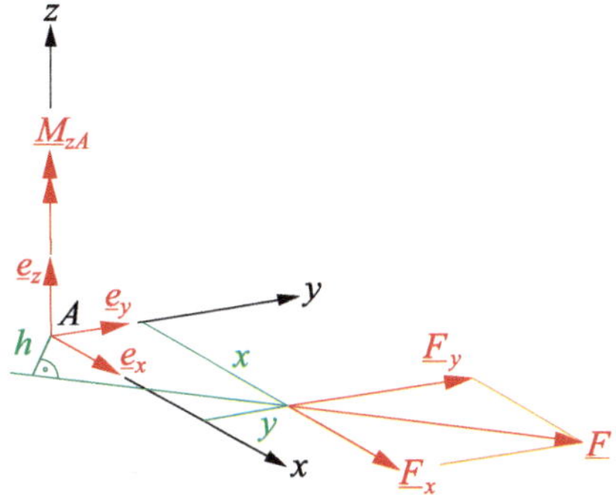

Fig. 2.35 Moment of a force

with respect to the coordinate origin A results from the two components of a resulting force R with the components R_y and R_x, multiplied by the respective lever arms x and y or from the force F with its lever arm h (Fig. 2.35):

$$M_{zA} = Fh = F_y x - F_x y. \tag{2.121}$$

A positive moment rotates in the sense of a clockwise screw around the reference axis, in this case the $z-$axis. It is therefore a moment that rotates around the z-axis, which is expressed by the index z.

A moment is a vector in space that is characterised by its magnitude and direction (i.e. its direction of rotation). A moment vector is represented graphically by a vector arrow with a double point, as shown in Fig. 2.35. The following applies to the moment vector $\underline{M}_{zA}$:

$$\underline{M}_{zA} = M_{zA}\underline{e}_z. \tag{2.122}$$

If there is an arbitrarily orientated moment vector $\underline{M}_A$ in space, it can be broken down into its components as follows:

$$\underline{M}_A = M_{xA}\underline{e}_x + M_{yA}\underline{e}_y + M_{zA}\underline{e}_z. \tag{2.123}$$

The components M_{xA}, M_{yA}, M_{zA} of this moment vector result for a force $\underline{F}$ oriented arbitrarily in space as (Fig. 2.36):

$$M_{xA} = F_z y - F_y z, \quad M_{yA} = F_x z - F_z x, \quad M_{zA} = F_y x - F_x y. \tag{2.124}$$

The magnitude of $\underline{M}_A$ follows as:

$$M_A = \left|\underline{M}_A\right| = \sqrt{M_{xA}^2 + M_{yA}^2 + M_{zA}^2}, \tag{2.125}$$

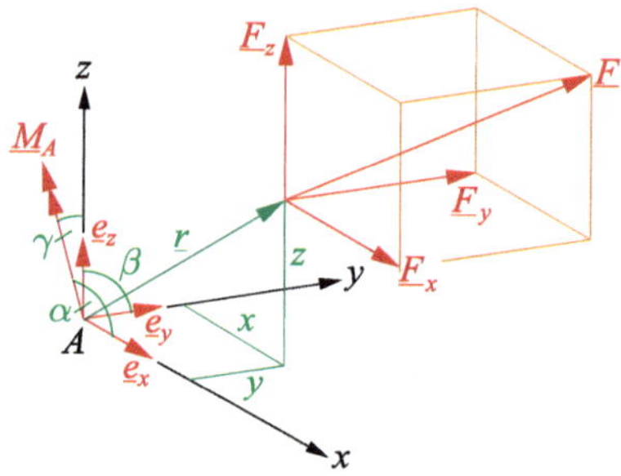

Fig. 2.36 Spatial moment vector

and its direction is defined by the angles α, β and γ as:

$$\cos\alpha = \frac{M_{xA}}{M_A}, \quad \cos\beta = \frac{M_{yA}}{M_A}, \quad \cos\gamma = \frac{M_{zA}}{M_A}. \tag{2.126}$$

It is easy to show that the moment $\underline{M}_A$ can also be represented as a vector product or cross product of the position vector $\underline{r}$ and the force vector $\underline{F}$ and that the representation (2.123) or (2.124) is obtained from this again:

$$\underline{M}_A = \underline{r} \times \underline{F}. \tag{2.127}$$

Accordingly, the moment vector $\underline{M}_A$ is orientated perpendicular to the plane spanned by the position vector $\underline{r}$ and force vector $\underline{F}$ (Fig. 2.37). The magnitude M_A of the moment vector $\underline{M}_A$ is then identical to the area of $\underline{r}$ and $\underline{F}$ of the resulting parallelogram:

$$M_A = Fr\sin\varphi = Fh. \tag{2.128}$$

The moment vector is a so-called free vector; it can be moved arbitrarily along its line of action and parallel to it.

If there are several spatial moment vectors $\underline{M}_i$, then the resulting moment vector $\underline{M}_R$ results from the sum of the moments:

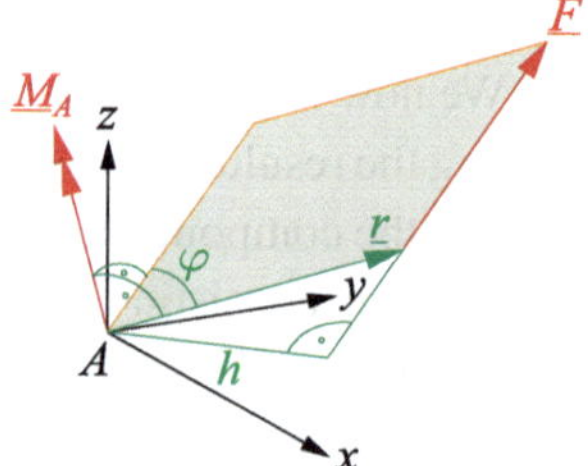

Fig. 2.37 Spatial moment vector

$$\underline{M}_R = \underline{M}_1 + \underline{M}_2 + ... + \underline{M}_n = \sum_{i=1}^{n} \underline{M}_i. \tag{2.129}$$

In a component notation we obtain:

$$M_{Rx} = \sum_{i=1}^{n} M_{ix}, \quad M_{Ry} = \sum_{i=1}^{n} M_{iy}, \quad M_{Rz} = \sum_{i=1}^{n} M_{iz}. \tag{2.130}$$

If the sum of the moments is zero, the resulting moment is identical to zero. The rigid body is then in moment equilibrium and there is no rotational effect. The following then applies:

$$\sum_{i=1}^{n} \underline{M}_i = \underline{0}, \tag{2.131}$$

or:

$$\sum_{i=1}^{n} M_{ix} = 0, \quad \sum_{i=1}^{n} M_{iy} = 0, \quad \sum_{i=1}^{n} M_{iz} = 0. \tag{2.132}$$

2.4.2 Equilibrium Conditions

Any non-central spatial force system consisting of the forces $\underline{F}_1, \underline{F}_2, ..., \underline{F}_n$ can be reduced to a resulting force vector and a moment vector with respect to a point A, analogous to the plane case. If we shift the lines of action of the forces $\underline{F}_1, \underline{F}_2, ..., \underline{F}_n$ so that they run through the point A and take into account the resulting moments $\underline{M}_{1A}, \underline{M}_{2A}, ..., \underline{M}_{nA}$, then a reduction to a resultant $\underline{R}$ and a resulting moment $\underline{M}_R$ can be carried out:

$$\underline{R} = \sum_{i=1}^{n} \underline{F}_i, \quad \underline{M}_R = \sum_{i=1}^{n} \underline{M}_i. \tag{2.133}$$

Analogous to the planar case, as many equilibrium conditions can be formulated in space as there are degrees of freedom. In the spatial case, there are therefore six equilibrium conditions, namely three force equilibrium conditions and three moment equilibrium conditions. For a spatial force group, it therefore applies that the resultant $\underline{R}$ and the resulting moment $\underline{M}_R$ disappear according to (2.133):

$$\sum_{i=1}^{n} \underline{F}_i = \underline{0}, \quad \sum_{i=1}^{n} M_{iA} = \underline{0}. \tag{2.134}$$

This means that the sum of the forces with respect to the $x-$, $y-$ and $z-$directions as well as the sums of moments around these axes with respect to a point A must disappear:

$$\sum_{i=1}^{n} F_{ix} = 0,$$
$$\sum_{i=1}^{n} F_{iy} = 0,$$
$$\sum_{i=1}^{n} F_{iz} = 0,$$
$$\sum_{i=1}^{n} M_{ixA} = 0,$$
$$\sum_{i=1}^{n} M_{iyA} = 0,$$
$$\sum_{i=1}^{n} M_{izA} = 0. \tag{2.135}$$

Example 2.15

Consider a rigid body with the dimensions shown in Fig. 2.38, which is loaded by the forces $F_1, F_2, ..., F_6$. The following applies: $F_1 = 2F_0$, $F_2 = F_0$, $F_3 = 3F_0$, $F_4 = F_0$, $F_5 = 2F_0$, $F_6 = F_0$ for a given F_0. We are looking for the resulting force and the moment with respect to the point A, which lies at the origin of the coordinates.

Solution:

We first determine the components of the resultant as follows:

$$\begin{aligned} R_x &= F_1 - F_2 = 2F_0 - F_0 = F_0, \\ R_y &= F_3 - F_4 = 3F_0 - F_0 = 2F_0, \\ R_z &= F_5 - F_6 = 2F_0 - F_0 = F_0. \end{aligned} \tag{2.136}$$

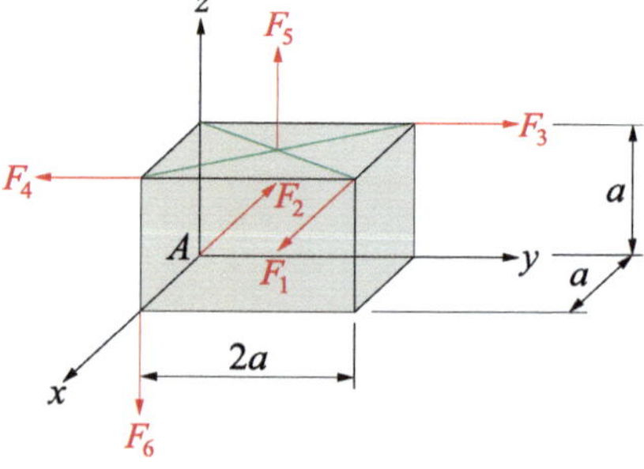

Fig. 2.38 Rigid body under load

The resultant can thus be specified as follows:

$$\underline{R} = R_x \underline{e}_x + R_y \underline{e}_y + R_z \underline{e}_z = \begin{pmatrix} R_x \\ R_y \\ R_z \end{pmatrix} = \begin{pmatrix} 1 \\ 2 \\ 1 \end{pmatrix} F_0. \tag{2.137}$$

Its magnitude is:

$$R = |\underline{R}| = \sqrt{R_x^2 + R_y^2 + R_z^2} = \sqrt{F_0^2 + (2F_0)^2 + F_0^2} = \sqrt{6}F_0. \tag{2.138}$$

The components of the resulting moment $\underline{M}_{RA}$ with respect to the point A (coordinate origin) are as follows:

$$\begin{aligned} M_{RxA} &= \sum_{i=1}^{n} M_{ixA} = -F_3 \cdot a + F_4 \cdot a + F_5 \cdot a = -3F_0 \cdot a + F_0 \cdot a + 2F_0 \cdot a = 0, \\ M_{RyA} &= \sum_{i=1}^{n} M_{iyA} = F_1 \cdot a - F_5 \cdot \frac{a}{2} + F_6 \cdot a = 2F_0 \cdot a - 2F_0 \cdot \frac{a}{2} + F_0 \cdot a = 2F_0 a, \\ M_{RzA} &= \sum_{i=1}^{n} M_{izA} = -F_1 \cdot 2a - F_4 \cdot a = -2F_0 \cdot 2a - F_0 \cdot a = -5F_0 a. \end{aligned} \tag{2.139}$$

This can be represented in vector form as:

$$\underline{M}_{RA} = M_{RxA} \underline{e}_x + M_{RyA} \underline{e}_y + M_{RzA} \underline{e}_z = \begin{pmatrix} M_{RxA} \\ M_{RyA} \\ M_{RzA} \end{pmatrix} = \begin{pmatrix} 0 \\ 2 \\ -5 \end{pmatrix} F_0 a. \tag{2.140}$$

The magnitude of the moment vector is

$$M_{RA} = |\underline{M}_{RA}| = \sqrt{M_{RxA}^2 + M_{RyA}^2 + M_{RzA}^2} = \sqrt{(2F_0 a)^2 + (-5F_0 a)^2} = \sqrt{29}F_0 a. \tag{2.141}$$

◀

3 Centre of Gravity

The concept of centre of gravity is of central importance in many technical applications. In this chapter, we will first look at the centre of gravity of groups of forces as well as surface and line loads before we turn our attention to determining the centres of gravity of bodies, areas and lines.

3.1 Centre of Gravity of a Parallel Force Group

We consider the situation of Fig. 3.1. Let a rigid massless beam carrying three weights with the masses m_1, m_2, m_3 at the points x_1, x_2, x_3 be given. This beam is to be supported at point A to be determined at the still unknown position x_A so that the beam remains in its horizontal position. The weight forces $m_1 g, m_2 g, m_3 g$ (g = acceleration due to gravity) are a group of parallel forces that can be reduced to a resultant force R. To determine the resultant R and the position x_A as well as the resulting support force A_V, we consider the free-body diagram of Fig. 3.1, middle. It is clear that the support position will be found exactly below the resultant R, as shown in Fig. 3.1, bottom.

We determine the magnitude of the support force A_V from the vertical force sum:

$$\overset{\uparrow}{\sum} V = 0: \quad A_V - m_1 g - m_2 g - m_3 g = 0 \quad \rightarrow \quad A_V = \sum_{i=1}^{3} m_i g. \tag{3.1}$$

The position of the support, so that equilibrium is guaranteed, is obtained from the sum of moments with respect to the coordinate origin $x = 0$:

C. Mittelstedt, *Engineering Mechanics 1: Statics*,
https://doi.org/10.1007/978-3-662-71852-0_3

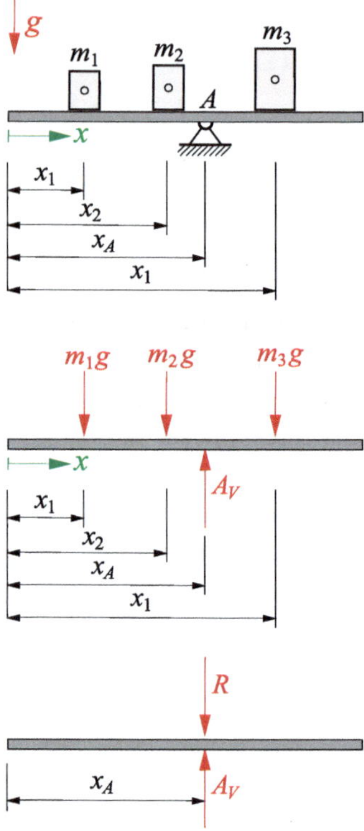

Fig. 3.1 Rigid massless beam with three weights (top), free-body diagram (centre), position of the resultant force (bottom)

$$\overset{\curvearrowleft}{\sum} M = 0: \quad A_V x_A - m_1 g x_1 - m_2 g x_2 - m_3 g x_3 = 0 \quad \rightarrow \quad x_A = \frac{\sum_{i=1}^{3} m_i g x_i}{\sum_{i=1}^{3} m_i g}. \tag{3.2}$$

The point A determined in this way is referred to as the centre of gravity S, wherein the designation x_S is often used for the position of the centre of gravity. In general, the following applies:

$$x_S = \frac{\sum_{i=1}^{n} F_i x_i}{\sum_{i=1}^{n} F_i}, \tag{3.3}$$

if n parallel forces F_1, F_2, ..., F_n are present.

We can proceed in a similar way to determine the resultant R and centre of gravity S of a spatial group of parallel forces (Fig. 3.2). Consider a rigid massless plate on which the

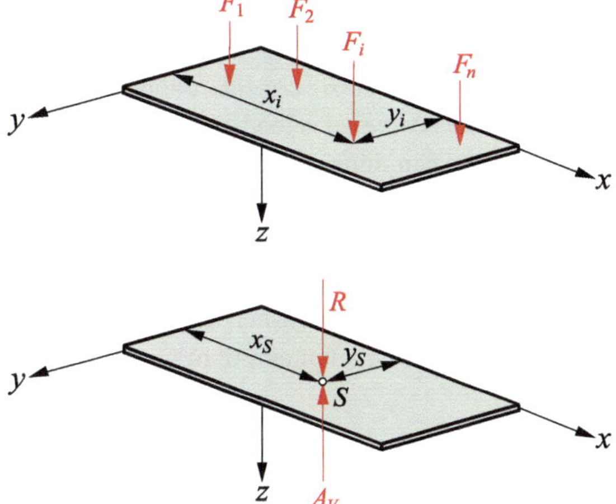

Fig. 3.2 Rigid plate with point forces, position of the centre of gravity

point forces $F_1, F_2, ..., F_i, ..., F_n$ act. The force F_i is applied at x_i and y_i $(i = 1, 2, ..., n)$. We want to determine the support force A_V and the coordinates x_S and y_S. We use the given Cartesian axis system x, y, z as the reference system, with the $x-$axis and the $y-$axis spanning the plate plane. The vertical force sum results in:

$$\overset{\uparrow}{\sum} V = 0: \quad A_V - F_1 - F_2 - ... - F_i - ... - F_n = 0 \quad \rightarrow \quad A_V = \sum_{i=1}^{n} F_i. \quad (3.4)$$

From the two sums of moments relating to the origin of the coordinates around the $x-$axis and the $y-$axis, we obtain

$$\sum M_x = 0: \quad -A_V y_S + F_1 y_1 + F_2 y_2 + ... + F_i y_i + ... + F_n y_n = 0,$$
$$\sum M_y = 0: \quad A_V x_S - F_1 x_1 - F_2 x_2 - ... - F_i x_i - ... - F_n x_n = 0. \quad (3.5)$$

We can solve these two equations for the centre of gravity coordinates x_S and y_S and obtain

$$x_S = \frac{\sum_{i=1}^{n} F_i x_i}{\sum_{i=1}^{n} F_i} = \frac{\sum_{i=1}^{n} F_i x_i}{A_V},$$
$$y_S = \frac{\sum_{i=1}^{n} F_i y_i}{\sum_{i=1}^{n} F_i} = \frac{\sum_{i=1}^{n} F_i y_i}{A_V}. \quad (3.6)$$

3.2 Centre of Gravity of Line Loads and Surface Loads

We can transfer the above considerations to the consideration of line loads. For this purpose, we consider a beam that is loaded by an arbitrary but continuously distributed line load $q(x)$ (Fig. 3.3). We can think of the continuously distributed line load $q(x)$ as an infinite number of infinitesimal individual forces $q\mathrm{d}x$, each acting on a partial length $\mathrm{d}x$ (Fig. 3.3, top). At this point, we can proceed in exactly the same way as in Sect. 3.1 and use Eq. (3.3), whereby we now replace the forces F_1, F_2, ..., F_n by the infinitesimal forces $q\mathrm{d}x$ and x_i by x and also take into account that when summing up an infinite number of infinitesimal forces, the summation turn into an integration:

$$x_S = \frac{\int q(x)x\mathrm{d}x}{\int q(x)\mathrm{d}x}. \tag{3.7}$$

Here, the expression in the denominator of (3.7) represents the resultant of the line load $q(x)$:

$$R = \int q(x)\mathrm{d}x. \tag{3.8}$$

If, for example, the line load acts over the entire length l of a beam, then the above integration must be carried out over the entire length of the beam:

$$x_S = \frac{\int_0^l q(x)x\mathrm{d}x}{\int_0^l q(x)\mathrm{d}x}. \tag{3.9}$$

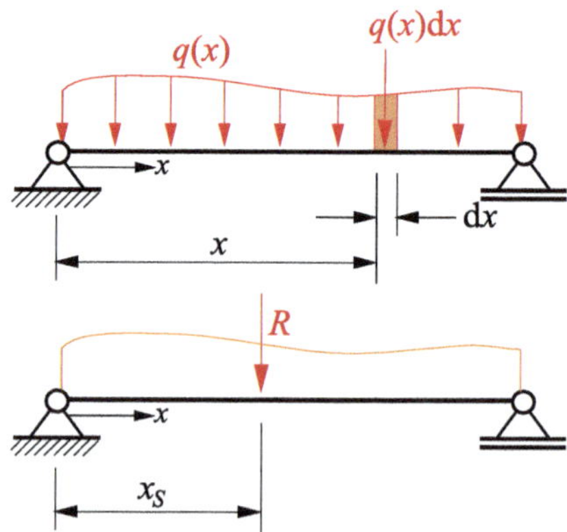

Fig. 3.3 Beam under line load $q(x)$ (top), resultant (bottom)

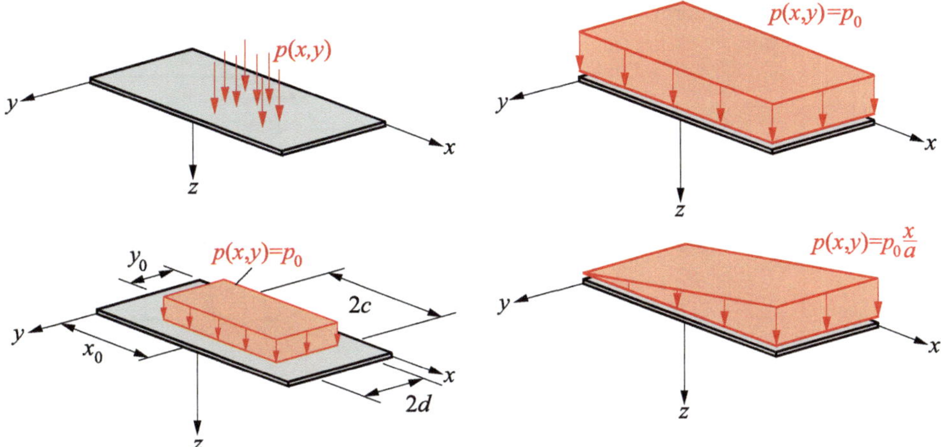

Fig. 3.4 Plate under surface load $p(x, y)$; examples of surface loads

Fig. 3.5 Determination of the centre of gravity S of the surface load $p(x, y)$

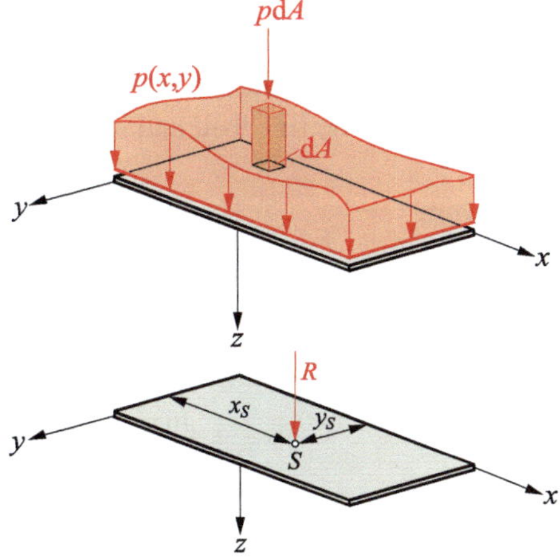

If a plate (i.e. a plane load-bearing structure that is loaded perpendicular to its plane by a surface load $p(x, y)$, Fig. 3.4) is considered, we can proceed in the same way to determine the centre of gravity and the resulting force. The position x_S, y_S of the centre of gravity S (Fig. 3.5) can then be specified as:

$$x_S = \frac{\int p(x, y)x\mathrm{d}A}{\int p(x, y)\mathrm{d}A}, \quad y_S = \frac{\int p(x, y)y\mathrm{d}A}{\int p(x, y)\mathrm{d}A}. \tag{3.10}$$

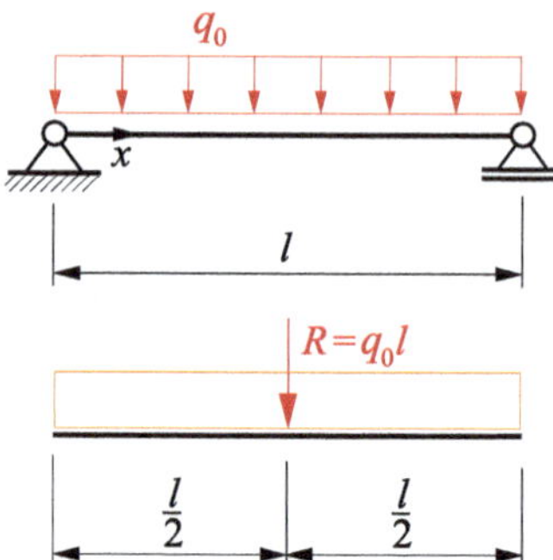

Fig. 3.6 Beam under uniform line load $q(x) = q_0$

Here, $\mathrm{d}A = \mathrm{d}x\mathrm{d}y$ is the surface element over which the integration is to be carried out. In (3.10), the integrations over $\mathrm{d}A$ mean integrations over the entire plate area A spanned by the x−axis and the y−axis.

Example 3.1

Consider a beam under the uniform line load $q(x) = q_0$ (Fig. 3.6). We are looking for the resultant R of the line load and its point of application x_S.

Solution:

We determine the position x_S of the centre of gravity according to (3.7):

$$x_S = \frac{\int_0^l q(x)x\mathrm{d}x}{\int_0^l q(x)\mathrm{d}x} = \frac{q_0 \int_0^l x\mathrm{d}x}{q_0 \int_0^l \mathrm{d}x} = \frac{\frac{1}{2}q_0 l^2}{q_0 l} = \frac{l}{2}. \tag{3.11}$$

The expression in the denominator of (3.11) represents the resultant R of the line load $q(x)$:

$$R = \int_0^l q(x)\mathrm{d}x = q_0 l. \tag{3.12}$$

Its location is shown in Fig. 3.6, bottom.

◀

Example 3.2

Consider a beam under the linear line load $q(x) = q_0 \frac{x}{l}$ (Fig. 3.7). Determine the resultant R of the line load and its point of application x_S.

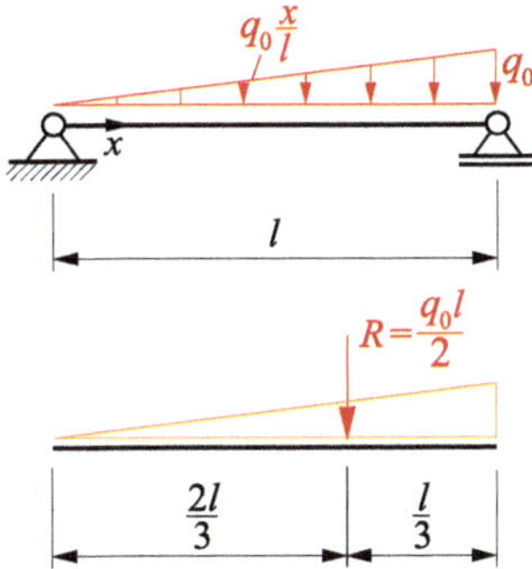

Fig. 3.7 Beam under linear line load $q(x) = q_0\frac{x}{l}$

Solution:

The position x_S of the centre of gravity S of the line load $q(x)$ follows from (3.7):

$$x_S = \frac{\int_0^l q(x)x\mathrm{d}x}{\int_0^l q(x)\mathrm{d}x} = \frac{\frac{q_0}{l}\int_0^l x^2\mathrm{d}x}{\frac{q_0}{l}\int_0^l x\mathrm{d}x} = \frac{\frac{q_0 l^2}{3}}{\frac{q_0 l}{2}} = \frac{2}{3}l. \tag{3.13}$$

The expression $\int_0^l q(x)\mathrm{d}x = \frac{q_0 l}{2}$ is the magnitude of the resultant R of the line load $q(x)$. It is shown in Fig. 3.7, bottom.

◀

3.3 Centre of Gravity of a Solid Body

Consider the three-dimensional body of Fig. 3.8, which has the volume V. Let the body be arbitrary in such a way that the density ρ can be a function of all three spatial directions: $\rho = \rho(x, y, z)$. An infinitesimal volume element $\mathrm{d}V$ then has the infinitesimal mass $\mathrm{d}m = \rho\mathrm{d}V$. We can imagine the body under consideration as an infinite number of such volume elements, and the infinitesimal weight forces $\mathrm{d}G = g\mathrm{d}m$ involved form a parallel group of forces. We can therefore use Eq. (3.3) to determine the position of the resultant, whereby we must note here that the summations turn into integrations when adding up the contributions of an infinite number of volume elements. We introduce an arbitrary coordinate system $\bar{x}, \bar{y}, \bar{z}$ for the calculation, and let the centre of gravity coordinate system be x, y, z. It follows for the centre of gravity coordinate $\bar{x}_S$:

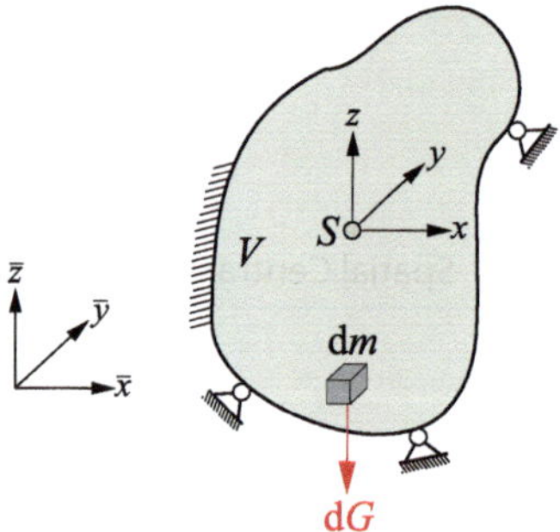

Fig. 3.8 Three-dimensional body with infinitesimal volume element

$$\bar{x}_S = \frac{\int_V \bar{x} \mathrm{d}G}{\int_V \mathrm{d}G}. \tag{3.14}$$

This involves integrating over the entire body volume V. We can make analogous considerations for the two remaining spatial directions and obtain the centre of gravity coordinates $\bar{y}_S$ and $\bar{z}_S$ of the three-dimensional inhomogeneous solid:

$$\bar{y}_S = \frac{\int_V \bar{y} \mathrm{d}G}{\int_V \mathrm{d}G}, \quad \bar{z}_S = \frac{\int_V \bar{z} \mathrm{d}G}{\int_V \mathrm{d}G}. \tag{3.15}$$

If we assume that the acceleration due to gravity g is constant over the body volume V (which is the case for almost all practically relevant technical purposes), then g can be eliminated from the above equations, and the coordinates of the centre of mass are:

$$\bar{x}_M = \frac{\int_V \bar{x} \mathrm{d}m}{\int_V \mathrm{d}m}, \quad \bar{y}_M = \frac{\int_V \bar{y} \mathrm{d}m}{\int_V \mathrm{d}m}, \quad \bar{z}_M = \frac{\int_V \bar{z} \mathrm{d}m}{\int_V \mathrm{d}m}. \tag{3.16}$$

If the density ρ is also constant, then the density ρ can be eliminated from the above equations with $\mathrm{d}m = \rho \mathrm{d}V$, and the centre of gravity of the now homogeneous body is obtained:

$$\bar{x}_V = \frac{\int_V \bar{x} \mathrm{d}V}{\int_V \mathrm{d}V}, \quad \bar{y}_V = \frac{\int_V \bar{y} \mathrm{d}V}{\int_V \mathrm{d}V}, \quad \bar{z}_V = \frac{\int_V \bar{z} \mathrm{d}V}{\int_V \mathrm{d}V}. \tag{3.17}$$

Example 3.3

For the homogeneous cuboid of Fig. 3.9 the position of the centre of gravity S is sought.

Solution:

To determine the centre of gravity S, we introduce a reference system $\bar{x}$, $\bar{y}$, $\bar{z}$ as shown in Fig. 3.9. To determine the centre of gravity coordinate $\bar{x}_V$ according to (3.17), we consider the infinitesimal volume element $\mathrm{d}V$ of Fig. 3.9, centre. With $\mathrm{d}V = hl\mathrm{d}\bar{x}$ we obtain:

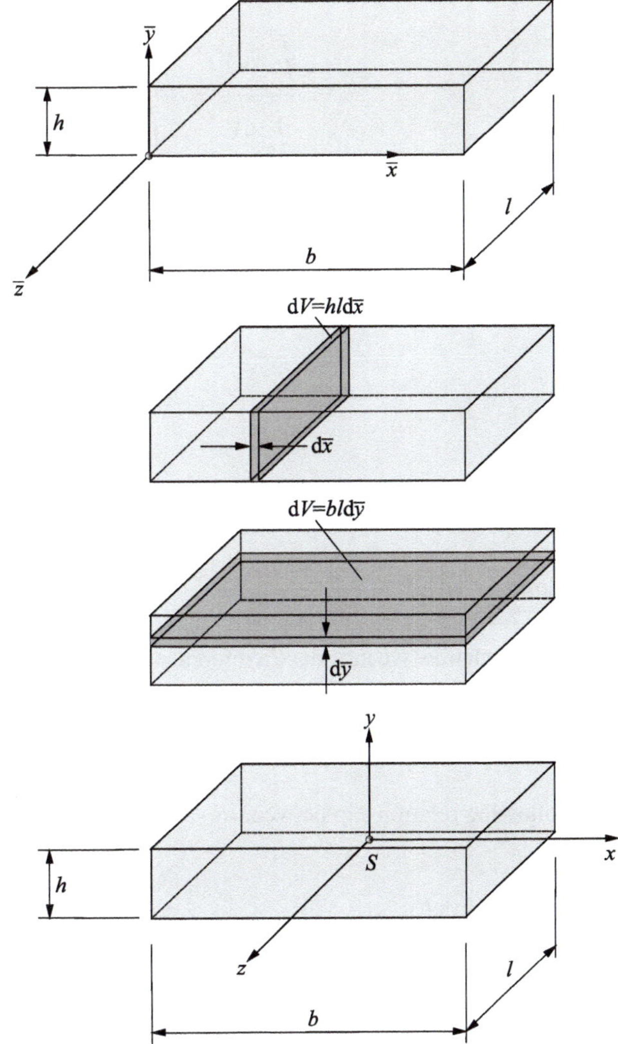

Fig. 3.9 Cuboid (top), infinitesimal volume elements (middle), centre of gravity (bottom)

$$\bar{x}_V = \frac{\int_V \bar{x} \mathrm{d}V}{\int_V \mathrm{d}V} = \frac{\int_0^b \bar{x} hl \mathrm{d}\bar{x}}{\int_0^b hl \mathrm{d}\bar{x}} = \frac{\frac{b^2 hl}{2}}{bhl} = \frac{b}{2}. \tag{3.18}$$

We proceed in the same way to determine y_V and consider the infinitesimal volume element $\mathrm{d}V = bl\mathrm{d}\bar{y}$ as shown in Fig. 3.9:

$$\bar{y}_V = \frac{\int_V \bar{y} \mathrm{d}V}{\int_V \mathrm{d}V} = \frac{\int_0^h \bar{y} bl \mathrm{d}\bar{y}}{\int_0^h bl \mathrm{d}\bar{y}} = \frac{\frac{bh^2 l}{2}}{bhl} = \frac{h}{2}. \tag{3.19}$$

Similarly, $\bar{z}_V = -\frac{l}{2}$ follows. The determination of this result is not shown here. The position of the centre of gravity S with the centre of gravity axis system x, y, z is shown in Fig. 3.9, bottom.

◀

Example 3.4

Determine the position of the centre of gravity S for a homogeneous cone with the radius R_0 (Fig. 3.10).

Solution:

For the calculation, we introduce a reference system $\bar{x}, \bar{y}, \bar{z}$ at the tip of the cone and consider the section element $\mathrm{d}V$ shown in Fig. 3.10, right. The following applies:

$$\mathrm{d}V = \pi r_0^2 \mathrm{d}\bar{x}. \tag{3.20}$$

The following relationship between R_0 and r_0 holds:

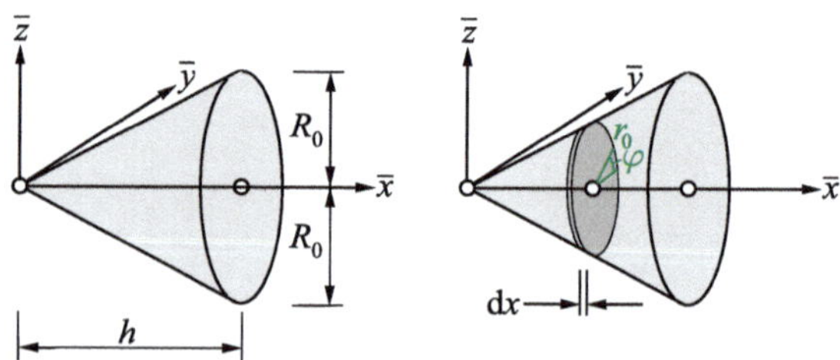

Fig. 3.10 Cone (left), infinitesimal sectional element (right)

$$r_0 = \frac{\bar{x}}{h} R_0. \tag{3.21}$$

Thus for $\mathrm{d}V$ follows:

$$\mathrm{d}V = \pi \frac{R_0^2}{h^2} \bar{x}^2 \mathrm{d}\bar{x}. \tag{3.22}$$

This allows the centre of gravity coordinate $\bar{x}_V$ to be determined as:

$$\bar{x}_V = \frac{\int_V \bar{x} \mathrm{d}V}{\int_V \mathrm{d}V} = \frac{\pi \frac{R_0^2}{h^2} \int_0^h \bar{x}^3 \mathrm{d}\bar{x}}{\pi \frac{R_0^2}{h^2} \int_0^h \bar{x}^2 \mathrm{d}\bar{x}} = \frac{3}{4} h. \tag{3.23}$$

The centre of gravity coordinates $\bar{y}_V$ and $\bar{z}_V$ follow without calculation due to the rotational symmetry of the cone:

$$\bar{y}_V = \bar{z}_V = 0. \tag{3.24}$$

◀

3.4 Centre of Gravity of an Area

We now consider the special case that the three-dimensional solid under consideration is very thin with respect to one coordinate direction (here the z-direction) and has a constant thickness h, which is very small compared to the dimensions with respect to x and y. We refer to such a body as a disk. Then the infinitesimal volume element $\mathrm{d}V$ can be written as $\mathrm{d}V = h\mathrm{d}A$ (Fig. 3.11). If such a disk is given, then the thickness h can be cancelled out from (3.17) and the coordinates of the area's centre of gravity are obtained as:

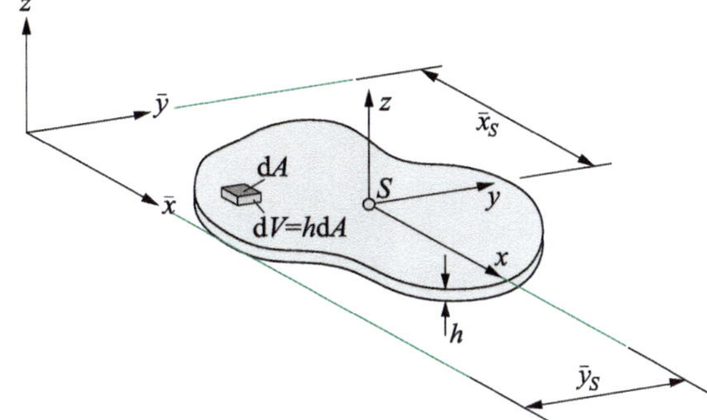

Fig. 3.11 Disk

$$\bar{x}_S = \frac{\int_A \bar{x} \mathrm{d}A}{\int_A \mathrm{d}A}, \quad \bar{y}_S = \frac{\int_A \bar{y} \mathrm{d}A}{\int_A \mathrm{d}A}, \tag{3.25}$$

whereby we assume that the centre of gravity is located exactly in the centre plane of the very thin disk under consideration. The expression

$$\int_A \mathrm{d}A = A \tag{3.26}$$

represents the area of the disk with respect to the xy−plane. The expressions

$$S_y = \int_A \bar{x} \mathrm{d}A = \bar{x}_S A, \quad S_x = \int_A \bar{y} \mathrm{d}A = \bar{y}_S A \tag{3.27}$$

are the so-called static moments or first moments of area of the area under consideration. According to the above definition, these disappear if the origin of the reference system x, y is located in the centre of gravity S of the area under consideration.

It is easy to show that in an area that has an axis of symmetry, the centre of gravity S lies exactly on this axis of symmetry. We consider the area of Fig. 3.12, where the y−axis is a symmetry axis. For each area element $\mathrm{d}A$ at a point x there is then an identical area element $\mathrm{d}A$ at the point $-x$, so that these are ultimately cancelled out in the Eq. (3.25) with regard to the calculation of x_S, the static moment S_y is identically zero:

$$S_y = \int_A x \mathrm{d}A = 0 \quad \rightarrow \quad x_S = \frac{S_y}{A} = 0. \tag{3.28}$$

The centre of gravity coordinate x_S is therefore identical to zero in such a case, and the centre of gravity is located on the y−axis. Symmetry axes of an area are therefore always centre of gravity axes.

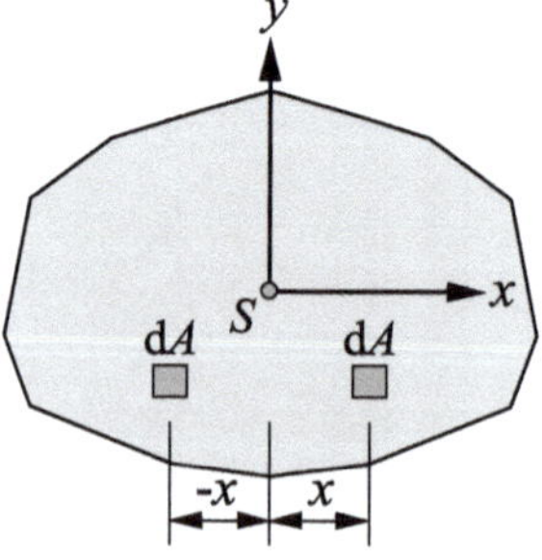

Fig. 3.12 Single symmetric surface

Example 3.5

For the rectangular area (width b, height h) of Fig. 3.13 the position of the centre of gravity S is searched for.

Solution:

We introduce a reference system $\bar{x}$, $\bar{y}$ as shown in Fig. 3.13 and determine the coordinates of the centre of gravity of the rectangular area according to (3.25):

$$\bar{x}_S = \frac{\int_A \bar{x}\mathrm{d}A}{\int_A \mathrm{d}A}, \quad \bar{y}_S = \frac{\int_A \bar{y}\mathrm{d}A}{\int_A \mathrm{d}A}. \tag{3.29}$$

For x_S we obtain with the area element $\mathrm{d}A = h\mathrm{d}\bar{x}$:

$$\bar{x}_S = \frac{h\int_0^b \bar{x}\mathrm{d}\bar{x}}{h\int_0^b \mathrm{d}\bar{x}} = \frac{\frac{1}{2}b^2h}{bh} = \frac{b}{2}. \tag{3.30}$$

For $\bar{y}_S$ we obtain with $\mathrm{d}A = b\mathrm{d}\bar{y}$:

$$\bar{y}_S = \frac{b\int_0^h \bar{y}\mathrm{d}\bar{y}}{b\int_0^h \mathrm{d}\bar{y}} = \frac{\frac{1}{2}bh^2}{bh} = \frac{h}{2}. \tag{3.31}$$

◀

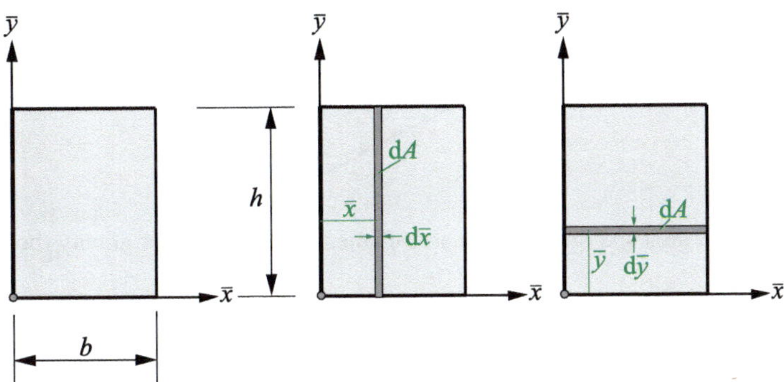

Fig. 3.13 Rectangular surface (left), infinitesimal area elements (right)

Example 3.6

For the trianglular area of Fig. 3.14 with the width b and the height h, the centre of gravity S is to be determined.
Solution:

We first determine the centre of gravity coordinate $\bar{x}_S$ according to

$$\bar{x}_S = \frac{\int_A \bar{x} \mathrm{d}A}{\int_A \mathrm{d}A}. \tag{3.32}$$

The infinitesimal area element dA is shown in Fig. 3.14, top left, and can be given as:

$$\mathrm{d}A = h(\bar{x})\mathrm{d}\bar{x}. \tag{3.33}$$

The height $h(\bar{x})$ is:

$$h(\bar{x})\bar{x}\frac{h}{b}. \tag{3.34}$$

Equation (3.32) can thus be evaluated, and we obtain:

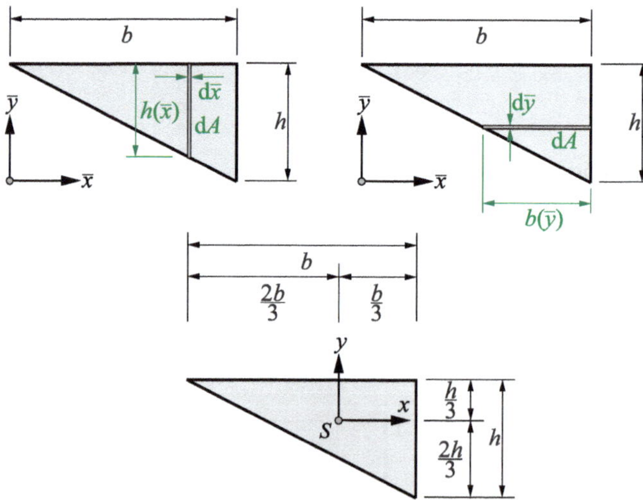

Fig. 3.14 Triangular area and infinitesimal area elements (top), centre of gravity (bottom)

$$\bar{x}_S = \frac{\frac{h}{b}\int_0^b \bar{x}^2 \mathrm{d}\bar{x}}{\frac{h}{b}\int_0^b \bar{x}\mathrm{d}\bar{x}} = \frac{\frac{1}{3}b^3}{\frac{1}{2}b^2} = \frac{2}{3}b. \tag{3.35}$$

We can proceed in the same way to determine $\bar{y}_S$. The following applies:

$$\bar{y}_S = \frac{\int_A \bar{y}\mathrm{d}A}{\int_A \mathrm{d}A}. \tag{3.36}$$

With the area element $\mathrm{d}A = b(\bar{y})\mathrm{d}\bar{y} = \frac{b}{h}\bar{y}\mathrm{d}\bar{y}$ follows:

$$\bar{y}_S = \frac{\frac{b}{h}\int_0^h \bar{y}^2 \mathrm{d}\bar{y}}{\frac{b}{h}\int_0^h \bar{y}\mathrm{d}\bar{y}} = \frac{\frac{1}{3}h^3}{\frac{1}{2}h^2} = \frac{2}{3}h. \tag{3.37}$$

The centre of gravity for the triangular area is shown in Fig. 3.14, bottom.

◀

Example 3.7

For the quarter circle area of Fig. 3.15 with the radius R the centre of gravity S is to be determined.

Solution:

We first determine the centre of gravity coordinate $\bar{x}_S$ using a vertical area element $\mathrm{d}A$ as shown in Fig. 3.15, middle. The area element $\mathrm{d}A$ results as:

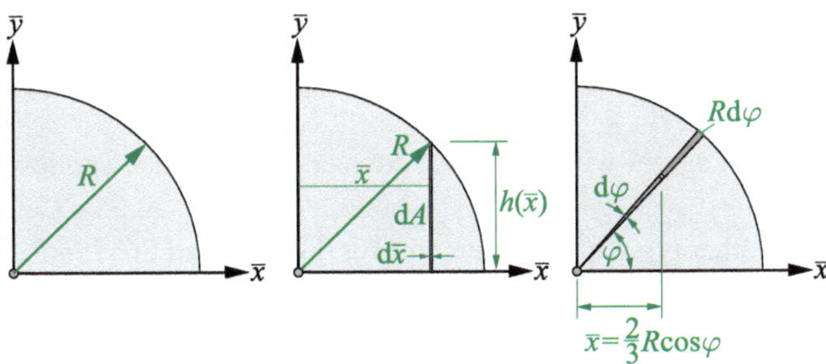

Fig. 3.15 Quarter circle surface and infinitesimal area elements

$$\mathrm{d}A = h(\bar{x})\mathrm{d}\bar{x}. \tag{3.38}$$

Using Pythagoras' theorem, we can express the height $h(\bar{x})$ of the area element as:

$$h = \sqrt{R^2 - \bar{x}^2}. \tag{3.39}$$

We then obtain the centre of gravity coordinate $\bar{x}_S$:

$$\begin{aligned}\bar{x}_S &= \frac{\displaystyle\int_A \bar{x}\mathrm{d}A}{\displaystyle\int_A \mathrm{d}A} = \frac{\displaystyle\int_0^R \bar{x}\sqrt{R^2 - \bar{x}^2}\mathrm{d}\bar{x}}{\displaystyle\int_0^R \sqrt{R^2 - \bar{x}^2}\mathrm{d}\bar{x}} \\ &= \frac{\frac{1}{3}\left.\left(R^2 - \bar{x}^2\right)^{\frac{3}{2}}\right|_0^R}{\frac{1}{2}\left.\left[\bar{x}\sqrt{R^2 - \bar{x}^2} + R^2 \arcsin\left(\frac{\bar{x}}{R}\right)\right]\right|_0^R} = \frac{\frac{1}{3}R^3}{\frac{\pi R^2}{4}} = \frac{4R}{3\pi}.\end{aligned} \tag{3.40}$$

For reasons of symmetry, the following applies:

$$\bar{y}_S = \frac{4R}{3\pi}. \tag{3.41}$$

An alternative way of calculation is to consider a circular area segment as shown in Fig. 3.15, right. The area of the element $\mathrm{d}A$ with the infinitesimal opening angle $\mathrm{d}\varphi$ is

$$\mathrm{d}A = \frac{1}{2}R^2\mathrm{d}\varphi, \tag{3.42}$$

whereby a triangular shape of the area element was assumed here. With the coordinate $\bar{x} = \frac{2}{3}R\cos\varphi$, the centre of gravity coordinate is $\bar{x}_S$:

$$\bar{x}_S = \frac{\displaystyle\int_A \bar{x}\mathrm{d}A}{\displaystyle\int_A \mathrm{d}A} = \frac{\displaystyle\int_0^{\frac{\pi}{2}} \frac{2}{3}R\cos\varphi \cdot \frac{1}{2}R^2\mathrm{d}\varphi}{\displaystyle\int_0^{\frac{\pi}{2}} \frac{1}{2}R^2\mathrm{d}\varphi} = \frac{\frac{1}{3}R^3}{\frac{\pi R^2}{4}} = \frac{4R}{3\pi}. \tag{3.43}$$

Apparently, this result agrees with (3.40). Of course, $\bar{y}_S = \frac{4R}{3\pi}$ also applies in this representation.

We can also use other types of area elements for this task, but this is not described further here.

◀

Example 3.8

For the area of Fig. 3.16 in the form of a quadratic parabola $\bar{y} = \frac{h}{a^2}\bar{x}^2$ the centre of gravity is to be determined.

Solution:

For reasons of symmetry, $\bar{x}_S = 0$, i.e. the centre of gravity is located on the $\bar{y}-$axis. To determine the centre of gravity coordinate $\bar{y}_S$, let us consider the area element $\mathrm{d}A = 2\bar{x}\mathrm{d}\bar{y}$ shown in Fig. 3.16. From the parabolic equation $\bar{y} = \frac{h}{a^2}\bar{x}^2$ we also obtain $\bar{x} = \sqrt{\frac{\bar{a}^2}{h}}$, so that we can determine $\bar{y}_S$ as follows:

$$\bar{y}_S = \frac{\int_A \bar{y}\mathrm{d}A}{\int_A \mathrm{d}A} = \frac{\int_0^h \bar{y}\cdot 2\sqrt{\frac{\bar{y}a^2}{h}}\mathrm{d}\bar{y}}{\int_0^h 2\sqrt{\frac{\bar{y}a^2}{h}}\mathrm{d}\bar{y}} = \frac{2\sqrt{\frac{a^2}{h}}\int_0^h \bar{y}^{\frac{3}{2}}\mathrm{d}\bar{y}}{2\sqrt{\frac{a^2}{h}}\int_0^h \bar{y}^{\frac{1}{2}}\mathrm{d}\bar{y}} = \frac{\frac{2}{5}\bar{y}^{\frac{5}{2}}\Big|_0^h}{\frac{2}{3}\bar{y}^{\frac{3}{2}}\Big|_0^h} = \frac{3}{5}h. \tag{3.44}$$

◀

Figure 3.17 contains a selection of areas and the position of the centres of gravity.

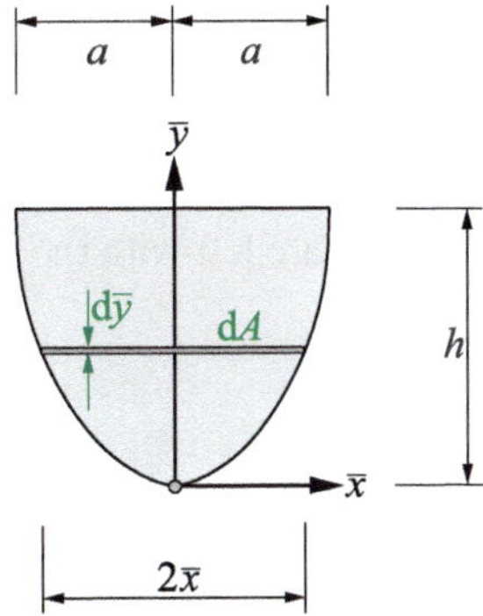

Fig. 3.16 Parabolic area and infinitesimal area element

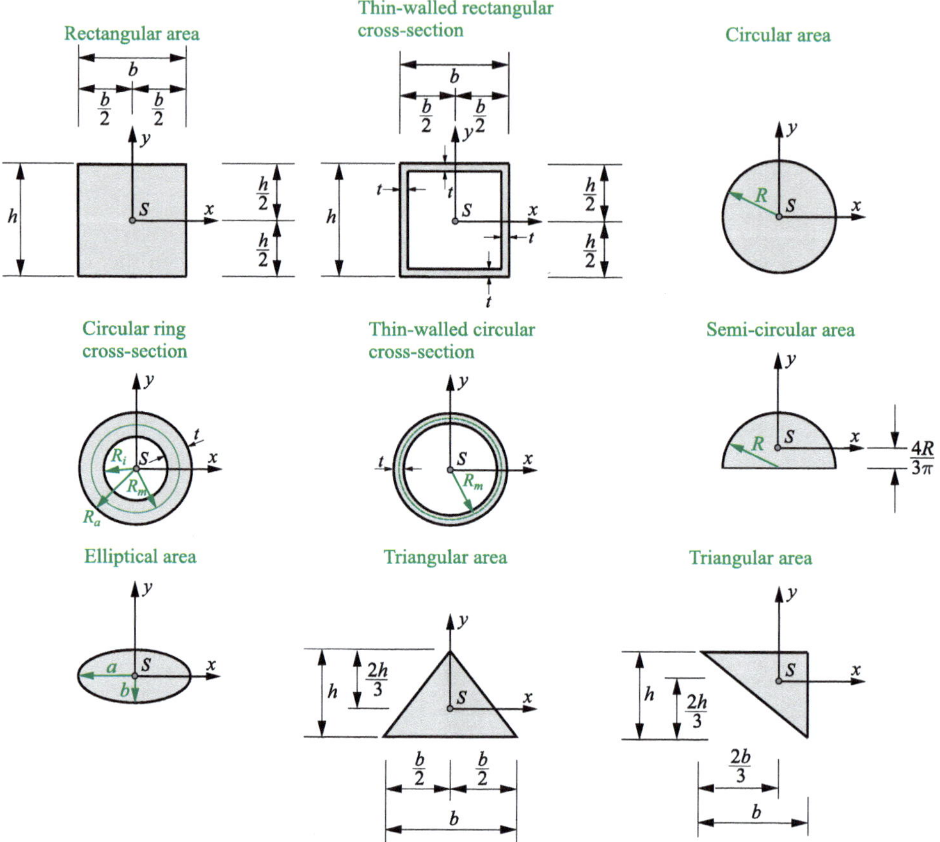

Fig. 3.17 Centres of gravity of selected surfaces

3.5 Line Centre of Gravity

If we consider a homogeneous body that is distributed along a line with the length l and the cross-section A (Fig. 3.18), then the volume of this body is $V = Al$. The infinitesimal volume element $\mathrm{d}V$ can then be written as $\mathrm{d}V = A\mathrm{d}l$, where $\mathrm{d}l$ is an infinitesimal line element. We can therefore eliminate A from Eq. (3.17) for the three-dimensional body A, and we are left with the line centre of gravity:

$$\bar{x}_L = \frac{\int_l \bar{x}\mathrm{d}l}{\int_l \mathrm{d}l}, \quad \bar{y}_L = \frac{\int_l \bar{y}\mathrm{d}l}{\int_l \mathrm{d}l}, \quad \bar{z}_L = \frac{\int_l \bar{z}\mathrm{d}l}{\int_l \mathrm{d}l}. \tag{3.45}$$

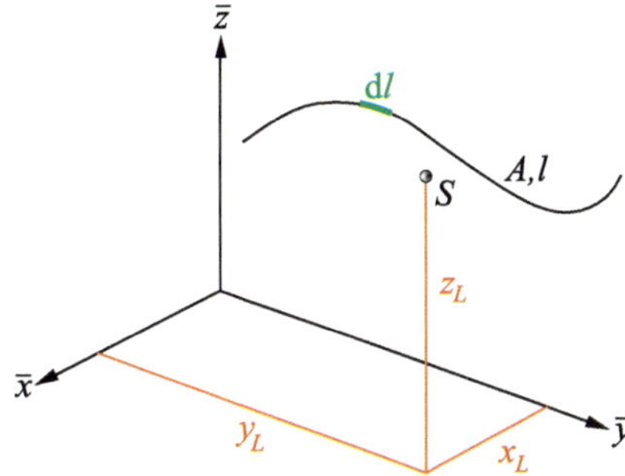

Fig. 3.18 Body distributed along a line

The expression

$$\int_l \mathrm{d}l = l \tag{3.46}$$

is the length l of the line under consideration. The line centre of gravity is generally not on the line under consideration.

Example 3.9

For the quarter circle arc of Fig. 3.19 with the radius R the line centre of gravity is to be determined.

Solution:

We determine the position x_l of the centre of gravity according to (3.45):

$$\bar{x}_L = \frac{\int_l \bar{x}\mathrm{d}l}{\int_l \mathrm{d}l}. \tag{3.47}$$

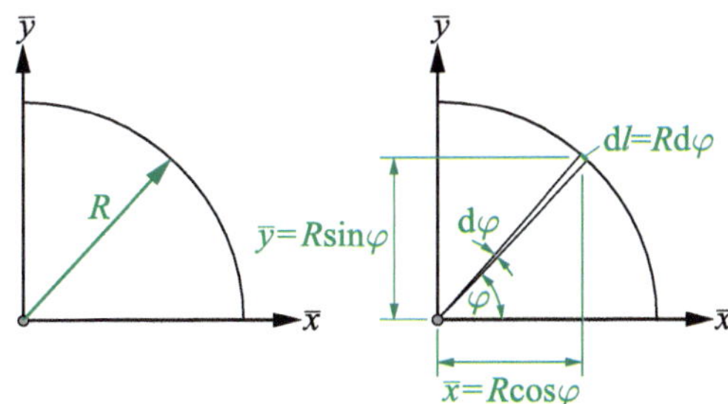

Fig. 3.19 Quarter circle arc and infinitesimal element

With $\bar{x} = R\cos\varphi$ and the infinitesimal line element $\mathrm{d}l = R\mathrm{d}\varphi$ follows:

$$\bar{x}_L = \frac{\int_0^{\frac{\pi}{2}} R\cos\varphi \cdot R\mathrm{d}\varphi}{\int_0^{\frac{\pi}{2}} R\mathrm{d}\varphi} = \frac{R^2 \sin\varphi|_0^{\frac{\pi}{2}}}{R\varphi|_0^{\frac{\pi}{2}}} = \frac{2R}{\pi}. \tag{3.48}$$

For the centre of gravity coordinate $\bar{y}_L$ follows with $\bar{y} = R\sin\varphi$:

$$\bar{y}_L = \frac{\int_l \bar{y}\mathrm{d}l}{\int_l \mathrm{d}l} = \frac{\int_0^{\frac{\pi}{2}} R\sin\varphi \cdot R\mathrm{d}\varphi}{\int_0^{\frac{\pi}{2}} R\mathrm{d}\varphi} = -\frac{R^2 \cos\varphi|_0^{\frac{\pi}{2}}}{R\varphi|_0^{\frac{\pi}{2}}} = \frac{2R}{\pi}. \tag{3.49}$$

◀

3.6 Composite Bodies

We consider a three-dimensional solid which is now composed of a number n of sub-solids V_i $(i = 1, 2, 3, \ldots, n)$, whereby the densities ρ_i of these sub-solids are constant in each case. The centre of gravity positions $\bar{x}_i$ and $\bar{y}_i$ of the sub-bodies are known. Such a composite body is shown as an example in Fig. 3.20. In such a case, the centre of gravity can be determined as follows. The mass m of the body is made up of the masses of the sub-bodies, i.e.:

$$m = \int_V \mathrm{d}m = \int_{V_1} \mathrm{d}m + \int_{V_2} \mathrm{d}m + \ldots + \int_{V_n} \mathrm{d}m = m_1 + m_2 + \ldots + m_n = \sum_{i=1}^{n} m_i, \tag{3.50}$$

where $m_1, m_2, \ldots, m_n$ are the masses of the sub-bodies. The $\bar{x}$−coordinate $\bar{x}_i$ of the centre of mass of sub-body i is calculated as:

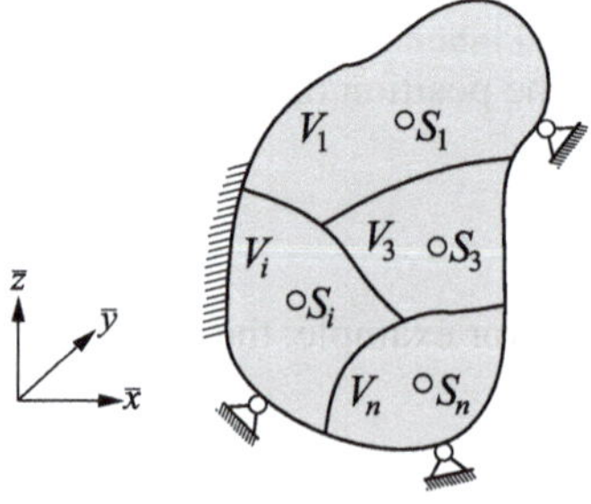

Fig. 3.20 Body composed of n sub-bodies V_i

$$\bar{x}_i = \frac{\int_{V_i} \bar{x} \mathrm{d}m}{\int_{V_i} \mathrm{d}m} \quad \rightarrow \quad \int_{V_i} \bar{x} \mathrm{d}m = \bar{x}_i m_i. \tag{3.51}$$

This results in the position of the centre of mass:

$$\bar{x}_M = \frac{\int_V \bar{x} \mathrm{d}m}{\int_V \mathrm{d}m} = \frac{1}{m} \left(\int_{V_1} \bar{x} \mathrm{d}m + \int_{V_2} \bar{x} \mathrm{d}m + \ldots + \int_{V_n} \bar{x} \mathrm{d}m \right) = \frac{\sum_{i=1}^{n} \bar{x}_i m_i}{\sum_{i=1}^{n} m_i}. \tag{3.52}$$

Similar results are obtained for the coordinates $\bar{y}_M$ and $\bar{z}_M$:

$$\bar{y}_M = \frac{\sum_{i=1}^{n} \bar{y}_i m_i}{\sum_{i=1}^{n} m_i}, \quad \bar{z}_M = \frac{\sum_{i=1}^{n} \bar{z}_i m_i}{\sum_{i=1}^{n} m_i}. \tag{3.53}$$

If the density is constant throughout the body under consideration, then the centre of gravity of the volume follows as:

$$\bar{x}_V = \frac{\sum_{i=1}^{n} \bar{x}_i V_i}{\sum_{i=1}^{n} V_i}, \quad \bar{y}_V = \frac{\sum_{i=1}^{n} \bar{y}_i V_i}{\sum_{i=1}^{n} V_i}, \quad \bar{z}_V = \frac{\sum_{i=1}^{n} \bar{z}_i V_i}{\sum_{i=1}^{n} V_i}, \tag{3.54}$$

where V_i is the volume of sub-body i. This also allows sections such as voids or holes to be taken into account by assigning negative volumes to the corresponding 'sub-bodies'.

Example 3.10

From the cuboid shown in Fig. 3.21 with the edge lengths $2a$, $3a$, $2a$, a cube with the edge length a was cut out at the position shown. Determine the position of the centre of gravity.

Solution:

We interpret the given structure as a cuboid with the volume $V_1 = 2a \cdot 3a \cdot 2a = 12a^3$ with a hole whose volume V_2 we assume to be negative with $V_2 = -a^3$. The centre of gravity positions of the two components of the structure under consideration are

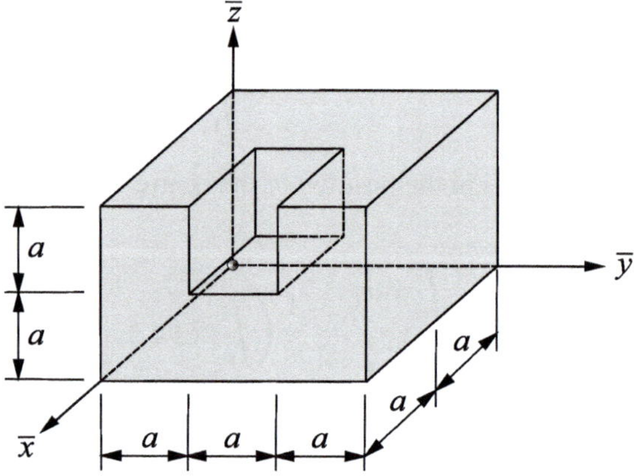

Fig. 3.21 Cuboid under consideration

$$\bar{x}_1 = a, \quad \bar{x}_2 = \frac{3}{2}a,$$
$$\bar{y}_1 = \frac{3}{2}a, \quad \bar{y}_2 = \frac{3}{2}a,$$
$$\bar{z}_1 = a, \quad \bar{z}_2 = \frac{3}{2}a. \tag{3.55}$$

We then obtain the coordinates of the centre of gravity as:

$$\bar{x}_V = \frac{\sum_{i=1}^{2} \bar{x}_i V_i}{\sum_{i=1}^{2} V_i} = \frac{a \cdot 12a^3 - \frac{3}{2}a \cdot a^3}{12a^3 - a^3} = \frac{21}{22}a,$$
$$\bar{y}_V = \frac{\sum_{i=1}^{2} \bar{y}_i V_i}{\sum_{i=1}^{2} V_i} = \frac{\frac{3}{2}a \cdot 12a^3 - \frac{3}{2}a \cdot a^3}{12a^3 - a^3} = \frac{3}{2}a,$$
$$\bar{z}_V = \frac{\sum_{i=1}^{2} \bar{z}_i V_i}{\sum_{i=1}^{2} V_i} = \frac{a \cdot 12a^3 - \frac{3}{2}a \cdot a^3}{12a^3 - a^3} = \frac{21}{22}a. \tag{3.56}$$

◀

3.7 Composite Areas

Analogous to the previous section, areas can also be treated that consist of a number n of partial areas whose centre of gravity coordinates are known. Then the following applies to the centre of gravity of the composite area:

$$\bar{x}_S = \frac{\sum_{i=1}^{n} \bar{x}_i A_i}{\sum_{i=1}^{n} A_i}, \quad \bar{y}_S = \frac{\sum_{i=1}^{n} \bar{y}_i A_i}{\sum_{i=1}^{n} A_i}. \tag{3.57}$$

Here, A_i is the area of the partial area i. In these equations, areas with cut-outs or holes can also be taken into account by treating the corresponding cut-outs as negative areas.

Example 3.11

For the area of Fig. 3.22 the position of the centre of gravity is to be determined.

Solution:

For the calculation, we divide the given area into two partial areas $A_1 = 3a^2$ and $A_2 = 3a^2$ as shown in Fig. 3.22, right. We introduce a coordinate system $\bar{x}$ and $\bar{y}$ as shown and can directly deduce $\bar{x}_S = 0$ from the symmetry of the area, so that the centre of gravity lies on the $\bar{y}$−axis. The positions $\bar{y}_1$ and $\bar{y}_2$ can be deduced as $\bar{y}_1 = \frac{7}{2}a$ and $\bar{y}_2 = \frac{3}{2}a$. The position of the centre of gravity is then given by

$$\bar{x}_S = 0, \quad \bar{y}_S = \frac{\bar{y}_1 A_1 + \bar{y}_2 A_2}{A_1 + A_2} = \frac{\frac{7}{2}a \cdot 3a^2 + \frac{3}{2}a \cdot 3a^2}{3a^2 + 3a^2} = \frac{5}{2}a. \tag{3.58}$$

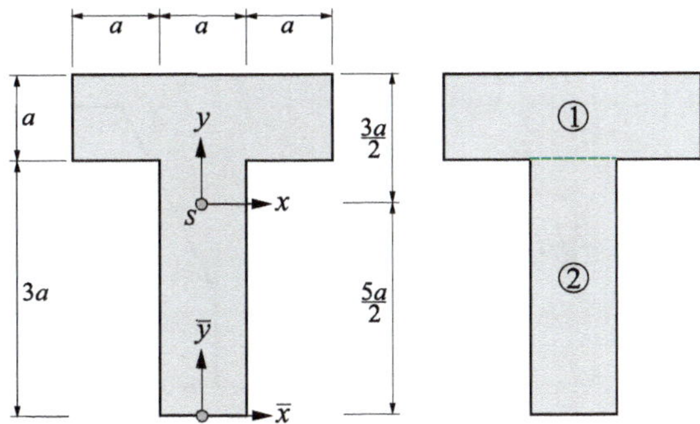

Fig. 3.22 Given area

The centre of gravity S along with the centre of gravity coordinate system x, y is shown in Fig. 3.22.

◀

Example 3.12

For the area of Fig. 3.23 the centre of gravity is to be determined.

Solution:

We divide the area into four sub-areas $A_1 = 80a^2$, $A_2 = 16a^2$, $A_3 = 16a^2$, $A_4 = 48a^2$ and define a reference system $\bar{x}$, $\bar{y}$ as shown in Fig. 3.23. Due to the symmetry of the area, $\bar{x}_S = 0$ follows immediately, so that only $\bar{y}_S$ needs to be determined. With the coordinates $\bar{y}_1 = 14a$, $\bar{y}_2 = \frac{28}{3}a$, $\bar{y}_3 = \frac{28}{3}a$, $\bar{y}_4 = 6a$, the positions of the centre of gravity are:

$$
\begin{aligned}
\bar{x}_S &= 0, \\
\bar{y}_S &= \frac{\bar{y}_1 A_1 + \bar{y}_2 A_2 + \bar{y}_3 A_3 + \bar{y}_4 A_4}{A_1 + A_2 + A_3 + A_4} \\
&= \frac{14a \cdot 80a^2 + \frac{28}{3}a \cdot 16a^2 + \frac{28}{3}a \cdot 16a^2 + 6a \cdot 48a^2}{80a^2 + 16a^2 + 16a^2 + 48a^2} = \frac{32}{3}a. \qquad (3.59)
\end{aligned}
$$

The centre of gravity S with the centre of gravity coordinate system x, y is depicted in Fig. 3.23.

◀

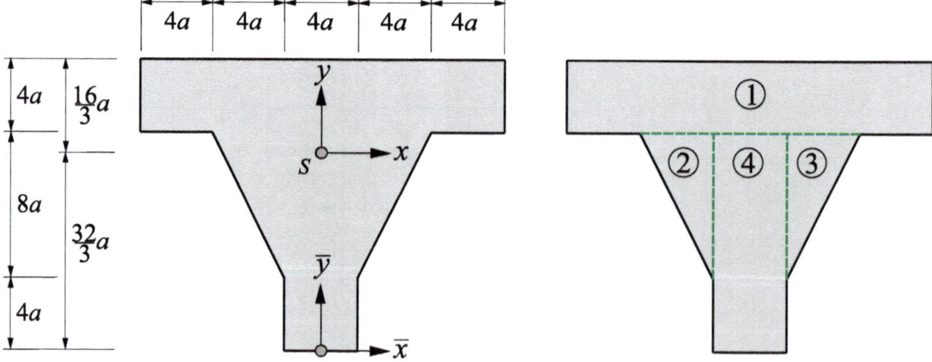

Fig. 3.23 Given area

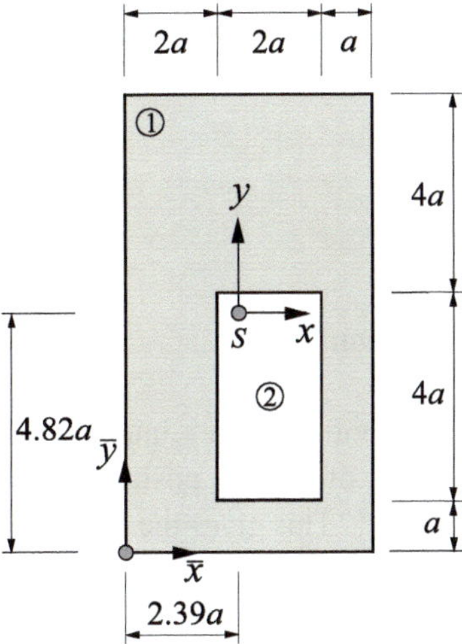

Fig. 3.24 Given area

Example 3.13

For the area of Fig. 3.24 the centre of gravity is to be determined.

Solution:

The area A_1 is $A_1 = 5a \cdot 9a = 45a^2$. The hole with the area $A_2 = -2a \cdot 4a = -8a^2$ must be subtracted from this. The centre of gravity positions of the two partial areas are $\bar{x}_1 = \frac{5}{2}a$, $\bar{x}_2 = 3a$ and $\bar{y}_1 = \frac{9}{2}a$, $\bar{y}_2 = 3a$. This allows the centre of gravity coordinates $\bar{x}_S$, $\bar{y}_S$ to be determined as:

$$\bar{x}_S = \frac{\bar{x}_1 A_1 + \bar{x}_2 A_2}{A_1 + A_2} = \frac{\dfrac{5}{2}a \cdot 45a^2 - 3a \cdot 8a^2}{45a^2 - 8a^2} = 2.39a,$$

$$\bar{y}_S = \frac{\bar{y}_1 A_1 + \bar{y}_2 A_2}{A_1 + A_2} = \frac{\dfrac{9}{2}a \cdot 45a^2 - 3a \cdot 8a^2}{45a^2 - 8a^2} = 4.82a. \tag{3.60}$$

The centre of gravity S with the centre of gravity axes x, y is shown in Fig. 3.24.

◀

Example 3.14

For the area of Fig. 3.25 the centre of gravity is to be determined.

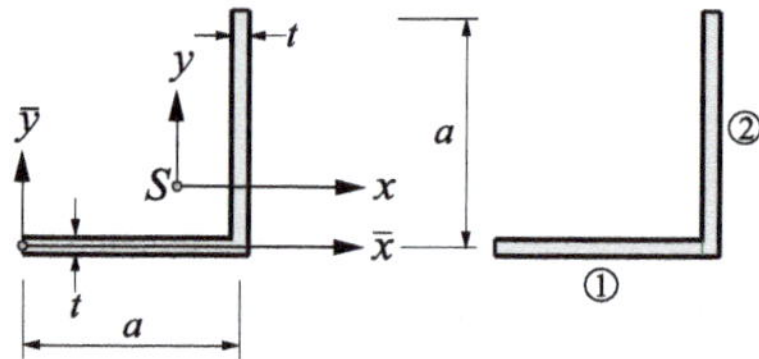

Fig. 3.25 Given area

Solution:

We divide the area into the two sub-areas $A_1 = (a - t)t$ and $A_2 = at$ as indicated. The centre of gravity positions of the sub-areas are $\bar{x}_1 = \frac{1}{2}(a - t)$, $\bar{x}_2 = a - \frac{t}{2}$ and $\bar{y}_1 = \frac{t}{2}$, $\bar{y}_2 = \frac{a}{2}$. This gives the coordinates of the centre of gravity S as:

$$\begin{aligned}\bar{x}_S &= \frac{\bar{x}_1 A_1 + \bar{x}_2 A_2}{A_1 + A_2} = \frac{\frac{1}{2}(a-t)(a-t)t + \left(a - \frac{t}{2}\right)at}{(a-t)t + at} \\ &= \frac{3}{4}a\frac{1 - \frac{t}{a} + \frac{1}{3}\frac{t^2}{a^2}}{1 - \frac{1}{2}\frac{t}{a}}, \\ \bar{y}_S &= \frac{\bar{y}_1 A_1 + \bar{y}_2 A_2}{A_1 + A_2} = \frac{\frac{t}{2}(a-t)t + \frac{a}{2}at}{(a-t)t + at} \\ &= \frac{1}{4}a\frac{1 + \frac{t}{a} - \frac{t^2}{a^2}}{1 - \frac{1}{2}\frac{t}{a}}.\end{aligned} \tag{3.61}$$

If we consider an area with $t \ll a$, then the expression $\frac{t}{a}$ is small compared to 1, and the terms $\frac{t}{a}$ and $\frac{t^2}{a^2}$ can be neglected. The following then remains:

$$\bar{x}_S = \frac{3}{4}a, \quad \bar{y}_S = \frac{1}{4}a. \tag{3.62}$$

◀

Figure 3.26 contains the centre of gravity positions for some technically relevant areas that are typical for cross-sections of beam and bar structures.

If we consider an area with $t \ll a$, as treated in Example 3.14, then such an area is considered mathematically using its skeleton line, i.e. the line that halves the thickness of the area at each point. This is a typical procedure for many thin-walled beam and bar cross-sections. Some exemplary thin-walled cross-sections are shown in Fig. 3.27. We consider

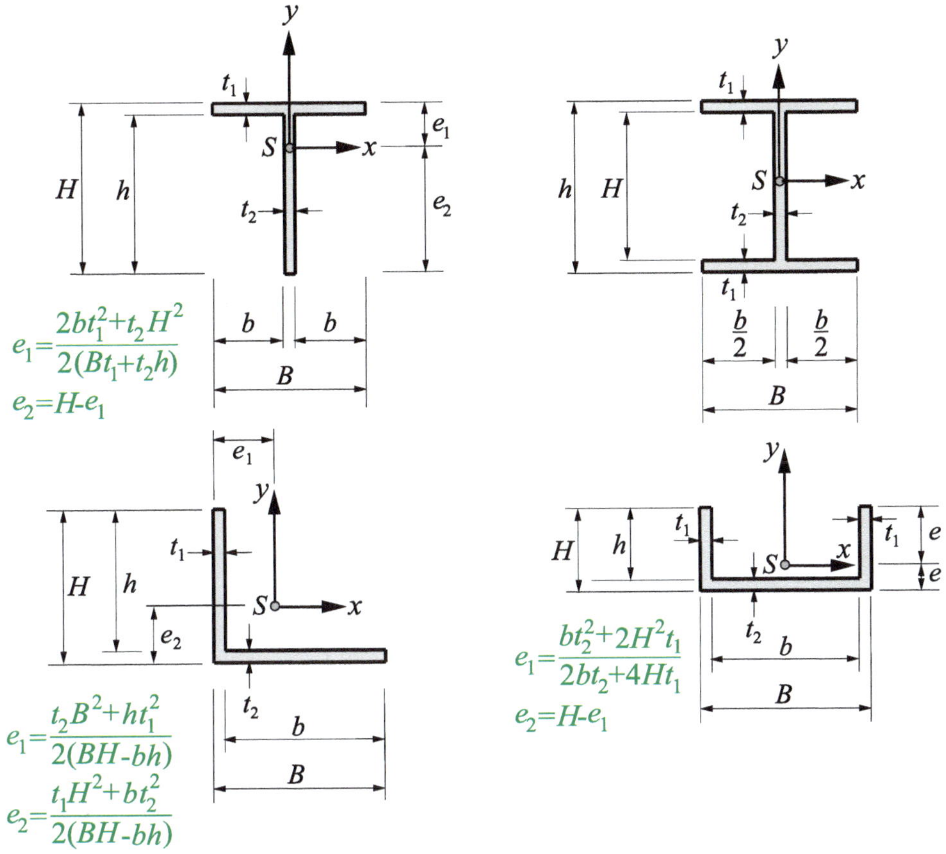

$$e_1=\frac{2bt_1^2+t_2H^2}{2(Bt_1+t_2h)}$$
$$e_2=H-e_1$$

$$e_1=\frac{t_2B^2+ht_1^2}{2(BH-bh)}$$
$$e_2=\frac{t_1H^2+bt_2^2}{2(BH-bh)}$$

$$e_1=\frac{bt_2^2+2H^2t_1}{2bt_2+4Ht_1}$$
$$e_2=H-e_1$$

Fig. 3.26 Centre of gravity positions for technically relevant composite areas

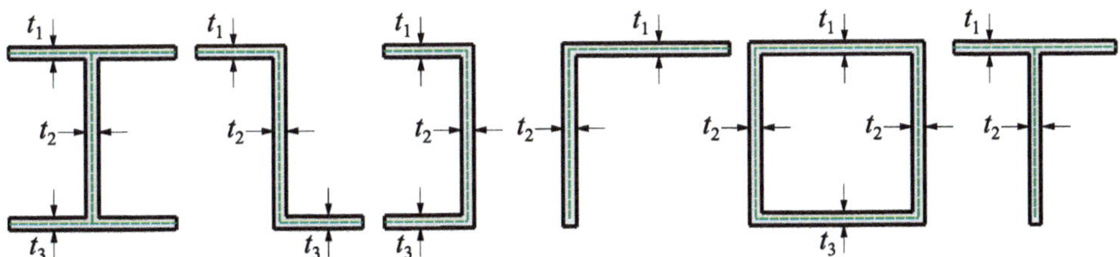

Fig. 3.27 Thin-walled cross-sections

the above example again and assume that a thin-walled cross-section is given, so that $t \ll a$ applies (Fig. 3.28). For the two subareas we have $A_1 = A_2 = at$, and the coordinates of the centres of gravity of the sub-areas are $\bar{x}_1 = \frac{a}{2}$, $\bar{x}_2 = a$ and $\bar{y}_1 = 0$, $\bar{y}_2 = \frac{a}{2}$. The position of the centre of gravity S can then be determined as:

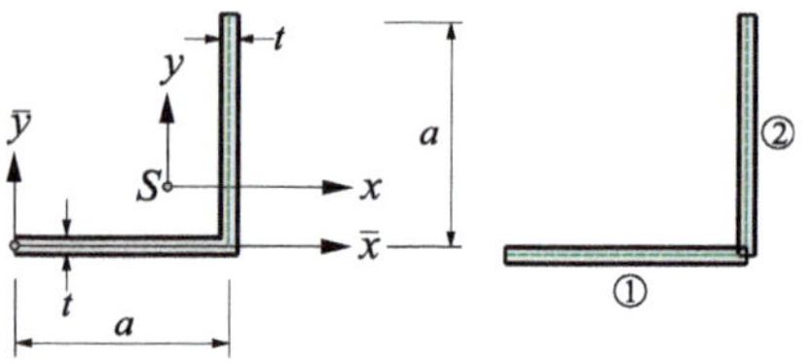

Fig. 3.28 Given thin-walled cross-section

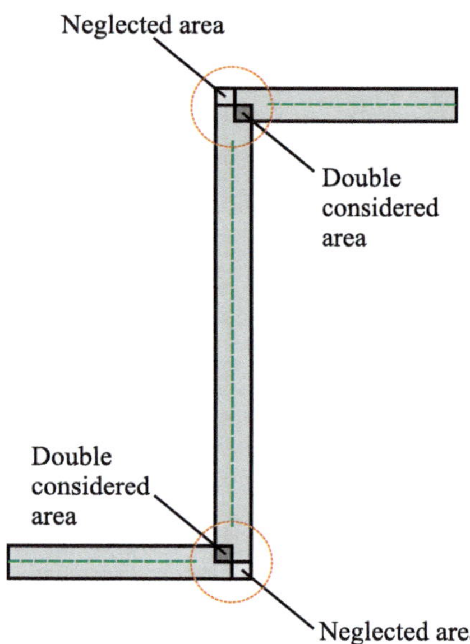

Fig. 3.29 Thin-walled Z cross-section

$$\bar{x}_S = \frac{\bar{x}_1 A_1 + \bar{x}_2 A_2}{A_1 + A_2} = \frac{\frac{a}{2} \cdot at + a \cdot at}{at + at} = \frac{3}{4}a,$$
$$\bar{y}_S = \frac{\bar{y}_1 A_1 + \bar{y}_2 A_2}{A_1 + A_2} = \frac{0 \cdot at + \frac{a}{2} \cdot at}{at + at} = \frac{1}{4}a. \qquad (3.63)$$

This result agrees with (3.62).

If the centre of gravity of a thin-walled cross-section is determined strictly along the skeleton line, then, as shown in Fig. 3.29 using a Z-cross-section, at the transition points between the flanges and the web this results in sub-areas that are not taken into account at all in the calculation, but also sub-areas that are included twice in the calculation. In the example of the Z cross-section shown, these are the points of contact between the horizontal segments and the vertical segment of the cross-section. Obviously, there is a small sub-area that is included twice in the calculation, but there is also a small sub-area that is not taken into account. This error is inevitable if we treat such a cross-section as a group of areas and orientate ourselves on the skeleton line. However, experience shows that the error made in this way remains within acceptable limits when dealing with very thin-walled cross-sections.

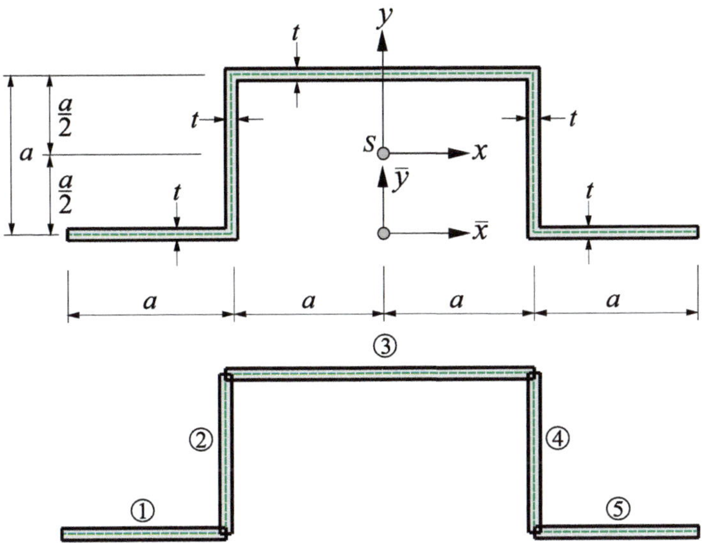

Fig. 3.30 Thin-walled cross-section

Example 3.15

For the area shown in Fig. 3.30, the position of the centre of gravity S is to be determined.

Solution:

We orientate ourselves in the treatment of the given thin-walled cross-section on the basis of the skeleton line and subdivide the cross-section into sub-areas as shown in Fig. 3.30, bottom. The centre of gravity coordinate $\bar{x}_S$ results as $\bar{x}_S = 0$ due to the symmetry of the cross-section. With the partial areas $A_1 = A_2 = A_4 = A_5 = at$, $A_3 = 2at$ and the coordinates $\bar{y}_1 = \bar{y}_5 = 0$, $\bar{y}_2 = \bar{y}_4 = \frac{a}{2}$, $\bar{y}_3 = a$, the centre of gravity coordinate $\bar{y}_S$ follows as:

$$\bar{y}_S = \frac{\bar{y}_1 A_1 + \bar{y}_2 A_2 + \bar{y}_3 A_3 + \bar{y}_4 A_4 + \bar{y}_5 A_5}{A_1 + A_2 + A_3 + A_4 + A_5} = \frac{1}{2}a. \tag{3.64}$$

◀

3.8 Composite Lines

In a similar way to the previous sections, lines composed of individual sub-segments can also be treated. The following then applies to the line centre of gravity:

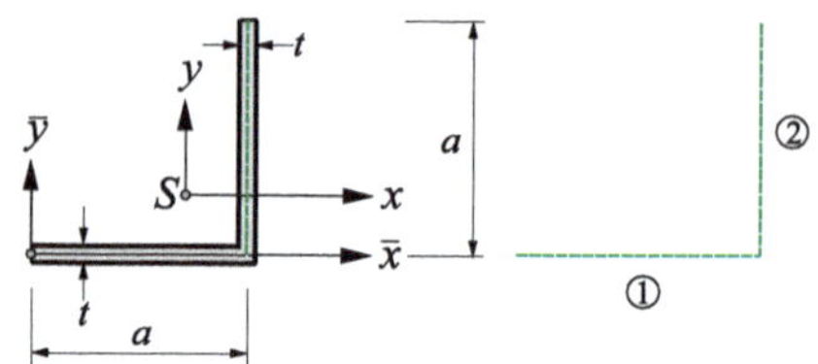

Fig. 3.31 Thin-walled cross-section

$$\bar{x}_L = \frac{\sum_{i=1}^{n} \bar{x}_i l_i}{\sum_{i=1}^{n} l_i}, \quad \bar{y}_L = \frac{\sum_{i=1}^{n} \bar{y}_i l_i}{\sum_{i=1}^{n} l_i}, \quad \bar{z}_L = \frac{\sum_{i=1}^{n} \bar{z}_i l_i}{\sum_{i=1}^{n} l_i}. \tag{3.65}$$

Therein, l_i is the length of the partial line i.

As an application, we again consider thin-walled cross-sections with $t \ll a$. If the wall thickness in each partial segment of a thin-walled cross-section is identical to the value t, then the centre of gravity S can be determined as the line centre of gravity. For the cross-section of Fig. 3.28, this procedure is permissible and we divide the skeleton line into two lines of lengths $l_1 = l_2 = a$ (Fig. 3.31). With $\bar{x}_1 = \frac{a}{2}$, $\bar{x}_2 = a$ and $\bar{y}_1 = 0$, $\bar{y}_2 = \frac{a}{2}$ the centre of gravity coordinates $\bar{x}_L$, $\bar{y}_L$ follow as:

$$\begin{aligned} \bar{x}_L &= \frac{\bar{x}_1 l_1 + \bar{x}_2 l_2}{l_1 + l_2} = \frac{\frac{3}{2}a^2}{2a} = \frac{3}{4}a, \\ \bar{y}_L &= \frac{\bar{y}_1 l_1 + \bar{y}_2 l_2}{l_1 + l_2} = \frac{\frac{1}{2}a^2}{2a} = \frac{1}{4}a. \end{aligned} \tag{3.66}$$

Support Reactions 4

In this chapter, we look at how the reaction forces and moments, i.e. the support reactions of static systems, can be determined. After some introductory remarks, we look at single-part and multi-part systems and conclude the chapter by discussing spatial structures.

4.1 Fundamentals

Components are connected to adjacent components or structures by suitable supports, and forces and/or moments arise in these supports under loads, depending on the type of support, which we summarise under the term support reactions and which, in addition to the geometry of the component and the type of load, also depend on the type of support. In this chapter, we will limit ourselves to the consideration of bars and beams, i.e. components whose cross-sectional dimensions are very small compared to their length and which are loaded in such a way that they are either stressed only in the direction of their axis (so-called bar) or perpendicular to it (so-called beam). In technical applications, curved beams or arches may occur as well, and if it we have an angled beam component, then this is often referred to as a frame. What all these classes of structures have in common is that they must be suitably supported so that they can fulfil their function. A spatial component has six degrees of freedom without support, three translations can occur in the three spatial directions x, y, z and three rotations around the three reference axes. The movement possibilities of components must therefore be restricted by adequate supports.

A very simple example of a supported component is shown in Fig. 4.1, in which a beam-like component is shown under individual forces $F_1, F_2, \ldots, F_n$ and individual moments $M_1, M_2, \ldots, M_n$. The beam rests on two supports and the contact points between the beam

C. Mittelstedt, *Engineering Mechanics 1: Statics*,
https://doi.org/10.1007/978-3-662-71852-0_4

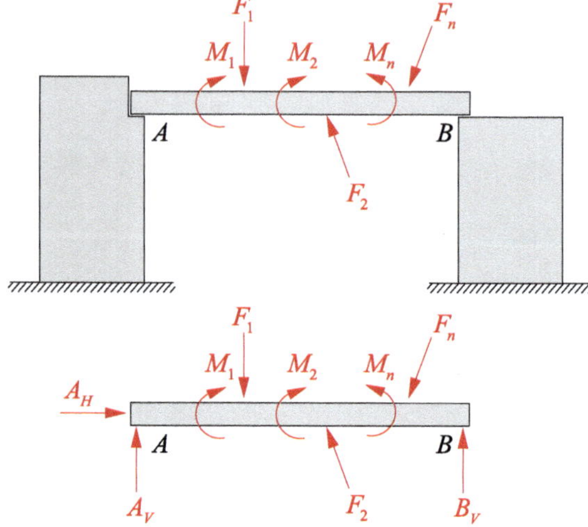

Fig. 4.1 Beam on two supports (top), free-body diagram (bottom)

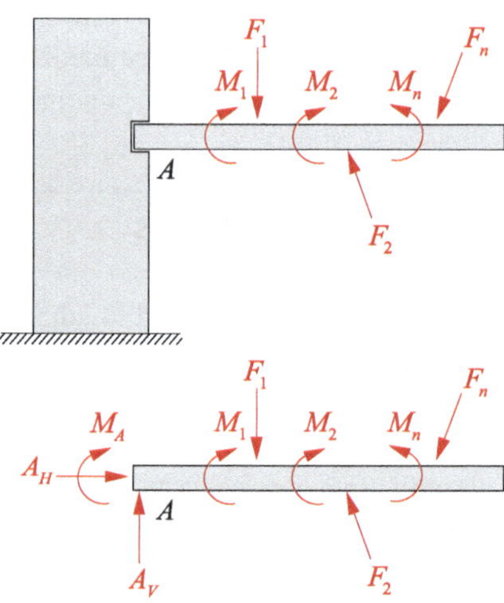

Fig. 4.2 Cantilever beam (top), free-body diagram (bottom)

and the supports are labelled A and B. We cut the beam free as shown in Fig. 4.1, bottom, and release the support reactions, here the two vertical support forces A_V and B_V, whereby it is very useful to label the support reactions released in this way according to their support points. The indices V indicate that the support reactions are vertical. In addition, a horizontal support force A_H will also occur at support point A, whereby the index H indicates the horizontal direction of action of this support force. The support forces A_V, A_H, B_V occurring here result from the fact that the beam at the two points A and B is prevented from moving in the form of translations in the vertical and horizontal directions by the two supports.

Another example is a cantilever beam that is free at its right end and fixed in a wall at its left end A (Fig. 4.2). With such a support, not only the two support forces A_V and A_H occur, but also a support moment or clamping moment M_A. The clamping moment M_A results from the fact that the beam is prevented from rotating in its plane at the clamping point A.

What all types of supports have in common is that they limit the motion possibilities of the bar or beam under consideration. To begin with, we will restrict ourselves to single bars and beams that are only loaded in their plane. We will deal with multi-part and three-dimensional structures later. If there is no support, a component has three degrees of freedom in the plane, namely two translations in the plane and a rotation around an axis that is orientated orthogonally to the plane. If a support is present, these degrees of freedom are restricted and support reactions are caused, as already shown in Figs. 4.1 and 4.2. If we take another look at the beam in Fig. 4.1, it can be seen that the two supports hinder the two translations in the horizontal and vertical directions. As a reaction to this, the support forces shown arise in the support points. This applies analogously to the cantilever beam of Fig. 4.2, where in addition to the two translations, the rotation in the plane is also restricted, which causes the two support forces shown as well as the depicted support moment. In general, the number f of degrees of freedom of a body in the plane is $f = 3$ if there is no support at all. Now let r be the number of support reactions. Then the number of degrees of freedom is defined as $f = 3 - r$. Using the examples in Figs. 4.1 and 4.2, $f = 3 - 3 = 0$ applies in each case, i.e. the beam can neither move nor can it rotate. In both cases, the horizontal immovability must be ensured by a suitable design of the support points. Furthermore, in the case of the beam in Fig. 4.1, the lifting of the beam in the vertical direction must be prevented by a suitable design of the support points A and B.

Supports are generally classified according to the number of support reactions they cause. A single-valued support (Fig. 4.3) refers to supports that are displaceable in one direction and that allow the beam to rotate and in which consequently only a single support force ($r = 1$) occurs. Examples of such supports are roller supports (Fig. 4.3, top left), rail bearings (Fig. 4.3, top centre) or supports by a bar (Fig. 4.3, top right). What they all have in common is that they are idealized by the support symbol shown in Fig. 4.3, bottom left, and cause a vertical support force A_V at the support point A. Such a single-valued support prevents translation of the beam in the vertical direction, whereas horizontal translation and rotation in the plane are still possible. If necessary, the lifting of the support point must be prevented by a suitable design for such a single-value support.

A two-valued support is a support that prevents both the vertical translation and the horizontal translation, but in which the rotation in the plane is still possible. Such a support is illustrated by the model shown in Fig. 4.4, left, and the two support forces A_V and A_H occur here at the support point A in the vertical and horizontal directions (Fig. 4.4, right), i.e. $r = 2$ applies. Since the rotation of the beam in its plane is not hindered by such a support, no support moment occurs.

The so-called sliding sleeve is a special case of a two-valued support (Fig. 4.5, left), in which horizontal displacement is possible, but the translation in the vertical direction and

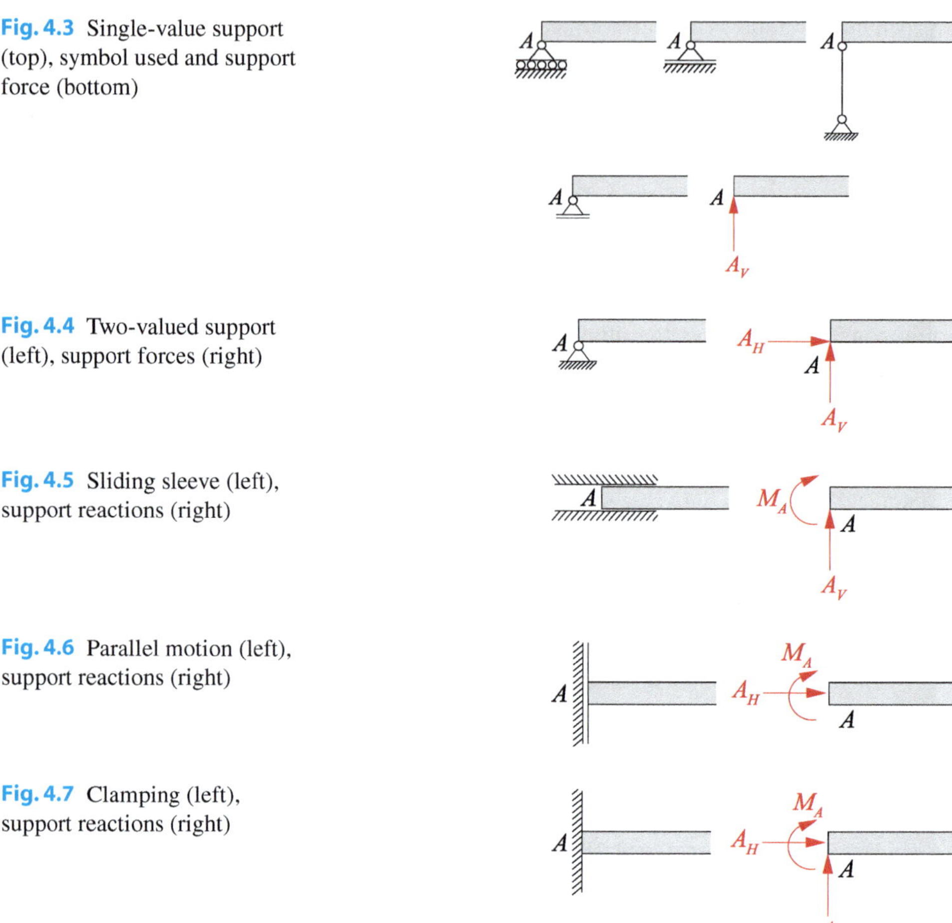

Fig. 4.3 Single-value support (top), symbol used and support force (bottom)

Fig. 4.4 Two-valued support (left), support forces (right)

Fig. 4.5 Sliding sleeve (left), support reactions (right)

Fig. 4.6 Parallel motion (left), support reactions (right)

Fig. 4.7 Clamping (left), support reactions (right)

the rotation in the plane are excluded. Accordingly, a vertical support force A_V and a support moment M_A occur here (Fig. 4.5, right).

Another special case of a two-valued support is the so-called parallel motion (Fig. 4.6, left). Here, both the horizontal displacement and the rotation in the plane are impeded, so that the horizontal support force A_H and the support moment M_A occur as support reactions (Fig. 4.6, right).

A clamping (Fig. 4.7, left) is a three-valued support ($r = 3$) in which, in addition to the two translations, the rotation in the plane is also prevented. Accordingly, in addition to the two support forces A_V and A_H, the support moment or clamping moment M_A also occurs (Fig. 4.7, right).

The determination of support reactions, i.e. support forces and support moments, is an important task for engineers, which is necessary for the design of bars and beams in general

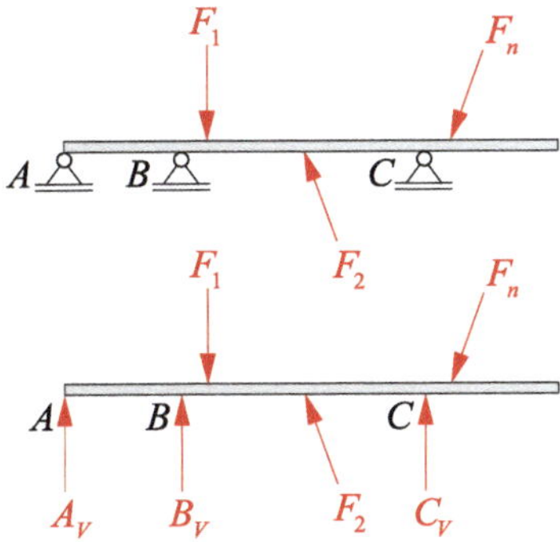

Fig. 4.8 Beam on three single-valued supports (top), free-body diagram (bottom)

and the support points in particular. In this chapter, we look at how to determine support reactions of bars and beams as well as in multi-part systems.

An important concept that should be highlighted at this point is the so-called static determinacy. A plane structure is labelled as statically determinate if its support reactions can be determined from the three plane equilibrium conditions. This means that a maximum of three unknown support reactions may occur in the plane. The beams of the Figs. 4.1 and 4.2 are therefore statically determinate: In both cases, three support reactions A_V, A_H, B_V or A_V, A_H, M_A occur, which can be determined from the equilibrium conditions. Both beams are completely restricted in their movement, $f = 3 - r = 3 - 3 = 0$ applies. In general, it can be stated that a structure is statically determinate if exactly three support reactions occur and the structure is immovably supported, whereby the support reactions are either three non-central forces that are not all parallel to each other, or two forces and a moment, whereby the forces are not parallel to each other. The fact that there are also exceptions to this is illustrated in Fig. 4.8, in which a beam is shown on three single-valued supports. The three vertically directed support forces A_V, B_V and C_V will occur on this beam, which cannot be determined from the equilibrium conditions despite $f = 3 - r = 3 - 3 = 0$, because the sum of the horizontal forces cannot be fulfilled. In addition, this beam is displaceable, as can be seen directly: The horizontal components of the acting forces will move the beam horizontally. Such a construction is unsuitable for technical application and must be avoided. We can convince ourselves of the mobility of a system, for example, by creating a so-called pole plan (see Sect. 7.4). If a pole plan can be produced without contradictions, then the system is movable and the construction is unusable. Such a system is called kinematically indeterminate.

If, on the other hand, a structure has more support reactions than equilibrium conditions, it is referred to as being statically indeterminate. The Fig. 4.9 shows two examples. For the beam of Fig. 4.9, top, $r = 4$ applies, i.e. there is one more support force than can be determined from the equilibrium conditions. It is then said that such a beam is simply statically indeterminately supported. For the beam of Fig. 4.9, bottom, $r = 6$ applies, so that one also speaks of a triple statically indeterminately supported system. We will not go into

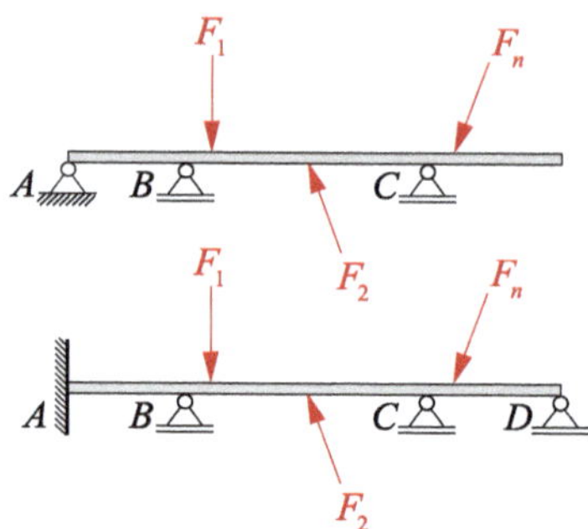

Fig. 4.9 Statically indeterminate beams

the treatment of statically indeterminate systems in this book; this topic is dealt with in detail in Volume 2 in Sects. 3.1.3 and 3.2.2 as well as 5.3 and 5.5.3 and in Chap. 8.

4.2 Determination of Support Reactions

To determine the support reactions of a bar or beam, we cut the structure under consideration free at the support points and thus expose the support reactions in the resulting free-body diagram. An example of this is shown in Fig. 4.10. We consider a beam of length l that is supported at its left end by a two-valued support and at its right end by a single-valued support. The beam is loaded by a vertical point force F in its centre. We now cut the beam free and thus set the support forces A_V, A_H and B_V free as shown in Fig. 4.10, bottom. The choice of the directions of action of the support reactions is arbitrary. Here, for example, we have assumed that the support force A_V acts positively upwards. The calculation result will later show whether this assumption was correct. If A_V is positive, then this support force is actually pointing upwards, whereas if it is negative, then this support force is actually pointing downwards. The equilibrium conditions in the plane are then formulated on the beam cut free in this way. These are the sum of the horizontal forces and the vertical forces as well as the sum of the moments with respect to a selected point. For the horizontal sum of forces follows:

$$\overset{\rightarrow}{\sum} H = 0: \quad A_H = 0. \tag{4.1}$$

The horizontal support force A_H is therefore identical to zero, which is a plausible result since there is no horizontal load applied on the beam.

The sum of the vertical forces results in:

$$\overset{\uparrow}{\sum} V = 0: \quad A_V + B_V - F = 0. \tag{4.2}$$

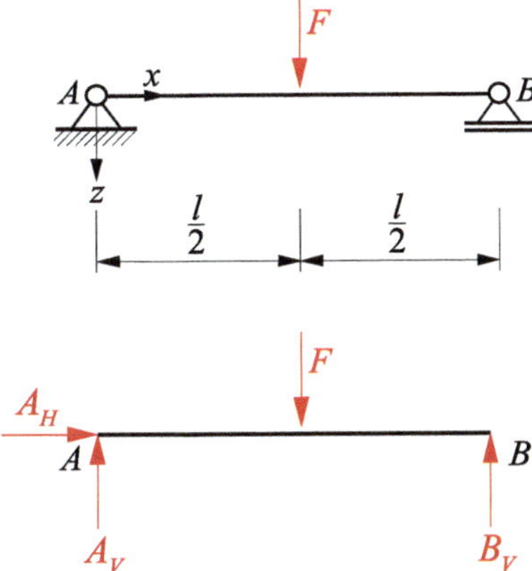

Fig. 4.10 Beam on two supports (top), free-body diagram (bottom)

This equation obviously contains two as yet unknown support forces, so that we will not consider this equation further for the time being, but will need it later.

The sum of the moments with respect to the support point A is as follows:

$$\overset{\curvearrowleft}{\sum} M_A = 0: \quad B_V \cdot l - F \cdot \frac{l}{2} = 0. \tag{4.3}$$

Only the support force B_V appears in this equation, so that this equation can be solved directly for B_V. It follows:

$$B_V = \frac{F}{2}. \tag{4.4}$$

Accordingly, the right support B carries exactly half of the applied force F. From Eq. (4.2), the support force A_V can then be determined immediately with known support force B_V, and it follows:

$$A_V = \frac{F}{2}. \tag{4.5}$$

Accordingly, the left support A also bears exactly half of the applied force F. The result $A_V = B_V = \frac{F}{2}$ agrees with our intuition for this elementary simple example: If the force is applied exactly in the centre of the beam, both supports bear exactly half of the force F.

Finally, the sum of the moments with respect to the support point B can be used for verification of our results:

$$\overset{\curvearrowright}{\sum} M_B = 0: \quad A_V \cdot l - F \cdot \frac{l}{2} = 0. \tag{4.6}$$

With $A_V = \frac{F}{2}$ this equation is obviously fulfilled identically.

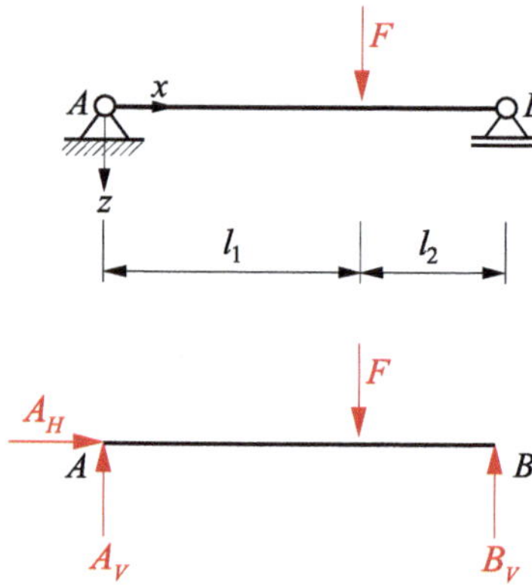

Fig. 4.11 Beam on two supports (top), free-body diagram (bottom)

Example 4.1

For the beam in Fig. 4.11 the support reactions are to be determined.

Solution:

We cut the beam free and apply the support reactions A_H, A_V and B_V as shown in Fig. 4.11, bottom. From the sum of the horizontal forces follows:

$$\overset{\rightarrow}{\sum} H = 0: \quad A_H = 0. \tag{4.7}$$

The sum of the moments with respect to the support A results in:

$$\overset{\curvearrowleft}{\sum} M_A = 0: \quad B_V(l_1 + l_2) - F l_1 = 0 \quad \rightarrow \quad B_V = F \frac{l_1}{l_1 + l_2}. \tag{4.8}$$

The support force A_V then results from the sum of the moments around the support point B:

$$\overset{\curvearrowright}{\sum} M_B = 0: \quad A_V(l_1 + l_2) - F l_2 = 0 \quad \rightarrow \quad A_V = F \frac{l_2}{l_1 + l_2}. \tag{4.9}$$

Finally, we use the sum of the vertical forces to check our results. It follows:

$$\overset{\uparrow}{\sum} V = 0: \quad A_V + B_V - F = 0 \quad \rightarrow \quad F \frac{l_2}{l_1 + l_2} + F \frac{l_1}{l_1 + l_2} - F = 0. \tag{4.10}$$

Obviously this equation is fulfilled.

◄

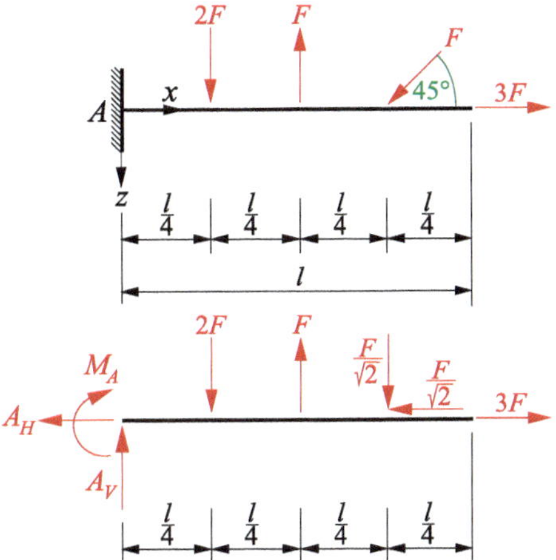

Fig. 4.12 Cantilever beam (top), free-body diagram (bottom)

Example 4.2

Consider a cantilever beam of length l, which is loaded by several forces (Fig. 4.12). We want to determine the support reactions.

Solution:

We determine the support reactions at point A and consider the free-body diagram in Fig. 4.12, bottom, in which we have applied the support reactions A_H, A_V and M_A and decomposed the force F acting at an angle of 45° into its horizontal and vertical components with $\sin 45° = \cos 45° = \frac{1}{\sqrt{2}}$. The equilibrium conditions are as follows:

$$\overset{\leftarrow}{\sum} H = 0: \quad A_H + \frac{F}{\sqrt{2}} - 3F = 0 \quad \rightarrow \quad A_H = \left(3 - \frac{1}{\sqrt{2}}\right) F,$$

$$\overset{\uparrow}{\sum} V = 0: \quad A_V - 2F + F - \frac{F}{\sqrt{2}} = 0 \quad \rightarrow \quad A_V = \left(1 + \frac{1}{\sqrt{2}}\right) F,$$

$$\overset{\curvearrowright}{\sum} M_A = 0: \quad M_A + 2F \cdot \frac{l}{4} - F \cdot \frac{l}{2} + \frac{F}{\sqrt{2}} \cdot \frac{3l}{4} = 0 \quad \rightarrow \quad M_A = -\frac{3Fl}{4\sqrt{2}} \tag{4.11}$$

◄

Example 4.3

We consider the example of Fig. 4.13. Consider a beam of length l (partial lengths l_1 and l_2) on two supports, which is loaded by a single moment M_0. The support forces are to be determined.

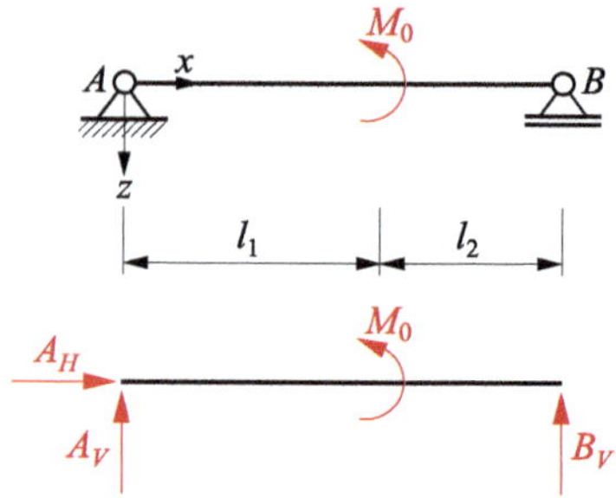

Fig. 4.13 Beam under single moment M_0

Solution:

The support reactions can be determined from the free-body diagram in Fig. 4.13, bottom. It follows:

$$\overset{\rightarrow}{\sum} H = 0: \quad A_H = 0,$$
$$\overset{\curvearrowright}{\sum} M_B = 0: \quad A_V l - M_0 = 0 \quad \rightarrow \quad A_V = \frac{M_0}{l},$$
$$\overset{\curvearrowleft}{\sum} M_A = 0: \quad B_V l + M_0 = 0 \quad \rightarrow \quad B_V = -\frac{M_0}{l}. \tag{4.12}$$

Obviously, the support forces A_V and B_V are independent of the location of the point of application of the moment. The sign of the support force B_V also indicates that this force is actually directed in exactly the opposite direction to that assumed at the beginning, i.e. it is pointing downwards.

◀

Example 4.4

Consider a beam of length l on two supports under the uniform line load q_0 (Fig. 4.14). We want to determine the support reactions.

Solution:

We cut the beam free at the support points A and B and draw the support forces A_H, A_V and B_V as well as the resultant $R = q_0 l$ of the uniform line load in the resulting free-body diagram (Fig. 4.14, bottom). The sum of the moments with respect to the support A results in:

$$\overset{\curvearrowleft}{\sum} M_A = 0: \quad B_V \cdot l - R \cdot \frac{l}{2} = 0 \quad \rightarrow \quad B_V = \frac{q_0 l}{2}. \tag{4.13}$$

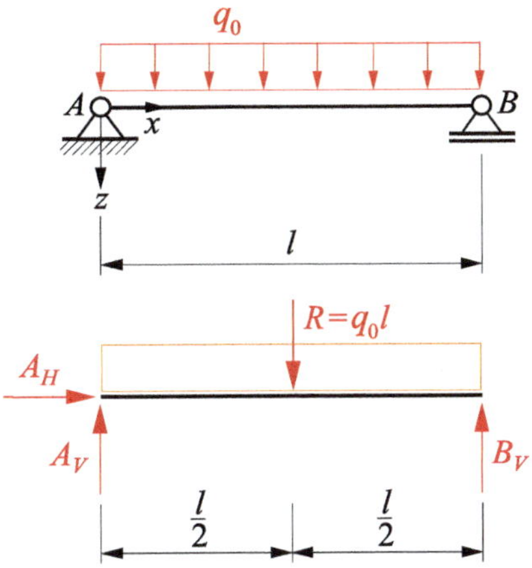

Fig. 4.14 Beam under uniform line load q_0

Fig. 4.15 Beam under linearly distributed line load $q = q_0 \frac{x}{l}$

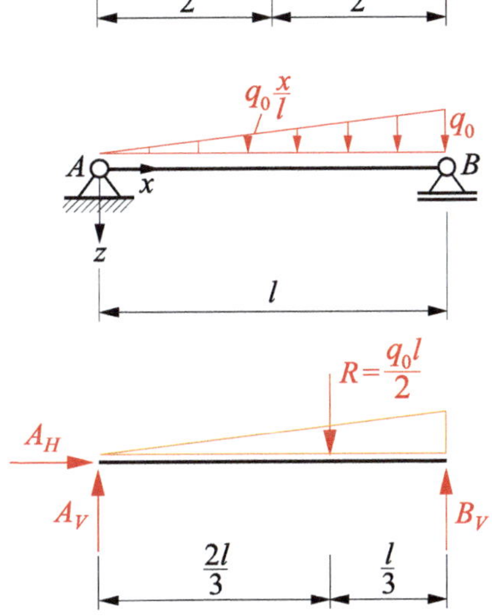

The support force A_V then results from the sum of the vertical forces:

$$\sum^{\uparrow} V = 0: \quad A_V + B_V - R = 0 \quad \rightarrow \quad A_V = \frac{q_0 l}{2}. \tag{4.14}$$

The support force A_H is zero for this example, there is no load in the horizontal direction.

◀

Example 4.5

Consider a beam of length l on two supports under the linear line load $q(x) = q_0 \frac{x}{l}$. We want to determine the support forces.

Solution:

We cut the beam free at the supports A and B. The support forces A_H, A_V and B_V released in this way and the resultant $R = \frac{1}{2}q_0 l$ of the line load are shown in the free-body diagram in Fig. 4.15, bottom. The support force B_V results from the sum of the moments with respect to the support A:

$$\overset{\curvearrowleft}{\sum} M_A = 0: \quad B_V \cdot l - R \cdot \frac{2l}{3} = 0 \quad \rightarrow \quad B_V = \frac{q_0 l}{3}. \tag{4.15}$$

The support force A_V can be determined from the sum of the vertical forces:

$$\overset{\uparrow}{\sum} V = 0: \quad A_V + B_V - R = 0 \quad \rightarrow \quad A_V = \frac{q_0 l}{6}. \tag{4.16}$$

The support force A_H is identical to zero.

◄

Example 4.6

For the angled beam of Fig. 4.16 the support forces are to be determined.

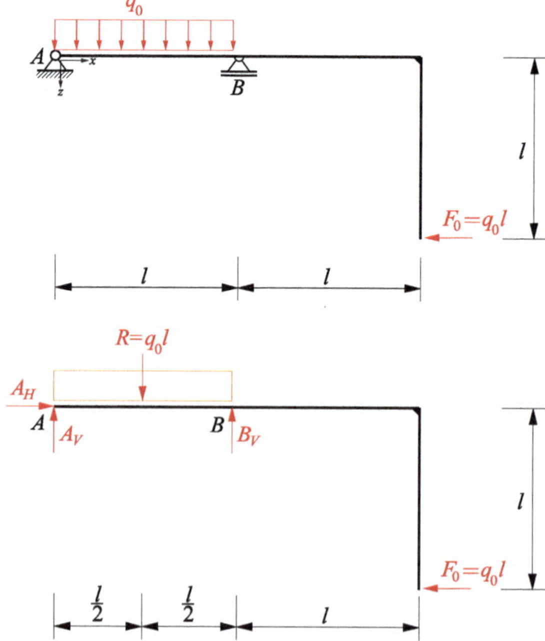

Fig. 4.16 Angled beam

Solution:

We cut the beam free at the support points A and B and also form the resultant $R = q_0 l$ of the line load (Fig. 4.16, bottom). The support force A_H results from the sum of the horizontal forces:

$$\overset{\rightarrow}{\sum} H = 0: \quad A_H - q_0 l = 0 \quad \rightarrow \quad A_H = q_0 l. \tag{4.17}$$

The sum of the moments with respect to the support point A leads to the support force B_V as follows:

$$\overset{\curvearrowleft}{\sum} M_A = 0: \quad B_V \cdot l - q_0 l \cdot \frac{l}{2} - q_0 l \cdot l = 0 \quad \rightarrow \quad B_V = \frac{3}{2} q_0 l. \tag{4.18}$$

Finally, the support force A_V results from the sum of the vertical forces:

$$\overset{\uparrow}{\sum} V = 0: \quad A_V + B_V - q_0 l = 0 \quad \rightarrow \quad A_V = -\frac{1}{2} q_0 l. \tag{4.19}$$

Accordingly, the actual direction of A_V is opposite to the direction shown in Fig. 4.16. We can use the sum of moments with respect to the support point B, for example, to check the result, but this is not shown here.

◀

Example 4.7

For the frame shown in Fig. 4.17 the support forces are to be determined.

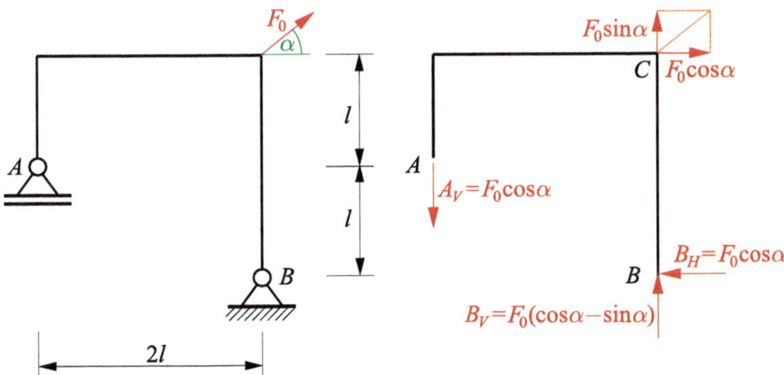

Fig. 4.17 Frame

Solution:

We cut the frame free at the support points A and B and divide the acting force F_0 into its horizontal and vertical components (Fig. 4.17, right). The support force B_H results from the sum of the horizontal forces:

$$\overset{\leftarrow}{\sum} H = 0: \quad B_H - F_0 \cos\alpha = 0 \quad \rightarrow \quad B_H = F_0 \cos\alpha. \tag{4.20}$$

We also form the sum of the moments with respect to the point C:

$$\overset{\curvearrowleft}{\sum} M_C = 0: \quad A_V \cdot 2l - B_H \cdot 2l = 0 \quad \rightarrow \quad A_V = F_0 \cos\alpha. \tag{4.21}$$

The sum of the vertical forces then results in the support force B_V:

$$\overset{\uparrow}{\sum} V = 0: \quad B_V - A_V + F_0 \sin\alpha = 0 \quad \rightarrow \quad B_V = F_0 (\cos\alpha - \sin\alpha). \tag{4.22}$$

◀

The superposition principle applies to the determination of support reactions. This means that if several loads are applied to a structure, the support reactions can be determined separately for each applied load and these can then be added up (superposed) to the actual support reactions. This is shown here for the beam in Fig. 4.18, which is loaded by two forces F_1 and F_2 as well as a uniform line load q. We release the support reactions A_H, A_V and M_A by cutting them free, whereby $A_H = 0$ results under the given loads, so that we do not consider this support reaction any further. We now consider the vertical support force A_V and the clamping moment M_A based on the individual loads and obtain

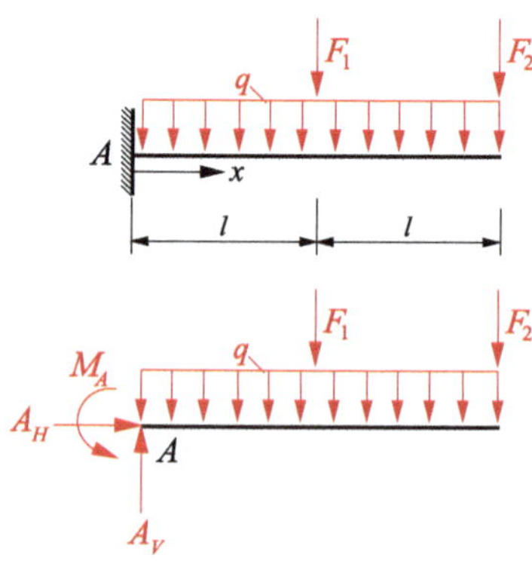

Fig. 4.18 Beam (top), free-body diagram (bottom)

$$
\begin{aligned}
A_{V,1} &= F_1, \quad A_{V,2} = F_2, \quad A_{V,q} = 2ql, \\
M_{A,1} &= F_1 l, \quad M_{A,2} = 2F_2 l, \quad M_{A,q} = 2ql^2.
\end{aligned} \tag{4.23}
$$

The support reactions A_V and M_A then result from summation:

$$
\begin{aligned}
A_V &= A_{V,1} + A_{V,2} + A_{V,q} = F_1 + F_2 + 2ql, \\
M_A &= M_{A,1} + M_{A,2} + M_{A,q} = F_1 l + 2F_2 l + 2ql^2.
\end{aligned} \tag{4.24}
$$

4.3 Multi-part Structures

In many technical applications, a beam consists not only of a single segment but of a large number of components. These components or parts are connected to each other using suitable structural measures. Forces and/or moments then occur at such connection points, which must be determined and which can be visualised in a free-body diagram by cutting them free, just like support reactions. Several different types can be considered as connecting elements between the segments of beams.

A hinge (Fig. 4.19) represents a type of connection through which any forces can be transmitted, but no bending moments can be transmitted due to the assumption of frictionless rotational behaviour. As a result, a hinge force occurs in such a hinge, which is often broken down into a vertical hinge force G_V and a horizontal hinge force G_H, as shown in Fig. 4.19, bottom. By cutting through the hinge, the hinge forces are visualised in the free-body diagram; they act in opposite directions on both beam segments due to the interaction principle of actio = reactio.

Connections can also be realised using bars (Fig. 4.20, top), which only transmit forces S in their own direction. With a parallel motion, on the other hand (Fig. 4.20, centre), rotation and translation in the direction of the connected beam segments is hindered, so that a horizontal force H and a moment M can be transmitted. In the same way, a moment M and a vertically acting force V are transmitted by a sliding sleeve (Fig. 4.20, bottom).

The forces and moments that are transmitted through the mentioned connecting elements are determined from the equilibrium conditions for the beam segments, whereby we can

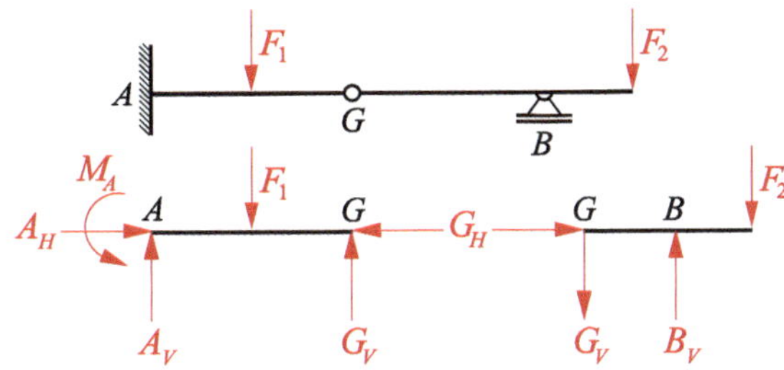

Fig. 4.19 Beam with hinge G (top), free-body diagram (bottom)

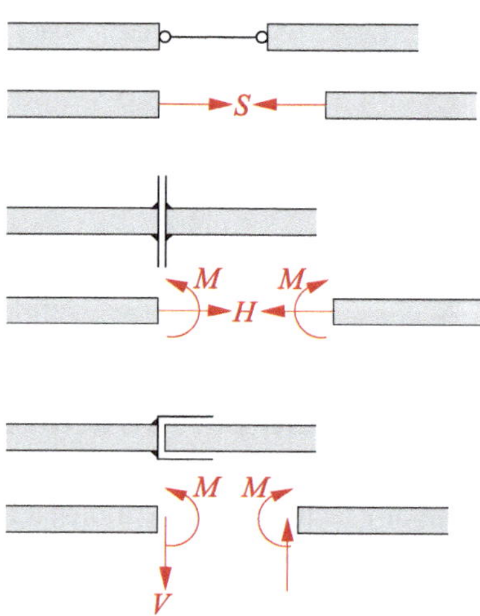

Fig. 4.20 Bar (top), parallel motion (centre), sliding sleeve (bottom)

use three equilibrium conditions for each beam segment. If n is the number of beam segments, then we can formulate a total of $3n$ equilibrium conditions. If there are also r support reactions and v connection forces and moments, then the structure under consideration is statically determinate if the $3n$ equilibrium conditions are sufficient to determine the occurring r support reactions and v connection forces and moments. The criterion $3n = r + v$ must therefore be fulfilled. If this is the case and the beam structure is rigid, i.e. non-displaceable, then the system is statically determinate. Accordingly, care must be taken to ensure that a system does not become moveable and therefore unusable due to the arrangement of connecting elements. This is illustrated using the example of Fig. 4.21 as an example. For both systems shown here, the criterion $3n = 6$ with $r + v = 4 + 2$ is fulfilled, and the system of Fig. 4.21, top, is non-displaceable and thus rigid and therefore suitable, whereas the system of Fig. 4.21, bottom, is displaceable in its right segment (rotation around the hinge point G is possible) and therefore kinematically indeterminate and unsuitable.

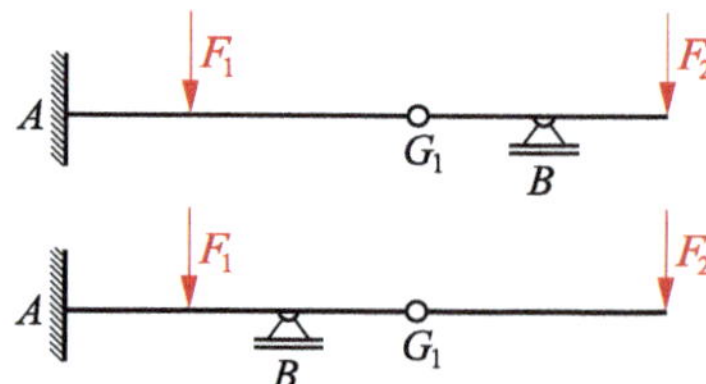

Fig. 4.21 Rigid system (top), kinematically indeterminate system (bottom)

Example 4.8

For the beam in Fig. 4.22 the support and hinge reactions are to be determined.

Solution:

With $n = 2$, $r = 4$ and $v = 2$, the criterion $3n = r + v$ is fulfilled. The beam is therefore statically determinate and also non-displaceable, i.e. rigid. We cut the beam free and look at the two free-body diagrams in Fig. 4.22, bottom, on which we have applied the support reactions A_H, A_V, M_A, B_V as well as the hinge forces G_H, G_V and the resultants of the two applied line loads. We first analyse the free body-diagram of the right beam segment and calculate the sum of the moments with respect to the hinge point G:

$$\overset{\curvearrowleft}{\sum} M_G = 0: \quad B_V \cdot l - F_0 \cdot \frac{l}{2} - \frac{1}{2} F_0 \cdot \frac{4}{3} l - F_0 \cdot 2l = 0 \quad \rightarrow \quad B_V = \frac{19}{6} F_0. \tag{4.25}$$

The sum of the vertical forces then results in the vertical hinge force G_V:

$$\overset{\uparrow}{\sum} V = 0: \quad G_V + \frac{19}{6} F_0 - F_0 - \frac{1}{2} F_0 - F_0 = 0 \quad \rightarrow \quad G_V = -\frac{2}{3} F_0. \tag{4.26}$$

The horizontal force sum leads to a vanishing horizontal hinge force G_H:

$$\overset{\rightarrow}{\sum} V = 0: \quad G_H = 0. \tag{4.27}$$

We now also consider at the free-body diagram of Fig. 4.22, bottom left, and obtain the support force A_V from the sum of the vertical forces:

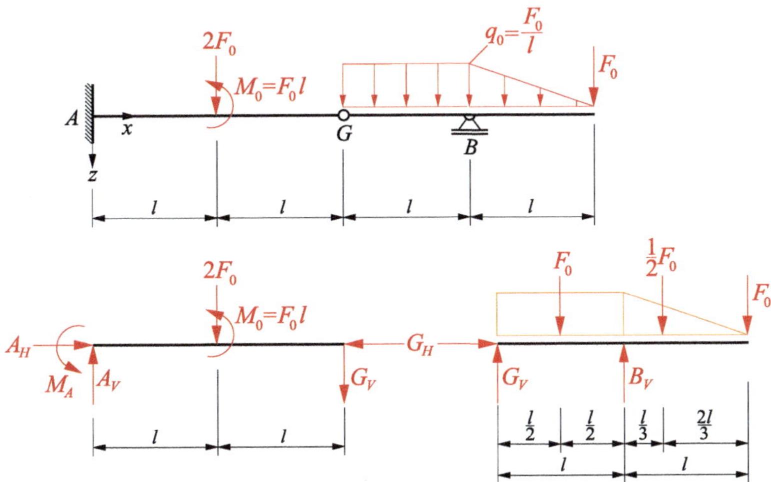

Fig. 4.22 Beam (top), determination of support and hinge reactions (bottom)

$$\overset{\uparrow}{\sum} V = 0: \quad A_V - 2F_0 + \frac{2}{3}F_0 = 0 \quad \rightarrow \quad A_V = \frac{4}{3}F_0. \tag{4.28}$$

The sum of the moments around the support point A results in the clamping moment M_A:

$$\overset{\curvearrowleft}{\sum} M_A = 0: \quad M_A + F_0 \cdot l - 2F. + \frac{2}{3}F_0 \cdot 2l = 0 \quad \rightarrow \quad M_A = -\frac{1}{3}F_0 l. \tag{4.29}$$

◀

Example 4.9

For the frame of Fig. 4.23 the support and hinge reactions are to be determined.

Solution:

With $n = 2, r = 4$ and $v = 2$, the criterion $3n = r + v$ is fulfilled and the frame is statically determinate. We cut the supports A and B free (see Fig. 4.23, centre) and calculate the moment sums around the two support points A and B:

$$\overset{\curvearrowright}{\sum} M_B = 0: \quad A_V \cdot l - ql \cdot \frac{l}{2} = 0 \quad \rightarrow \quad A_V = \frac{q_0 l}{2},$$
$$\overset{\curvearrowleft}{\sum} M_A = 0: \quad B_V \cdot l - ql \cdot \frac{l}{2} = 0 \quad \rightarrow \quad B_V = \frac{q_0 l}{2}. \tag{4.30}$$

The vertical force sum can be used to check the results for A_V and B_V:

$$\overset{\uparrow}{\sum} V = 0: \quad A_V + B_V - ql = 0. \tag{4.31}$$

It can be seen that the support forces $A_V = B_V = \frac{ql}{2}$ fulfil this condition.

To determine the remaining support forces A_H and B_H as well as the hinge reactions G_H and G_V, we now also cut through the hinge point G and thus obtain the two free-body diagrams in Fig. 4.23, bottom. We form the sum of the moments around the hinge point G at the left free body image:

$$\overset{\curvearrowleft}{\sum} M_G = 0: \quad A_H \cdot h - A_V \cdot \frac{l}{2} + \frac{ql}{2} \cdot \frac{l}{4} = 0 \quad \rightarrow \quad A_H = \frac{ql^2}{8h}. \tag{4.32}$$

The horizontal joint force G_H can then be determined from the sum of the horizontal forces:

$$\overset{\leftarrow}{\sum} H = 0: \quad G_H - A_H = 0 \quad \rightarrow \quad G_H = \frac{ql^2}{8h}. \tag{4.33}$$

The support force B_H can then be determined in the same way on the right-hand free-body diagram:

$$\overset{\leftarrow}{\sum} H = 0: \quad B_H - G_H = 0 \quad \rightarrow \quad B_H = \frac{ql^2}{8h}. \tag{4.34}$$

Finally, from one of the two free-body diagrams in Fig. 4.23, bottom, the hinge force G_V can be determined. If we look at the left free-body diagram, the vertical force sum results in:

$$\overset{\uparrow}{\sum} H = 0: \quad G_V + A_V - \frac{ql}{2} = 0 \quad \rightarrow \quad G_V = 0. \tag{4.35}$$

◀

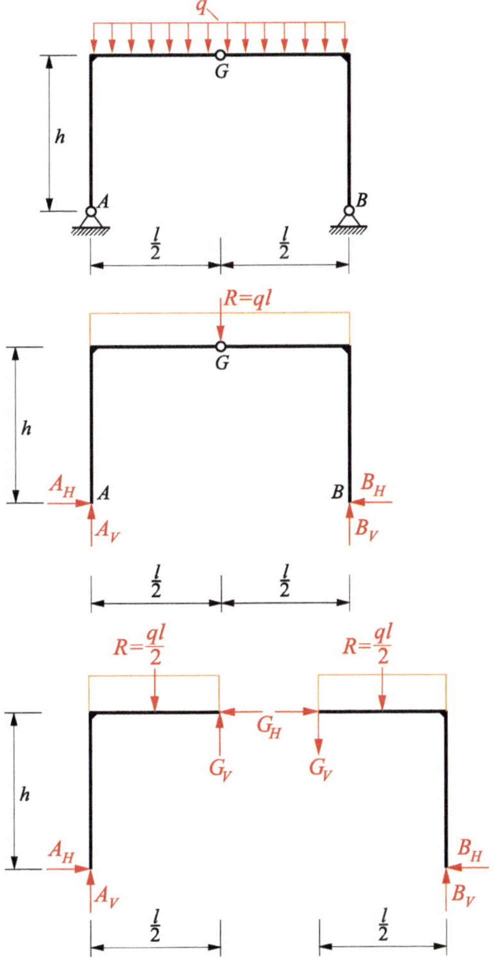

Fig. 4.23 Frame (top), free-body diagram (centre), section through the hinge (bottom)

Example 4.10

For the frame of Fig. 4.24, the support and hinge reactions are to be determined.

Solution:

We first cut the two supports A and B free and look at the free-body diagram in Fig. 4.24, bottom left. The vertical support force A_V can be obtained from the sum of the moments around the support B:

$$\overset{\curvearrowright}{\sum} M_B = 0: \quad A_V \cdot 2l + F_2 \cdot l + F_1 \cdot l - 2q_0 l \cdot l = 0 \quad \rightarrow \quad A_V = -\frac{q_0 l}{2}. \quad (4.36)$$

Accordingly, A_V actually points downwards. The vertical support force B_V then follows from the sum of the vertical forces:

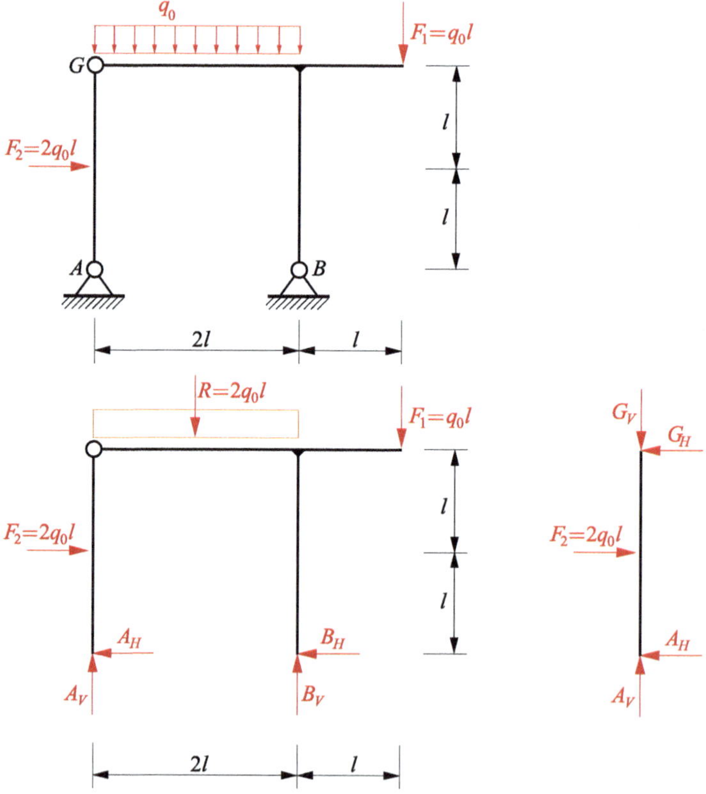

Fig. 4.24 Frame (top), free-body diagram (bottom left), section through the hinge (bottom right)

$$\overset{\uparrow}{\sum} V = 0: \quad B_V + A_V - F_1 - 2q_0 l = 0 \quad \rightarrow \quad B_V = \frac{7q_0 l}{2}. \tag{4.37}$$

The results for A_V and B_V can be put to the test by forming the sum of moments with respect to the support point A:

$$\overset{\curvearrowleft}{\sum} M_A = 0: \quad B_V \cdot 2l - F_2 \cdot l - F_1 \cdot 3l - 2q_0 l \cdot l = 0. \tag{4.38}$$

It can be seen that this equilibrium condition is fulfilled identically.

We now look at the free-body diagram in Fig. 4.24, bottom right, in which we have cut out the beam between support A and hinge G. From the vertical force sum we obtain the vertical joint force G_V as:

$$\overset{\downarrow}{\sum} V = 0: \quad G_V - A_V = 0 \quad \rightarrow \quad G_V = -\frac{q_0 l}{2}. \tag{4.39}$$

We obtain the horizontal support force A_H from the sum of the moments with respect to the hinge G:

$$\overset{\curvearrowright}{\sum} M_G = 0: \quad A_H \cdot 2l - F_2 \cdot l = 0 \quad \rightarrow \quad A_H = q_0 l. \tag{4.40}$$

The sum of the horizontal forces then results in the horizontal hinge force G_H as:

$$\overset{\leftarrow}{\sum} H = 0: \quad G_H + A_H - F_2 = 0 \quad \rightarrow \quad G_H = q_0 l. \tag{4.41}$$

Let us now look again at the free-body diagram in Fig. 4.24, bottom left. We can now finally determine the horizontal support force B_H from the sum of the horizontal forces:

$$\overset{\leftarrow}{\sum} H = 0: \quad B_H + A_H - F_2 = 0 \quad \rightarrow \quad B_H = q_0 l. \tag{4.42}$$

◀

4.4 Spatial Structures

The procedure for determining support reactions of spatial structures is analogous to the previously considered planar structures. It should be noted that a body in space has a total of six degrees of freedom, namely three translations and three rotations, which are partially or completely hindered by supports, so that support reactions are caused.

As in the plane case, the support types are differentiated according to the number of support reactions they cause. A clamped support (Fig. 4.25) restrains all translations as well as all rotations, so that a total of six support reactions are caused, namely three support forces in the direction of the reference axes and three clamping moments about the three spatial axes. A hinged support (see Fig. 4.26, support point A) has three support forces. Such a support

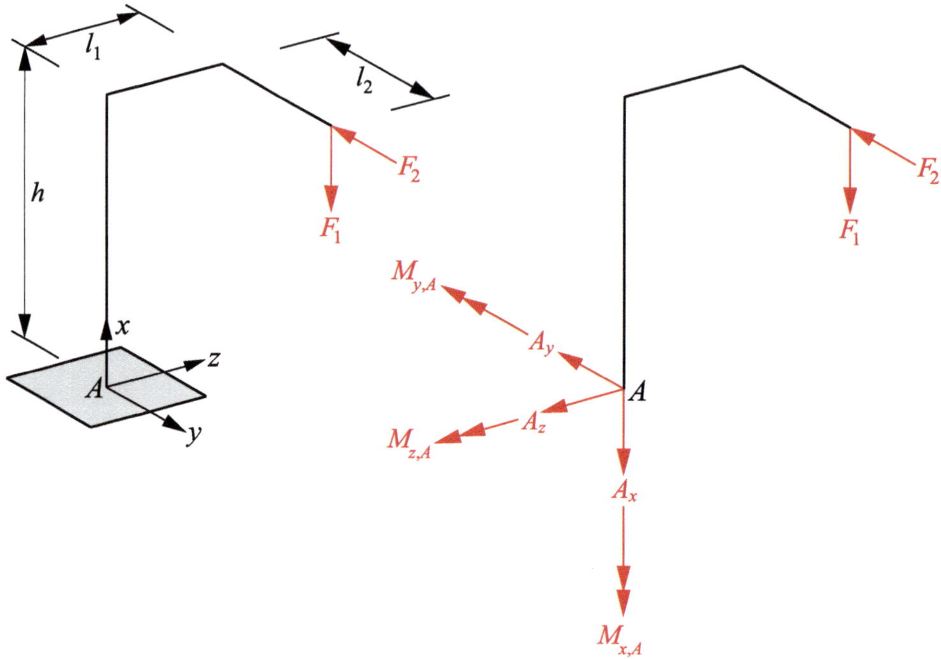

Fig. 4.25 Frame (left), free-body diagram (right)

hinders the three spatial translations, but allows all three rotations, so that three support forces occur, but no support moments. A bar, on the other hand, only absorbs one force in its own direction, so that at the support point B of Fig. 4.26 two support forces occur due to the two bars present there. Similarly, a support as in point C of Fig. 4.26 is single-valued, it restricts only one direction of movement and allows all other translational and rotational degrees of freedom, so that only a single support force occurs here. A spatial structure is statically determinate if all support reactions can be determined from the six spatial equilibrium conditions.

Example 4.11

For the clamped angled spatial frame of Fig. 4.25 the support reactions are to be determined.

Solution:

We cut the frame free at the clamping point A and draw the spatial support forces A_x, A_y, A_z and support moments $M_{x,A}$, $M_{y,A}$, $M_{z,A}$ in the resulting free-body diagram (Fig. 4.25, right). The equilibrium conditions here lead to the following results:

$$\sum F_x = 0: \quad A_x + F_1 = 0 \quad \rightarrow \quad A_x = -F_1,$$
$$\sum F_y = 0: \quad A_y + F_2 = 0 \quad \rightarrow \quad A_y = -F_2,$$
$$\sum F_z = 0: \quad A_z = 0,$$
$$\sum M_{x,A} = 0: \quad M_{x,A} - F_2 l1 = 0 \quad \rightarrow \quad M_{x,A} = F_2 l_1,$$
$$\sum M_{y,A} = 0: \quad M_{y,A} + F_1 l1 = 0 \quad \rightarrow \quad M_{y,A} = -F_1 l_1,$$
$$\sum M_{z,A} = 0: \quad M_{z,A} + F_2 h - F_1 l_2 = 0 \quad \rightarrow \quad M_{z,A} = F_1 l_2 - F_2 h. \quad (4.43)$$

◀

Example 4.12

For the system of Fig. 4.26 the support reactions are to be determined.

Solution:

We cut the static system free at the support points A, B and C and draw the spatial support forces A_x, A_y, A_z, B_x, B_z, C_z in the free-body diagram. The equilibrium conditions are then:

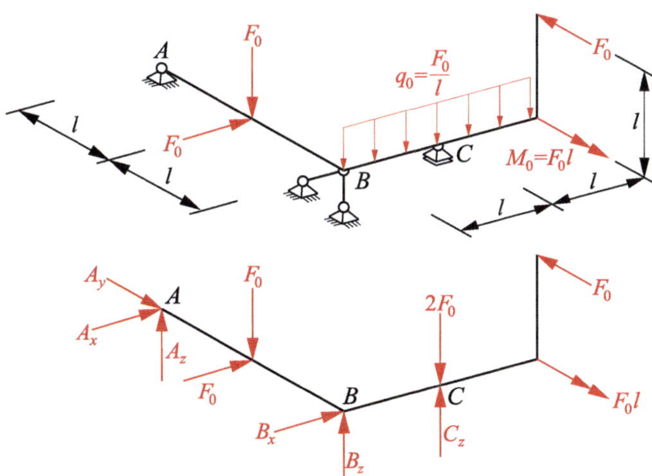

Fig. 4.26 Static system (top), free-body diagram (bottom)

$$\begin{aligned}
&\sum F_y = 0: \quad -A_y + F_0 = 0 \quad \rightarrow \quad A_y = F_0,\\
&\sum M_{y,A} = 0: \quad -C_z l + 2F_0 l - F_0 l = 0 \quad \rightarrow \quad C_z = F_0,\\
&\sum M_{x,A} = 0: \quad -B_z 2l - F_0 2l + F_0 l + 2F_0 2l - F_0 l = 0 \quad \rightarrow \quad B_z = F_0,\\
&\sum F_z = 0: \quad A_z - F_0 + F_0 + F_0 - 2F_0 = 0 \quad \rightarrow \quad A_z = F_0,\\
&\sum M_{z,A} = 0: \quad B_x 2l + F_0 l + F_0 2l = 0 \quad \rightarrow \quad B_x = -\frac{3}{2} F_0,\\
&\sum F_x = 0: \quad A_x + F_0 - \frac{3}{2} F_0 = 0 \quad \rightarrow \quad A_x = \frac{1}{2} F_0.
\end{aligned} \tag{4.44}$$

◀

Trusses 5

In this chapter, we consider plane load-bearing structures that consist exclusively of straight bars that are connected to each other by ideal hinges and that are loaded exclusively by point forces acting in the hinges. Such structures are known as trusses. We will investigate how to determine the forces in the members of trusses and will familiarise ourselves with both analytical procedures and a purely graphical method.

5.1 Fundamentals

A truss is a structure that consists exclusively of straight bars that are connected to each other in ideal and frictionless hinges (the so-called nodes). The bars are assumed to be massless and the load consists exclusively of point forces that act in the nodes but not on the members of the truss itself. We assume that the forces acting in a truss are ideally centred on the bars, so that the bars of the truss are exclusively under tensile or compressive normal forces. An exemplary truss is shown in Fig. 5.1, top. This truss under two point forces consists of nine bars that are connected to each other in a total of six nodes, whereby those nodes that are subject to a bearing are counted as well. At this point, we want to agree that we number the members consecutively with Arabic numerals, whereas we want to use Roman numerals for the nodes. Accordingly, the truss in Fig. 5.1 has bars 1–9, which are connected to each other in nodes I-VI.

A truss is said to be statically determinate if the equilibrium conditions in the plane are sufficient to determine the support reactions. Accordingly, the truss of Fig. 5.1 is statically determinate. By cutting it free as shown in Fig. 5.1, bottom, we visualise the support forces A_H, A_V in node I and B_V in node III. To determine the support forces, the truss can be assumed to be a rigid body (solidification principle). We first calculate the sum of the

C. Mittelstedt, *Engineering Mechanics 1: Statics*,
https://doi.org/10.1007/978-3-662-71852-0_5

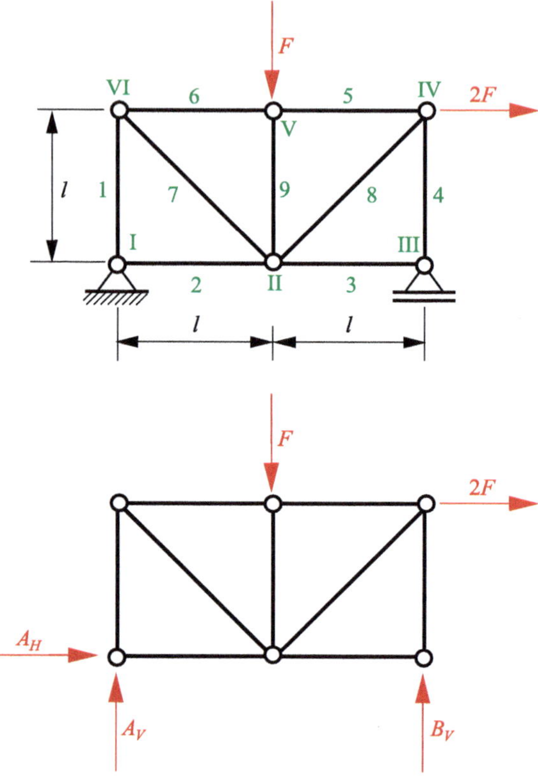

Fig. 5.1 Truss consisting of nine bars connected to each other in six nodes

moments with respect to support node I and obtain:

$$\sum^{\curvearrowleft} M_B = 0: \quad B_V \cdot 2l - F \cdot l - 2F \cdot l = 0, \tag{5.1}$$

from which the support force B_V can be determined as:

$$B_V = \frac{3}{2} F. \tag{5.2}$$

We obtain the sum of the vertical forces as follows:

$$\sum^{\uparrow} V = 0: \quad A_V + B_V - F = 0, \tag{5.3}$$

from which we obtain the still unknown support force A_V as:

$$A_V = -\frac{1}{2} F. \tag{5.4}$$

Accordingly, A_V actually points downwards. Finally, the still unknown support force A_H results from the sum of the horizontal forces:

$$\sum^{\rightarrow} H = 0: \quad A_H + 2F = 0, \tag{5.5}$$

from which A_H follows as:

$$A_H = -2F. \tag{5.6}$$

Accordingly, this support force actually points to the left.

The truss of Fig. 5.2, on the other hand, is statically indeterminate - here the equilibrium conditions are not sufficient to determine the support reactions.

The bar forces of a truss are determined mathematically by cutting the nodes free (so-called nodal section method, see Sect. 5.2) or by deliberately performing sections through selected bars (the so-called Ritter method or method of sections, see Sect. 5.3). A truss is described as internally statically determinate if all bar forces of a truss can be determined by using the equilibrium conditions. This is the case for the truss of Fig. 5.1, this truss is both externally statically determinate and internally statically determinate. The truss in Fig. 5.2 is externally statically indeterminate, but internally statically determinate. Once the support reactions are available, the bar forces can be determined using the equilibrium conditions in the plane. The Fig. 5.3 shows a truss that is both externally and internally statically indeterminate, which is derived from the truss in Fig. 5.2 by adding the two additional diagonal bars 10 and 11. In this case, both the support forces and the member forces can no longer be determined from the equilibrium conditions. In this chapter, we will deal exclusively with such plane trusses that are both externally and internally statically determinate. Statically indeterminate systems are discussed in detail in Volume 2.

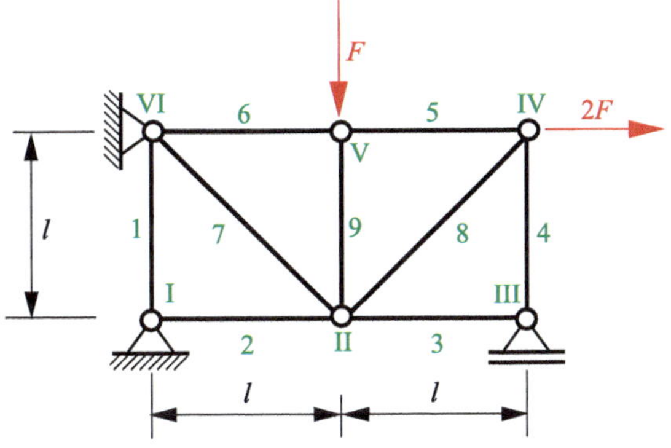

Fig. 5.2 Statically indeterminate truss

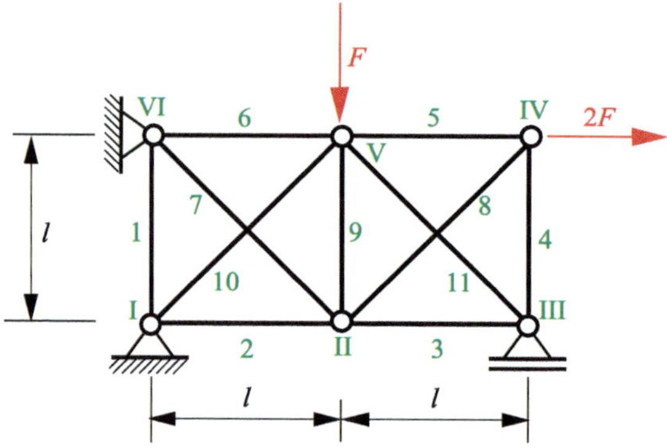

Fig. 5.3 Externally and internally statically indeterminate truss

If the nodal section method is used to determine the bar forces, i.e. if each node is cut free and the two available force equilibrium conditions are considered there, then $2i$ equations for the $j + r$ unknown member forces and support reactions can be given for a truss with i nodes, j members and r support forces. In order for a truss to be statically determinate, the following criterion must be fulfilled:

$$2i = j + r. \tag{5.7}$$

Using the example of the truss in Fig. 5.1 with six nodes, $2i = 2 \cdot 6 = 12$ results with $j = 9$ bars and $r = 3$ support forces results in static determinacy, whereas the two trusses in Figs. 5.2 and 5.3 are statically indeterminate.

A truss must always be constructed in such a way that it cannot move in whole or in part. If this is not guaranteed, it is also referred to as a kinematically indeterminate truss. If a truss is immovable, it is kinematically determinate. The truss in Fig. 5.4 is created from the truss in Fig. 5.1 by removing bars 7 and 8. This obviously makes the truss displaceable as indicated and therefore kinematically indeterminate. The criterion (5.7) as a necessary condition for kinematic and static determinacy is no longer fulfilled. Another example is shown in Fig. 5.5. This truss is created from the truss in Fig. 5.2 by removing the bar 9. Even if this truss is externally statically indeterminate, the removal of bar 9 will cause the two bars 5 and 6 to rotate around nodes IV and VI, with node V moving downwards. However, in contrast to the truss in Fig. 5.4, this is a displacement 'on a small scale', i.e. there is only an infinitesimally small deflection. Nevertheless, such a truss is unsuitable for practical use and must be avoided.

Trusses are also categorised according to the type of structure. A so-called simple truss exists if it is made up of a number of bar triangles. This applies, for example, to the truss shown in Fig. 5.1, which is made up of a total of four bar triangles. Such a truss is always internally statically and also kinematically determinate. Whether it is also externally statically determinate depends on the number of unknown support reactions. However, care must be taken with the arrangement of the nodes - the three nodes of two neighbouring members

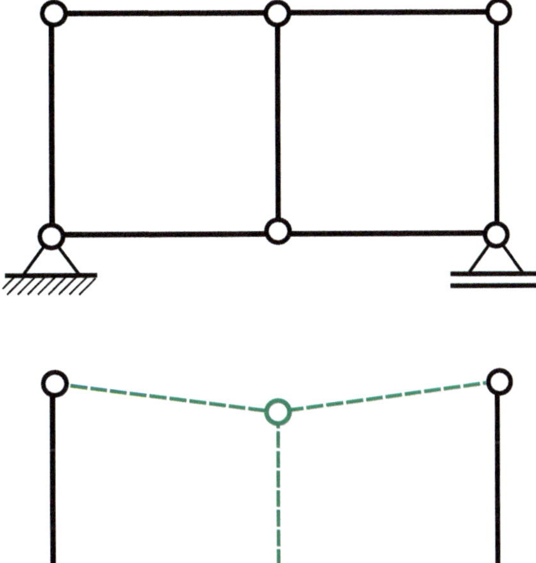

Fig. 5.4 Kinematically indeterminate truss

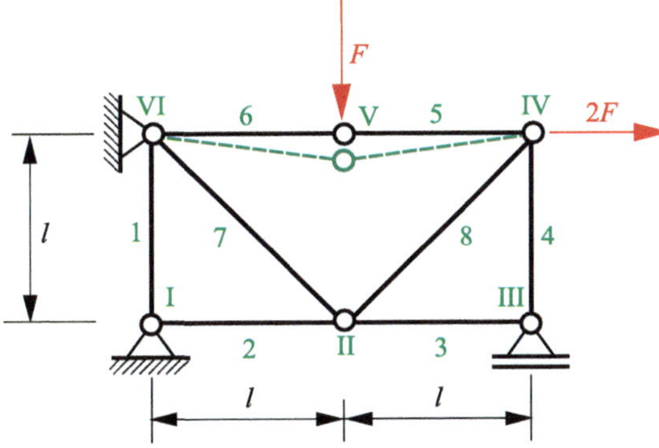

Fig. 5.5 Kinematically indeterminate truss

must not lie on a line, as the truss would otherwise become locally displaceable. An example is shown in Fig. 5.5, in which the three joints of bars 5 and 6 lie on a line so that node V can move in the vertical direction.

Trusses can also be constructed in such a way that two or more system components, which in turn all consist of bar triangles (i.e. represent simple partial trusses in themselves), are connected to each other by three individual members. However, these individual bars must not all be parallel to each other. An example is shown in Fig. 5.6, top. Here, two simple

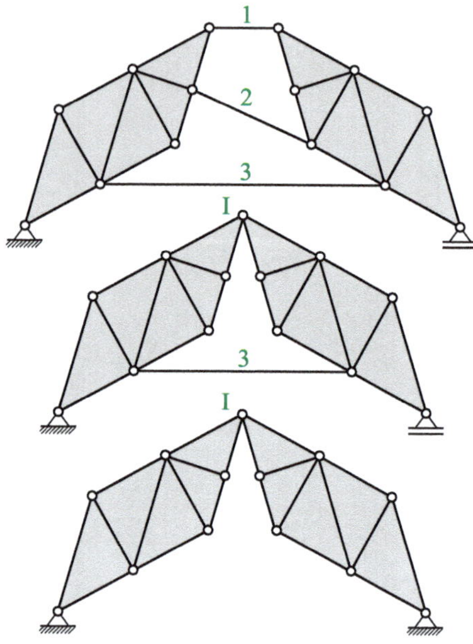

Fig. 5.6 Truss consisting of two simple partial trusses

truss structures (highlighted in grey) have been connected by the three members 1, 2 and 3. Another option for connecting the two trusses shown here is to merge the two trusses at node I and only add member 3 (Fig. 5.6, centre). The resulting overall truss is statically and kinematically determinate. Member 3 can be omitted if the right single-valued support is replaced by a two-valued support (Fig. 5.6, bottom) to prevent kinematic indeterminacy.

5.2 Nodal Section Method

5.2.1 Procedure

The so-called nodal section method is the common procedure for determining bar forces in trusses. Each node of the truss (including support nodes) is cut free and the two available force equilibrium conditions are set up at each of the i nodes so that the j member forces can be determined. In many cases, the support forces of a given truss will be determined in advance so that they do not have to be determined from node sections. The member forces acting at the truss nodes cut out in this way are applied, whereby we agree that positive member forces are tensile forces, i.e. they point away from the node in the free-body diagrams. The subsequent calculation will show whether this assumption was correct or not. If the result is a bar force with a positive sign, then the bar under consideration is a tensile bar. If, on the other hand, it has a negative sign, then it is a compressive member and the member

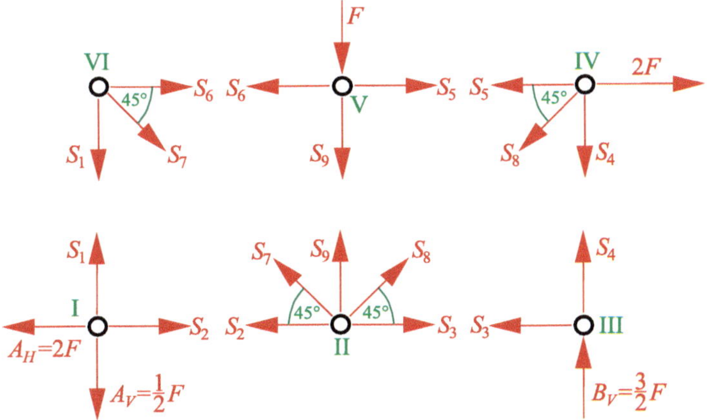

Fig. 5.7 Nodal section method

force points to the node under consideration. Since a bar is always connected to two nodes, the bar force of a bar must also appear in the free-body diagrams of both nodes due to the interaction principle 'actio = reactio'.

To illustrate this procedure, we look at the truss in Fig. 5.1 again and first draw the free-body diagrams of the individual nodes. This is shown in Fig. 5.7, where we have applied the previously calculated support forces A_H, A_V and B_V with their magnitudes and actual directions of action.

We will now consider node by node and start our considerations with node I. From the vertical and horizontal force sum we obtain

$$\overset{\rightarrow}{\sum} H = 0: \quad S_2 - 2F = 0 \quad \rightarrow \quad S_2 = 2F,$$
$$\overset{\uparrow}{\sum} V = 0: \quad S_1 - \frac{1}{2}F = 0 \quad \rightarrow \quad S_1 = \frac{1}{2}F. \tag{5.8}$$

This determines the two bar forces S_1 and S_2, both of which are tensile forces, so that bars 1 and 2 are tensile members.

We can proceed analogously for node II, whereby we find that with known S_2 with the bars S_3, S_7, S_8, S_9 there are a total of four unknown bar forces at this node at this point in the calculation, for which we only have two equilibrium conditions available. Therefore, the consideration of this node is not practical at this point of the calculation, so we will continue with node III and consider node II later.

For node III, the vertical and horizontal force sums result in:

$$\overset{\leftarrow}{\sum} H = 0: \quad S_3 = 0,$$

$$\overset{\uparrow}{\sum} V = 0: \quad S_4 + \frac{3}{2} F = 0 \quad \rightarrow \quad S_4 = -\frac{3}{2} F. \tag{5.9}$$

Obviously, bar 3 is a so-called zero bar (for the determination of zero bars, some explanations will follow later), whereas bar 4 is a compression bar—the bar force S_4 results with a negative sign.

We continue with the considerations at node IV and first look at the vertical force sum:

$$\overset{\downarrow}{\sum} V = 0: \quad S_4 + S_8 \sin 45° = 0 \quad \rightarrow S_8 = -\frac{S_4}{\sin 45°}. \tag{5.10}$$

If we insert the previously determined result $S_4 = -\frac{3}{2} F$ and also $\sin 45° = \frac{1}{\sqrt{2}}$, then it follows that bar 8 is a tensile bar with the bar force S_8 as follows:

$$S_8 = \frac{3\sqrt{2}}{2} F = \frac{3}{\sqrt{2}} F. \tag{5.11}$$

We now also consider the horizontal force sum at node IV and obtain:

$$\overset{\leftarrow}{\sum} H = 0: \quad S_5 + S_8 \cos 45° - 2F = 0 \quad \rightarrow \quad S_5 = \frac{1}{2} F. \tag{5.12}$$

Hence, bar 5 is a tensile bar.

At node V, the following result is obtained:

$$\overset{\leftarrow}{\sum} H = 0: \quad S_6 - S_5 = 0 \quad \rightarrow \quad S_6 = \frac{1}{2} F,$$

$$\overset{\downarrow}{\sum} V = 0: \quad S_9 + F = 0 \quad \rightarrow \quad S_9 = -F. \tag{5.13}$$

Finally, we look at node VI and obtain the horizontal sum of forces as:

$$\overset{\rightarrow}{\sum} H = 0: \quad S_6 + S_7 \cos 45° = 0. \tag{5.14}$$

With $S_6 = \frac{1}{2} F$ and $\cos 45° = \frac{1}{\sqrt{2}}$ we obtain:

$$S_7 = -\frac{\sqrt{2}}{2} F = -\frac{1}{\sqrt{2}} F. \tag{5.15}$$

From the vertical force sum we get the possibility to check our result for the bar force S_1. The following applies:

$$\overset{\downarrow}{\sum} V = 0: \quad S_1 + S_7 \sin 45^\circ = 0. \tag{5.16}$$

With $S_7 = -\frac{1}{\sqrt{2}}F$ and $\sin 45^\circ = \frac{1}{\sqrt{2}}$ it follows:

$$S_1 = \frac{1}{2}F. \tag{5.17}$$

This corresponds to the result (5.8).

We can now use the force sums at node II to check our results. From the vertical force sum follows:

$$\overset{\uparrow}{\sum} V = 0: \quad S_7 \sin 45^\circ + S_9 + S_8 \sin 45^\circ = 0. \tag{5.18}$$

With the bar forces S_7, S_8 and S_9 already determined, it can be seen that the left side of the vertical force sum becomes zero—equilibrium is therefore guaranteed. For the horizontal force sum, we obtain

$$\overset{\leftarrow}{\sum} H = 0: \quad S_2 + S_7 \cos 45^\circ - S_8 \cos 45^\circ - S_3 = 0. \tag{5.19}$$

With the already calculated bar forces S_2, S_7, S_8 and S_3, the left side of the equilibrium condition is also zero here, so that the horizontal force equilibrium is also fulfilled.

The results for the bar forces are summarised in Table 5.1.

Table 5.1 Bar forces S_1–S_9

Bar force	
S_1	$\frac{1}{2}F$
S_2	$2F$
S_3	0
S_4	$-\frac{3}{2}F$
S_5	$\frac{1}{2}F$
S_6	$\frac{1}{2}F$
S_7	$-\frac{1}{\sqrt{2}}F$
S_8	$\frac{3}{\sqrt{2}}F$
S_9	$-F$

Example 5.1

Consider the truss shown in Fig. 5.8, which consists of 11 bars. The bar forces are searched for using the nodal section method.

Solution:

We cut the nodes of the truss free and obtain the free-body diagram of Fig. 5.9. This truss is an example that can be solved without first determining the support forces A_H, B_H and B_V. We start the calculations by looking at node IV. The vertical force sum follows as:

$$\overset{\uparrow}{\sum} V = 0: \quad S_4 \sin 45^\circ - F = 0 \quad \rightarrow \quad S_4 = \frac{F}{\sin 45^\circ} = \sqrt{2}F. \tag{5.20}$$

Bar 4 is therefore a tensile bar. The horizontal force sum yields:

$$\overset{\leftarrow}{\sum} H = 0: \quad S_3 + S_4 \cos 45^\circ = 0 \quad \rightarrow \quad S_3 = -S_4 \cos 45^\circ = -F. \tag{5.21}$$

Bar 3 is therefore a compressive bar.

We now consider node V. From the vertical force sum we obtain:

$$\overset{\downarrow}{\sum} V = 0: \quad S_4 \sin 45^\circ + S_{11} = 0 \quad \rightarrow \quad S_{11} = -S_4 \sin 45^\circ = -F. \tag{5.22}$$

This means that bar 11 is a compressive bar. The sum of the horizontal forces is as follows:

$$\overset{\leftarrow}{\sum} H = 0: \quad S_5 - S_4 \cos 45^\circ = 0 \quad \rightarrow \quad S_5 = S_4 \cos 45^\circ = F. \tag{5.23}$$

Bar 5 is therefore a tensile bar.

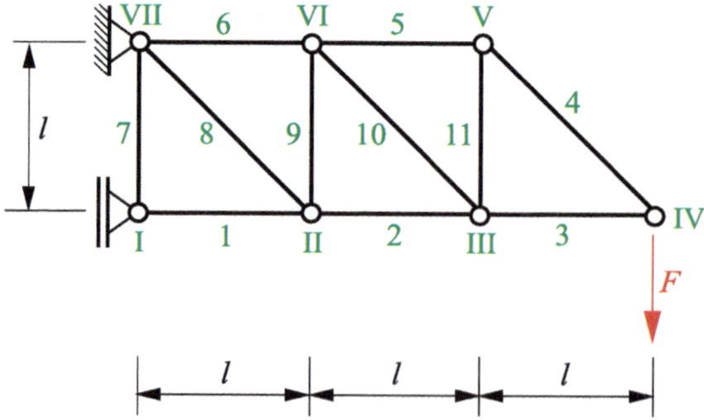

Fig. 5.8 Given truss with 11 bars

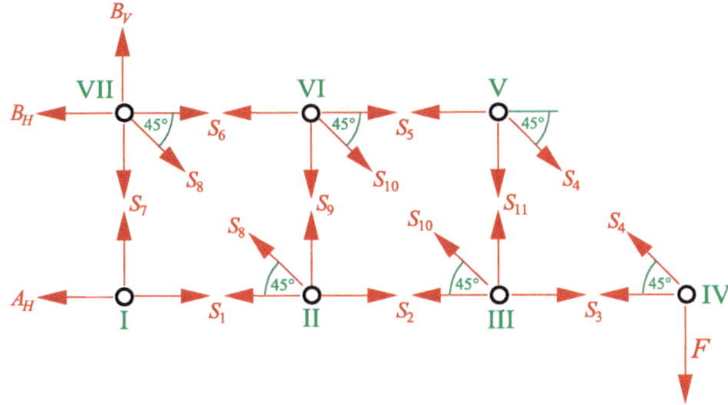

Fig. 5.9 Free-body diagram

We continue with node III and obtain:

$$\overset{\uparrow}{\sum} V = 0: \quad S_{11} + S_{10} \sin 45° = 0 \quad \rightarrow \quad S_{10} = -\frac{S_{11}}{\sin 45°} = \sqrt{2}F,$$
$$\overset{\leftarrow}{\sum} H = 0: \quad S_2 + S_{10} \cos 45° - S_3 = 0 \quad \rightarrow \quad S_2 = -S_{10} \cos 45° + S_3 = -2F. \tag{5.24}$$

The results at node VI are:

$$\overset{\downarrow}{\sum} V = 0: \quad S_9 + S_{10} \sin 45° = 0 \quad \rightarrow \quad S_9 = -F,$$
$$\overset{\leftarrow}{\sum} H = 0: \quad S_6 - S_{10} \cos 45° - S_5 = 0 \quad \rightarrow \quad S_6 = S_{10} \cos 45° + S_5 = 2F. \tag{5.25}$$

We continue with node II and calculate the bar forces S_1 and S_8:

$$\overset{\uparrow}{\sum} V = 0: \quad S_9 + S_8 \sin 45° = 0 \quad \rightarrow \quad S_8 = -\frac{S_9}{\sin 45°} = \sqrt{2}F,$$
$$\overset{\leftarrow}{\sum} H = 0: \quad S_1 + S_8 \cos 45° - S_2 = 0 \quad \rightarrow \quad S_1 = -S_8 \cos 45° + S_2 = -3F. \tag{5.26}$$

Both the bar force S_7 and the support force A_H can then be determined at node I. It follows:

$$\overset{\uparrow}{\sum} V = 0: \quad S_7 = 0,$$
$$\overset{\leftarrow}{\sum} H = 0: \quad A_H - S_1 = 0 \quad \rightarrow \quad A_H = S_1 = -3F. \tag{5.27}$$

Accordingly, bar 7 is a so-called zero-force bar. The support force A_H results here with a negative sign, hence it is actually directed to the right and exerts pressure on node I.

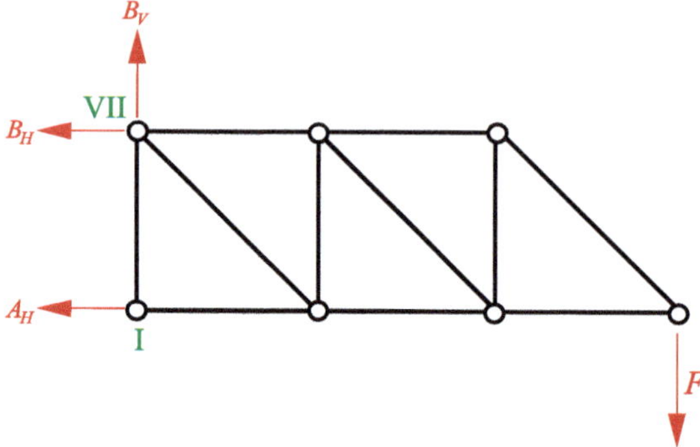

Fig. 5.10 Free-body diagram for the entire truss to determine the support forces

Finally, we consider the force sums at node VII in order to determine the support reactions B_H and B_V occurring there. It follows:

$$\overset{\uparrow}{\sum} V = 0: \quad B_V - S_7 - S_8 \sin 45^\circ = 0 \quad \rightarrow \quad B_V = S_7 + S_8 \sin 45^\circ = F,$$

$$\overset{\leftarrow}{\sum} H = 0: \quad B_H - S_6 - S_8 \cos 45^\circ = 0 \quad \rightarrow \quad B_H = S_6 + S_8 \cos 45^\circ = 3F \tag{5.28}$$

For these two support forces, the assumed direction of action was therefore correct, both support forces exert tension on node VII.

It is easy to see that the same support reactions A_H, B_H and B_V are obtained if the equilibrium is considered for the entire truss (Fig. 5.10). Firstly, we consider the sum of the moments with respect to node VII:

$$\overset{\curvearrowright}{\sum} M = 0: \quad A_H \cdot l + F \cdot 3l = 0 \quad \rightarrow \quad A_H = -3F. \tag{5.29}$$

The two additional support forces B_H and B_V then follow from the sum of the forces in the vertical and horizontal directions:

$$\overset{\uparrow}{\sum} V = 0: \quad B_V - F = 0 \quad \rightarrow \quad B_V = F,$$

$$\overset{\leftarrow}{\sum} H = 0: \quad B_H + A_H = 0 \quad \rightarrow \quad B_H = -A_H = 3F. \tag{5.30}$$

The calculated bar forces for this example are summarised in Table 5.2.

◀

Table 5.2 Bar forces S_1-S_{11}

Bar force	
S_1	$-3F$
S_2	$-2F$
S_3	$-F$
S_4	$\sqrt{2}F$
S_5	F
S_6	$2F$
S_7	0
S_8	$\sqrt{2}F$
S_9	$-F$
S_{10}	$\sqrt{2}F$
S_{11}	$-F$

Example 5.2

Consider the truss of Fig. 5.11. The bar forces are to be determined using the nodal section method.

Solution:

Firstly, we determine the support forces of the truss and consider the free-body diagram of the entire truss, as shown in Fig. 5.12. The sum of the vertical forces results in the support force A_V:

$$\overset{\uparrow}{\sum} V = 0: \quad A_V - 3F - 5F - 8F = 0 \quad \rightarrow \quad A_V = 16F. \tag{5.31}$$

The sum of the moments with respect to the support point B results in:

$$\overset{\curvearrowleft}{\sum} M_B = 0: \quad A_H \cdot 6l - 5F \cdot 5l - 8F \cdot 10l = 0 \quad \rightarrow \quad A_H = 17.5F. \tag{5.32}$$

Finally, the support force B_H follows from the horizontal force sum:

$$\overset{\leftarrow}{\sum} V = 0: \quad B_H - A_H = 0 \quad \rightarrow \quad B_H = A_H = 17.5F. \tag{5.33}$$

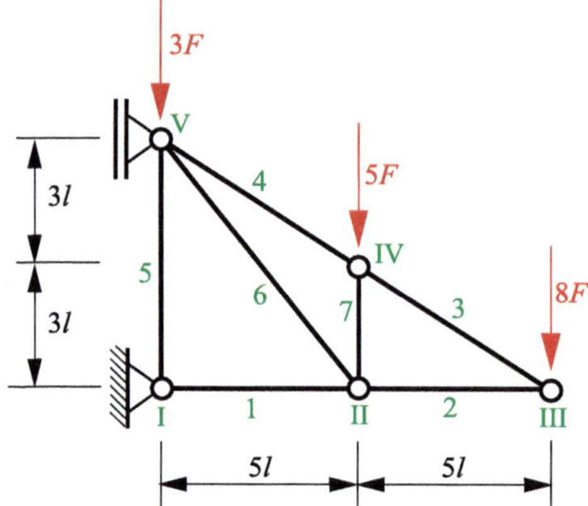

Fig. 5.11 Given truss with seven bars

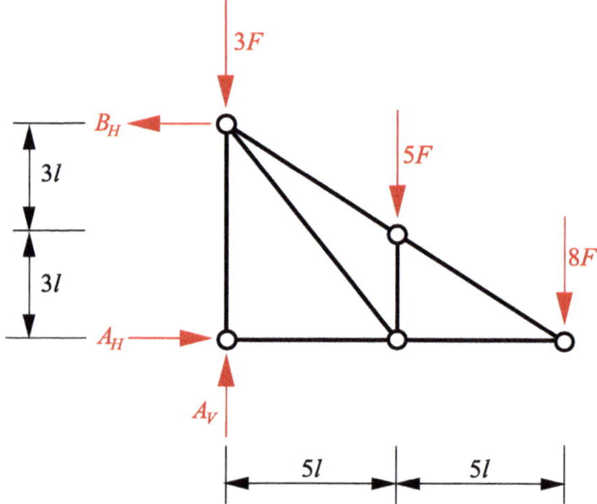

Fig. 5.12 Determination of the support forces

In the next step, we consider the free-body diagram of all truss nodes as shown in Fig. 5.13. The bar forces S_2 and S_3 can be obtained from the equilibrium at node III. We first form the equilibrium of the vertical forces and obtain

$$\overset{\uparrow}{\sum} V = 0: \quad S_3 \sin\alpha - 8F = 0. \tag{5.34}$$

With $\alpha = 30.96°$ it follows:

$$S_3 = \frac{8F}{\sin\alpha} = 15.55F. \tag{5.35}$$

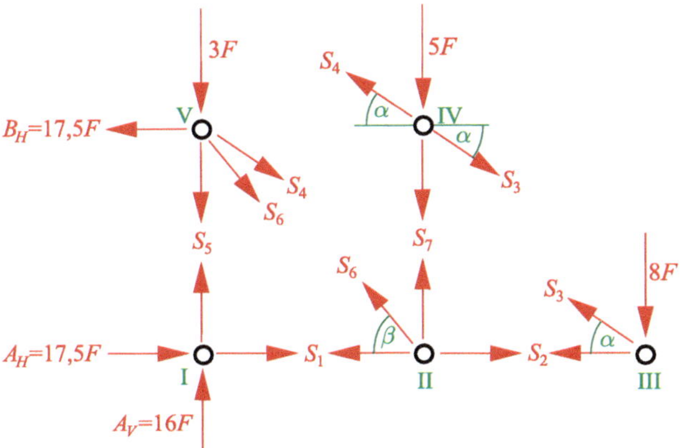

Fig. 5.13 Free-body diagrams of the truss nodes

The sum of horizontal forces results in:

$$\overset{\leftarrow}{\sum} H = 0: \quad S_2 + S_3 \cos\alpha = 0 \quad \rightarrow \quad S_2 = -S_3 \cos\alpha = -13.33F. \tag{5.36}$$

The two bar forces S_1 and S_5 can be determined from the equilibrium for node I:

$$\overset{\rightarrow}{\sum} H = 0: \quad S_1 + A_H = 0 \quad \rightarrow \quad S_1 = -A_H = -17.5F,$$

$$\overset{\uparrow}{\sum} V = 0: \quad S_5 + A_V = 0 \quad \rightarrow \quad S_5 = -A_V = -16F. \tag{5.37}$$

We also consider the force equilibrium at node II. The still unknown bar force S_6 results from the horizontal force sum:

$$\overset{\rightarrow}{\sum} H = 0: \quad S_6 \cos\beta + S_1 - S_2 = 0. \tag{5.38}$$

With $\beta = 50.19°$ it follows:

$$S_6 = \frac{-S_1 + S_2}{\cos 50.19°} = 6.51F. \tag{5.39}$$

From the sum of forces in the vertical direction follows:

$$\overset{\uparrow}{\sum} V = 0: \quad S_7 + S_6 \sin\beta = 0 \quad \rightarrow \quad S_7 = -S_6 \sin\beta = -5F. \tag{5.40}$$

The missing bar force S_4 can be determined from the sum of forces at node IV. We also have the opportunity to check our result for the bar force S_7 at this nodal section. The sum of the horizontal forces results in

$$\overset{\leftarrow}{\sum} H = 0: \quad S_4 \cos\alpha - S_3 \cos\alpha = 0 \quad \rightarrow \quad S_4 = S_3 = 15.55F. \tag{5.41}$$

This determines the last remaining bar force.

From the vertical force sum follows:

$$\overset{\downarrow}{\sum} V = 0: \quad S_7 + 5F = 0 \quad \rightarrow \quad S_7 = -5F. \tag{5.42}$$

This result agrees with the result that we have already determined for nodal section II.

The nodal section V offers us the opportunity to check our calculation results for the bar forces S_4, S_5 and S_6. We first calculate the horizontal force sum and obtain

$$\overset{\rightarrow}{\sum} H = 0: \quad S_4 \cos\alpha + S_6 \cos\beta - 17.5F = 0. \tag{5.43}$$

If the results already determined for S_4 and S_6 are inserted here, it can be seen that the left-hand side of the above equation gives exactly the value zero. The equilibrium condition is therefore fulfilled.

We also consider the sum of the vertical forces. The following applies:

$$\overset{\downarrow}{\sum} V = 0: \quad S_5 + S_6 \sin\beta + S_4 \sin\alpha + 3F = 0. \tag{5.44}$$

Here, too, it can be seen that the left-hand side of this equilibrium condition becomes zero after inserting S_4, S_5 and S_6.

The determined member forces are summarised in Table 5.3.

◀

Table 5.3 Bar forces S_1–S_7

Bar force	
S_1	$-17.5F$
S_2	$-13.33F$
S_3	$15.55F$
S_4	$15.55F$
S_5	$-16F$
S_6	$6.51F$
S_7	$-5F$

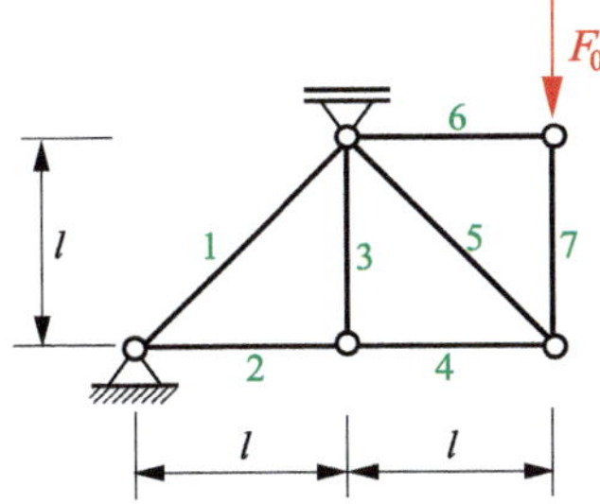

Fig. 5.14 Given truss with seven bars

Table 5.4 Bar forces S_1-S_7

Bar force	
S_1	$\sqrt{2}F_0$
S_2	$-F_0$
S_3	0
S_4	$-F_0$
S_5	$\sqrt{2}F_0$
S_6	0
S_7	$-F_0$

Example 5.3

For the truss shown in Fig. 5.14 the bar forces are to be determined.

Solution:

We will not describe the solution at this point; this is left to the reader for practice purposes. At this point, we only report the results for the bar forces (Table 5.4).

◀

5.2.2 General Rules

In many trusses, there are some bars that do not require any calculation, but from which we can draw general conclusions about bar forces simply by taking a look at them. One such general rule is recognising so-called obvious zero-force bars. The search for such obvious zero bars should always be at the beginning of a truss calculation in order to avoid unnecessary work. The following rules can be used to determine obvious zero-force bars that can be recognised without calculation.

If there is a node on a truss where two members are connected to each other in an unloaded node, these two members are zero-force bars. This can be seen in the truss of Fig. 5.15, top, for node I with the two bars 1 and 2, as can be seen immediately by cutting this node free and setting up the vertical and horizontal force sums.

If we examine a node on which two bars act and on which a force also points in the direction of one of the two bars, then the other bar is a zero bar and the applied force is fully supported by the bar whose direction corresponds to the direction of action of the force. This is shown on the truss in Fig. 5.15, centre, in the case of node I, where bar 1 is a zero-force member ($S_1 = 0$) and bar S_2 fully transmits the force F (i.e. $S_2 = F$). If the applied force F were to point to the left, i.e. if it exerts pressure on node I, bar 2 would be a compressions bar with $S_2 = -F$.

At an unloaded truss node to which a single-valued support is attached and at which two bars meet, one of which is aligned parallel to the support force and the other perpendicular to it, the bar perpendicular to the support reaction is a zero-force member. This applies to the two trusses in Fig. 5.15, top and centre, for bar 3, which is a zero-force bar in both cases.

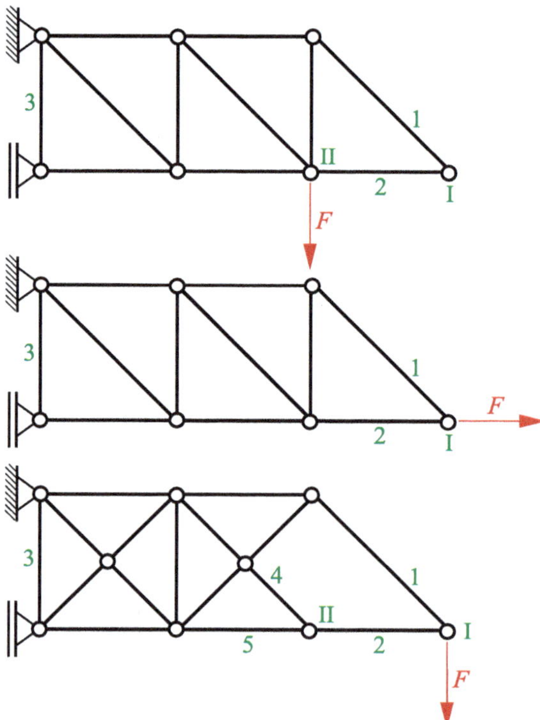

Fig. 5.15 Obvious zero-force bars

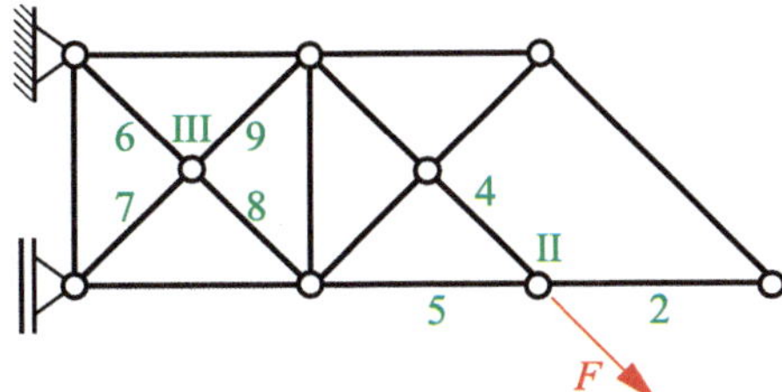

Fig. 5.16 Recognising obvious bar forces

However, care must be taken with the truss in Fig. 5.15, bottom. Bar 3 is not a zero-force member here due to the diagonal bar attached to this node.

We now also consider a truss node that is free of any load and to which two parallel bars are attached and another member that has a different direction to the first two. At such a node, the latter bar is then a zero-force bar and the two parallel bars have identical bar forces. Such a node is located with node II on the truss of Fig. 5.15, bottom, where bar 4 is a zero-force bar ($S_4 = 0$) and the two members 2 and 5 have identical bar forces ($S_2 = S_5$).

Further obvious correlations can be identified, which we can illustrate using Fig. 5.16.

At a loaded truss node with two parallel members and a third member orientated at an arbitrary angle to it, at which a force points exactly in the direction of the third bar, the bar forces of the two parallel members are identical and the bar force of the third member corresponds to the acting force. Such a node is located on the truss in Fig. 5.16 with node II, where the two member forces of members 2 and 5 are identical ($S_2 = S_5$) and the member force S_4 corresponds exactly to the applied force F ($S_4 = F$).

At an unloaded truss node with four members acting on it, two of which are parallel to each other, the member forces of the parallel members are identical. This is the case with the truss in Fig. 5.16 with node III, and the identities $S_6 = S_8$ and $S_7 = S_9$ apply to the bars at this node. Furthermore, this general rule applies not only to nodes at which the bars are orientated at right angles to each other, but also to two pairs of parallel bars that are orientated at arbitrary angles to each other.

Example 5.4

Consider the truss of Fig. 5.17. We want to determine all obvious zero-force members and equality relationships between bar forces. Finally, we want to determine the bar forces.

Solution:

At node II, bar 6 is obviously a zero-force member. This node is unloaded and bar 6 has a different direction than bars 1 and 2, whose directions match. Accordingly, the bar forces S_1 and S_2 must also be identical, as can be seen from the force equilibrium at this node:

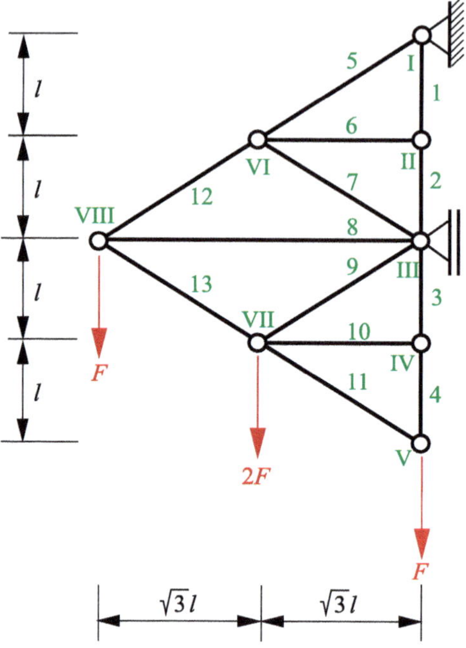

Fig. 5.17 Given truss with 13 bars

$$S_1 = S_2, \quad S_6 = 0. \tag{5.45}$$

If bar 6 is a zero-force member, it is immediately apparent from the observation of node VI that bar 7 is also a zero-force member. This node is also unloaded, and bar 7 has a different direction than bars 5 and 12, whose directions match. Accordingly, the bar forces S_5 and S_{12} must also be identical. The following therefore applies:

$$S_5 = S_{12}, \quad S_7 = 0. \tag{5.46}$$

Node IV also immediately shows that bar 10 is a zero-force bar. In addition, the member forces S_3 and S_4 are identical:

$$S_3 = S_4, \quad S_{10} = 0. \tag{5.47}$$

Looking at node V also shows that the force F applied there is borne exclusively by bar 4. Bar 11 is therefore a zero-force member. Hence:

$$S_4 = F, \quad S_{11} = 0. \tag{5.48}$$

The bar force S_3 therefore results as

$$S_3 = F. \tag{5.49}$$

Once bars 10 and 11 have been identified as zero-force members, it can be immediately concluded from the observation of node VII that the two bar forces S_9 and S_{13} must be identical. They act on this node at identical angles to the horizontal, and node VII is exclusively loaded by a vertical force $2F$:

$$S_9 = S_{13}. \tag{5.50}$$

We now determine the not yet known bar forces with the help of the nodal section method and start the considerations at node VII (see Fig. 5.18). With the angle $\alpha = 30°$ it follows from the sum of the vertical forces:

$$\overset{\uparrow}{\sum} V = 0: \quad S_9 \sin 30° + S_{13} \sin 30° - 2F = 0. \tag{5.51}$$

With the already established equality of the bar forces S_9 and S_{13} it follows:

$$S_9 = S_{13} = \frac{F}{\sin 30°} = 2F. \tag{5.52}$$

The two bar forces S_8 and S_{12} can be determined considering node VIII. The corresponding free-body diagram is shown in Fig. 5.19. We obtain from the sum of the vertical forces:

$$\overset{\uparrow}{\sum} V = 0: \quad S_{12} \sin\alpha - S_{13} \sin\alpha - F = 0 \quad \rightarrow \quad S_{12} = S_{13} + \frac{F}{\sin\alpha} = 4F. \tag{5.53}$$

The bar force S_5 can thus also be determined as

$$S_5 = S_{12} = 4F. \tag{5.54}$$

We determine the bar force S_8 by considering the horizontal force sum. The result is

$$\overset{\rightarrow}{\sum} H = 0: \quad S_8 + S_{12} \cos\alpha + S_{13} \cos\alpha = 0 \quad \rightarrow \quad S_8 = -4F \cos 30° - 2F \cos 30° = -3\sqrt{3}F. \tag{5.55}$$

The still unknown bar force S_2 can be determined from investigating node III. To do this, we consider the free-body diagram in Fig. 5.20. From the vertical force sum follows:

$$\overset{\uparrow}{\sum} V = 0: \quad S_2 - S_9 \sin\alpha - S_3 = 0 \quad \rightarrow \quad S_2 = S_9 \sin\alpha + S_3 = 2F \sin 30° + F = 2F. \tag{5.56}$$

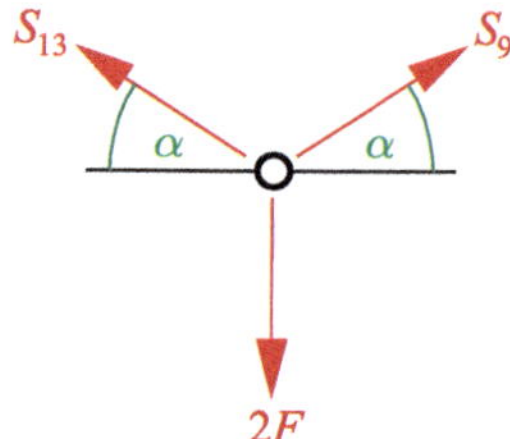

Fig. 5.18 Free-body diagram of node VII

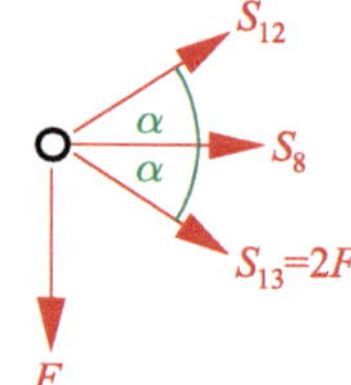

Fig. 5.19 Free-body diagram of node VIII

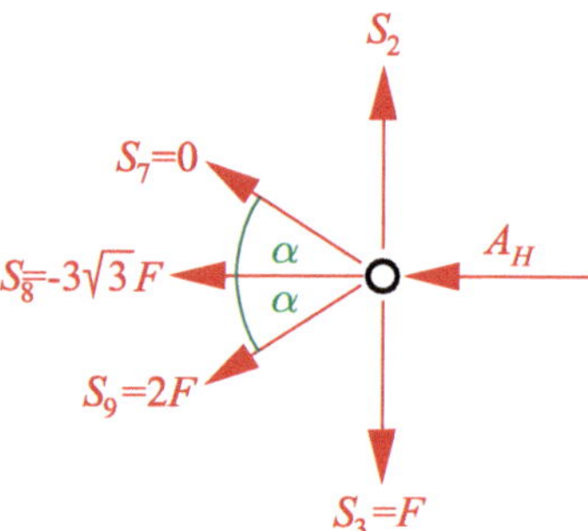

Fig. 5.20 Free-body diagram of node III

The last remaining member force S_1 is thus also immediately known as:

$$S_1 = 2F. \tag{5.57}$$

It should also be noted that all member forces could be determined without determining the support reactions in advance. If we are interested in the support reactions, these can be determined by analysing the nodal sections I and III. Alternatively, these can be determined using the equilibrium of the entire truss.

The member forces for the truss under consideration are summarised in Table 5.5.

◀

Table 5.5 Bar forces S_1–S_{13}

Bar force	
S_1	$2F$
S_2	$2F$
S_3	F
S_4	F
S_5	$4F$
S_6	0
S_7	0
S_8	$-3\sqrt{3}F$
S_9	$2F$
S_{10}	0
S_{11}	0
S_{12}	$4F$
S_{13}	$2F$

5.3 The Ritter Method

In some cases it is useful to use a strategy other than the nodal intersection method when determining member forces in trusses. One such alternative approach is the so-called Ritter method[1] or method of sections. The Ritter method consists of dividing the truss into two parts by means of a cut, whereby this cut consists of cutting three bars simultaneously. These bars must not all belong to the same node. Alternatively, we can also cut through a node and a bar. Suitable equilibrium conditions are then formulated on the resulting subsystem, from which the thus released bar forces can be determined. The choice of which of the two subsystems created by the cut is used to determine the member forces is made according to considerations of expediency. Ritter's method is particularly suitable if we are only interested in some individual bar forces and not in all bar forces of a truss. If necessary, the support reactions must be determined in advance.

We illustrate the procedure using the truss in Fig. 5.1 and want to determine the member forces 2, 6 and 7 with the help of the Ritter method. To do this, we cut through the relevant members and obtain the free-body diagram in Fig. 5.21, whereby we have applied the previously determined support forces with their respective directions of action.

We now consider the free-body diagram of Fig. 5.21, left, and first calculate the sum of moments around the point at which the lines of action of the member forces S_2 and S_7 intersect, i.e. around node II, which we have also included here. By choosing this reference

[1] Georg Dietrich August Ritter, 1826–1908, German physicist.

point, the bar forces S_2 and S_7 do not have a lever arm and do not contribute to the moment balance. The moment equilibrium results in:

$$\overset{\curvearrowright}{\sum} M_{II} = 0: \quad S_6 \cdot l - A_V \cdot l = 0 \quad \rightarrow \quad S_6 = A_V = \frac{1}{2}F. \tag{5.58}$$

Obviously, this result agrees with the member force as already determined from the nodal section method with (5.13).

We now have several options for determining the two remaining bar forces S_2 and S_7. We calculate the sum of the moments around node VI and obtain

$$\overset{\curvearrowleft}{\sum} M_{VI} = 0: \quad S_2 \cdot l - A_H \cdot l = 0 \quad \rightarrow \quad S_2 = A_H = 2F. \tag{5.59}$$

This result was also already determined using the nodal section method.

Finally, we determine the member force S_7 from the sum of the horizontal forces as follows:

$$\overset{\rightarrow}{\sum} H = 0: \quad S_6 + S_7 \cos 45° + S_2 - A_H = 0 \quad \rightarrow \quad S_7 = \frac{1}{\cos 45°}(-S_6 - S_2 + A_H). \tag{5.60}$$

With the bar forces already determined and $\cos 45° = \frac{1}{\sqrt{2}}$, it follows:

$$S_7 = \sqrt{2}(-\frac{1}{2}F - 2F + 2F) = -\frac{1}{\sqrt{2}}F. \tag{5.61}$$

This result was also determined with (5.15) using the nodal section method.

We can check our result for the member force S_7 on the current free-body diagram with the sum of the vertical forces. We obtain:

$$\overset{\downarrow}{\sum} V = 0: \quad S_7 \sin 45° + A_V = 0. \tag{5.62}$$

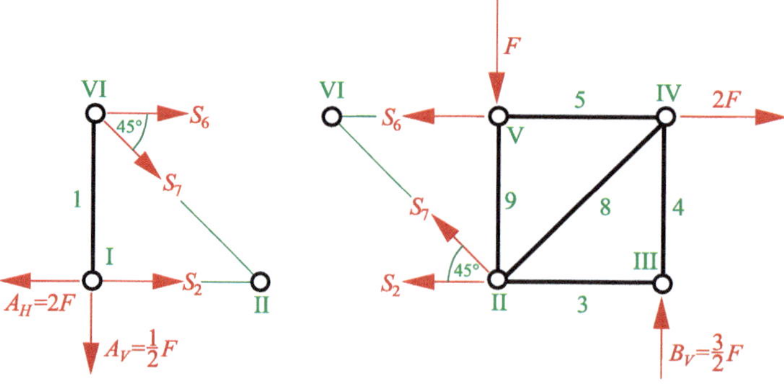

Fig. 5.21 Application of the Ritter method

If the member force S_7 and the support force A_V are inserted here, it can be seen that the left side of the equilibrium condition becomes zero, equilibrium is therefore guaranteed.

The same results are also obtained for the bar forces if we look at the right-hand side of the free-body diagram in Fig. 5.21. The sum of the moments around node VI results in

$$\overset{\curvearrowright}{\sum} M_{VI} = 0: \quad S_2 \cdot l + F \cdot l - B_V \cdot 2l = 0 \quad \rightarrow \quad S_2 = -F + 2B_V = 2F. \tag{5.63}$$

From the sum of moments around node II we obtain

$$\overset{\curvearrowleft}{\sum} M_{II} = 0: \quad S_6 \cdot l - 2F \cdot l + B_V \cdot l = 0 \quad \rightarrow \quad S_6 = 2F - B_V = \frac{1}{2} F. \tag{5.64}$$

Finally, we look at the sum of the horizontal forces. It follows:

$$\overset{\leftarrow}{\sum} H = 0: \quad S_6 + S_7 \cos 45^\circ + S_2 - 2F = 0 \quad \rightarrow \quad S_7 = \frac{1}{\cos 45^\circ}(-S_6 - S_2 + 2F) = -\frac{1}{\sqrt{2}} F \tag{5.65}$$

The bar forces determined on the right-hand free-body diagram are therefore identical to those determined on the left-hand free-body diagram.

Example 5.5

We consider the truss of Example 5.1 and now want to determine the bar forces 2, 5 and 10 with the help of the Ritter method.

Solution:

We cut the truss by making sections through members 2, 5 and 10 (Fig. 5.22), wherein we have also plotted the support reactions with magnitude and direction of action. This example is a truss for which it is not necessary to determine the support reactions in advance if we want to determine selected member forces using the Ritter method. This becomes immediately apparent if we look at the free-body diagram in Fig. 5.22, right, from which the required bar forces S_2, S_5 and S_{10} can be determined.

We calculate the sum of the moments around node VI, which is located at the intersection of the lines of action of the bar forces S_5 and S_{10}. It follows:

$$\overset{\curvearrowright}{\sum} M_{VI} = 0: \quad S_2 \cdot l + F \cdot 2l = 0 \quad \rightarrow \quad S_2 = -2F. \tag{5.66}$$

This result was already determined in Example 5.1 using the nodal section method.

We also calculate the sum of the moments around node III and obtain

$$\overset{\curvearrowleft}{\sum} M_{III} = 0: \quad S_5 \cdot l - F \cdot l = 0 \quad \rightarrow \quad S_5 = F. \tag{5.67}$$

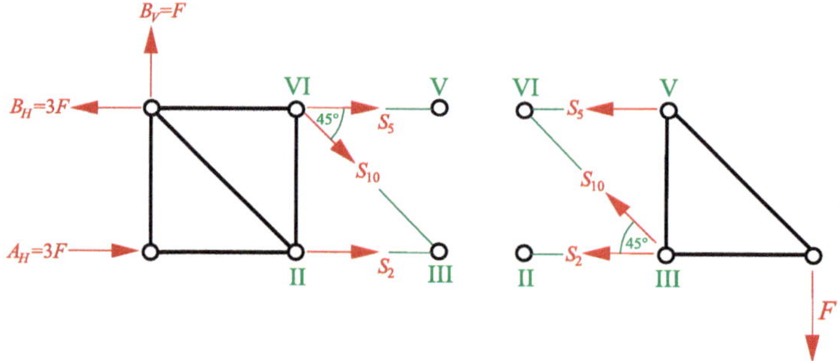

Fig. 5.22 Free-body diagram resulting from a cut through bars 2, 5 and 10

This result was also already provided in Example 5.1.

Finally, we look at the sum of the vertical forces:

$$\sum^{\uparrow} H = 0: \quad S_{10} \sin 45° - F = 0 \quad \rightarrow \quad S_{10} = \frac{F}{\sin 45°} = \sqrt{2}F. \tag{5.68}$$

This member force was also determined in Example 5.1.

As a verification, we also consider the sum of the horizontal forces and obtain

$$\sum^{\leftarrow} H = 0: \quad S_2 + S_5 + S_{10} \cos 45° = 0. \tag{5.69}$$

If the already determined memebr forces S_2, S_5 and S_{10} are inserted here, it can be seen that the left side of the force equilibrium condition becomes zero. Equilibrium is therefore guaranteed.

It can also be shown for this example that investigating the free-body diagram of Fig. 5.22, left, gives the same results for the member forces.

◀

Example 5.6

We consider again the truss from Example 5.4 and now want to determine the bar forces S_3, S_9 and S_{13} using the Ritter method.

Solution:

We make a section through bars 3, 9 and 13 and obtain the free-body diagram shown in Fig. 5.23. From the moment equilibrium with respect to node VII we obtain:

$$\sum^{\curvearrowleft} M_{VII} = 0: \quad S_3 \cdot \sqrt{3}l - F \cdot \sqrt{3}l = 0 \quad \rightarrow \quad S_3 = F. \tag{5.70}$$

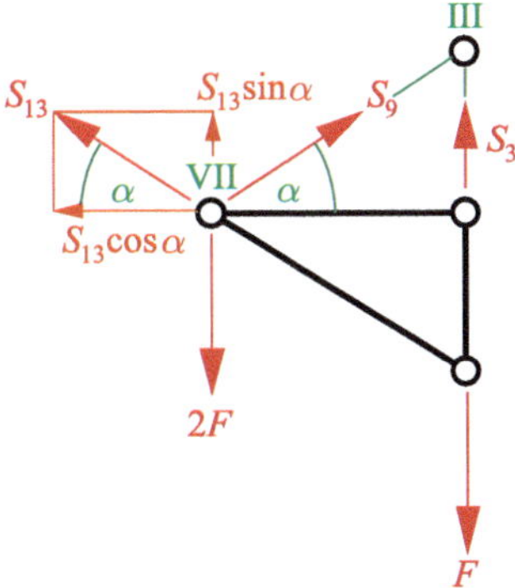

Fig. 5.23 Free-body diagram resulting from cutting through bars 3, 9 and 13

This result was also determined in Example 5.4.

We also consider the sum of the moments with respect to node III, which represents the intersection of the lines of action of the member forces S_3 and S_9. To do this, we divide the bar force S_{13} into its horizontal and vertical components. The result is as follows:

$$\sum^{\curvearrowright} M_{III} = 0: \quad S_{13}\sin\alpha \cdot \sqrt{3}l + S_{13}\cos\alpha \cdot l - 2F \cdot \sqrt{3}l = 0. \tag{5.71}$$

This leads to the result already known from Example 5.4

$$S_{13} = 2F. \tag{5.72}$$

The remaining member force S_9 results from the horizontal force sum:

$$\sum^{\rightarrow} H = 0: \quad S_9\cos\alpha - S_{13}\cos\alpha = 0, \tag{5.73}$$

from which the equality of the bar forces S_9 and S_{13} already determined in Example 5.4 immediately follows:

$$S_9 = 2F. \tag{5.74}$$

◄

5.4 The Cremona Plan

In addition to the explained options for the mathematical determination of member forces in plane trusses, it is also possible to determine the bar forces graphically. This is referred to as the Cremona plan,[2] which is explained in this section.

To graphically determine the member forces of a plane truss, the first step is to determine the support forces from the equilibrium conditions on the entire truss. The Cremona plan

[2] Antonio Luigi Gaudenzio Giuseppe Cremona, 1830–1903, Italian scientist.

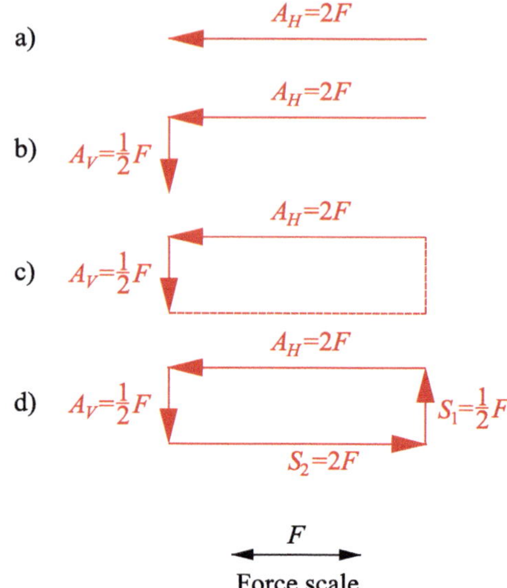

Fig. 5.24 Force polygonal for node I

utilises the fact that the equilibrium conditions must be fulfilled at each node of a truss, so that a closed force polygonal can be drawn for each truss node. These force polygonals are then later combined to form the Cremona plan, so that we obtain a graphic representation of the member forces together with support forces and other forces in one diagram.

To illustrate the Cremona plan, let us look again at the truss in Fig. 5.1. The support forces are known as $A_H = 2F$ (i.e. actually pointing to the left), $A_V = -\frac{1}{2}F$ (pointing downwards) and $B_V = \frac{3}{2}F$ (pointing upwards). A positive direction of rotation is now defined at the beginning for the truss nodes, whereby this direction is arbitrary but must be identical for all nodes. The direction of rotation is important for the treatment of the equilibrium at the truss nodes, and we agree on a positive anti-clockwise direction of rotation for this truss in all further explanations.

The resulting closed force polygonals are now constructed for each node, starting with the consideration of node I (Fig. 5.24). We first define a scale for the forces and start by considering the support force A_H, which we apply according to the scale with its actual direction of action (Fig. 5.24a). In accordance with the agreed direction of rotation, we now apply the support force A_V with its actual direction of action (Fig. 5.24b). Now we consider the member force S_2, whose horizontal line of action must be tangent to the force arrow of A_V. We also apply the line of action of S_1, which must be connected to the force arrow of the horizontal support force A_H (Fig. 5.24c). The actual length of the force arrows of S_1 and S_2 is limited on the one hand by the force arrows of the support reactions A_H and A_V and on the other hand by the intersection point between the lines of action of S_1 and S_2 (Fig. 5.24d). This completes the closed force polygonal for node I, and the magnitudes of the forces S_1 and S_2 and their directions of action can be seen from the sketch in Fig. 5.24d. For the force

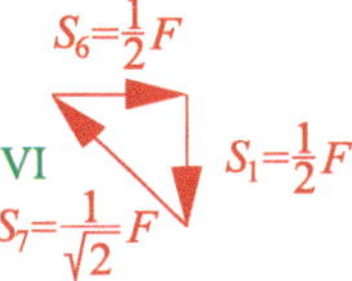

Fig. 5.25 Force polygonal for node VI

polygonal to be closed, $S_1 = \frac{1}{2}F$ must point upwards and $S_2 = 2F$ to the right. Obviously this agrees with the previously determined results for S_1 and S_2.

We proceed in the same way for the other force polygonals of this truss, whereby it is advisable to choose the order of the nodes to be treated in such a way that the sequence is as simple as possible and thus less error-prone. For example, let us now consider node VI. At this node, the member force $S_1 = \frac{1}{2}F$ is already known as a tensile force, whereby it must be ensured that this force is drawn exactly opposite to the image of node I due to the interaction principle. If we now draw the closed force polygonal, we obtain the image in Fig. 5.25.

The same procedure is followed for the remaining nodes of the truss, whereby we could use the node sequence I, VI, V, IV, III and finally node II remains for verifying the results. The individual force polygonals are shown in Fig. 5.26, where the actual sign of a bar force can be deduced from the direction of the force arrow. For example, the member force S_9 at node V is a compressive force and therefore negative; it points towards the node and exerts

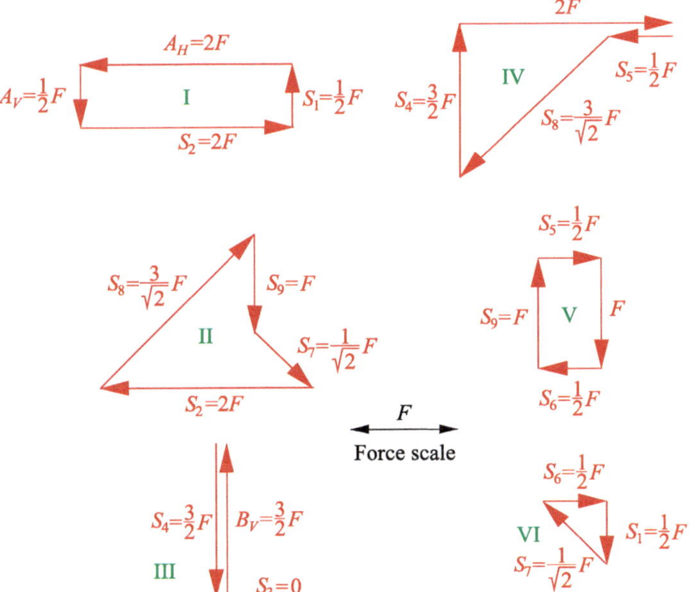

Fig. 5.26 Force polygonals for nodes I-VI

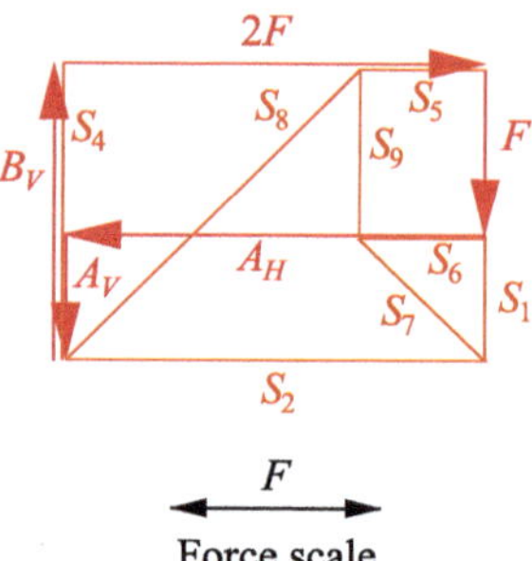

Fig. 5.27 Cremona plan

pressure on it. It should also be noted that the force polygonal degenerates into a line in the case of node III, bar 3 is therefore a zero-force bar.

The Cremona plan is now created by combining the previously created force polygonals in such a way that each bar force only needs to be drawn once. This is shown in Fig. 5.27. We have drawn the bar forces without a directional arrow, as they appear twice in each of the force polygonals involved here. The bar forces can be measured at this resultant force polygonal, which is now valid for the entire truss. In addition, the applied forces and the support reactions in themselves also result in a closed force polygonal (highlighted in red).

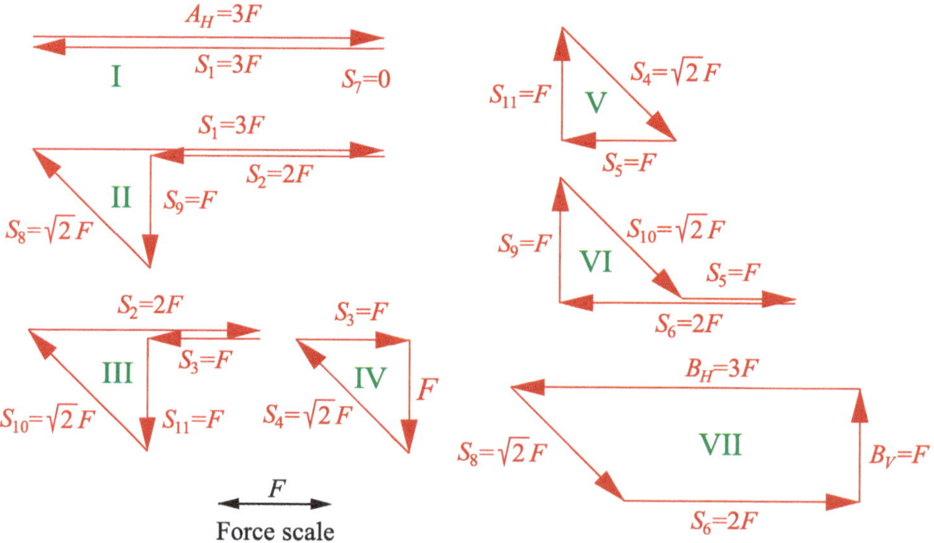

Fig. 5.28 Force polygonals

Example 5.7

The Cremona plan is to be determined for the truss from Example 5.1.

Solution:

The force polygonals of the individual nodes of the truss are shown in Fig. 5.28. These were created in the order IV, V, III, VI, II, I, VII with an anti-clockwise direction of rotation. This truss is again an example in which the support reactions do not have to be determined in advance, but can be determined from the force polygonals. The Cremona plan that can be created from this is shown in Fig. 5.29.

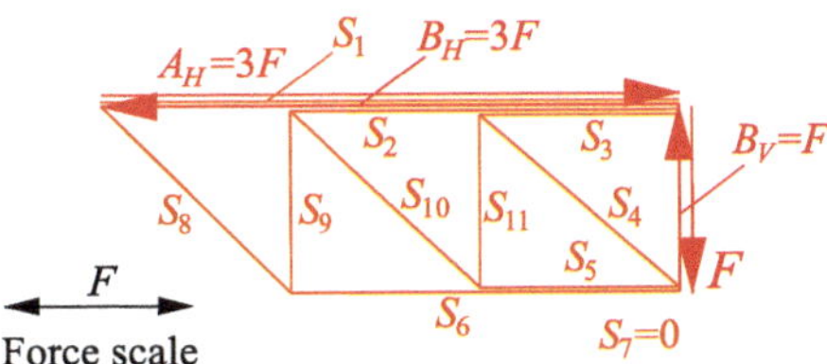

Fig. 5.29 Cremona plan

◀

Beams

6

This chapter is dedicated to the determination of internal forces of plane and spatial straight beams as well as curved beams. After a basic introduction, internal forces, i.e. normal and shear forces as well as bending moments, are determined on straight beams under concentrated forces and moments as well as line loads. In addition, a basic relationship between load and internal forces is derived and used to determine internal forces and moments by integration. We also deal with multi-field problems and thus enable the determination of internal forces for more complex systems and also consider angled beams, frames and curved beams. The chapter concludes with the treatment of spatial beam structures.

6.1 Fundamentals

Internal forces and moments or so-called stress resultants will develop in a beam or a system of beams under applied loads (Fig. 6.1). It is possible to relate these forces to an area, so that one speaks of the so-called stresses inside a beam. How to calculate these stresses and how they are distributed over the cross-section of a beam is discussed in detail in Volume 2 and cannot be dealt with in this book. In this chapter, we will therefore limit ourselves exclusively to the determination of these internal forces and moments, which are also referred to as the so-called stress resultants and which are visualised by cutting a beam and drawing a free-body diagram. If, for example, a beam is loaded by the two line loads q_y and q_z as well as by individual forces F_y and F_z, then there are generally six different stress resultants. These are the so-called normal force N, which points in the longitudinal direction of the beam, and the two so-called transverse shear forces Q_z and Q_z, which are orientated tangentially to the cross-section. In addition to these internal forces, there are also three internal moments.

C. Mittelstedt, *Engineering Mechanics 1: Statics*,
https://doi.org/10.1007/978-3-662-71852-0_6

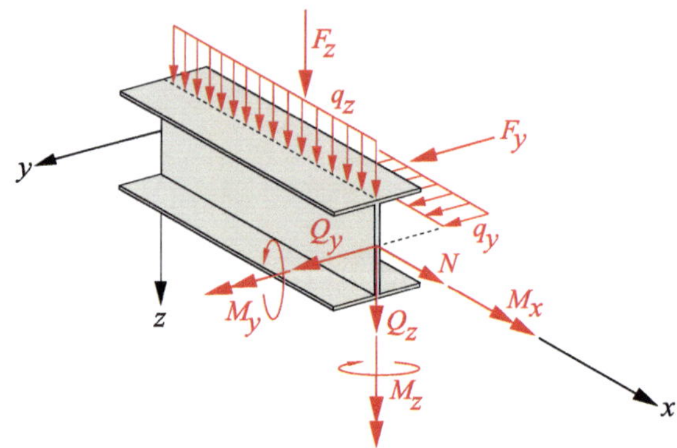

Fig. 6.1 Internal forces and moments in a beam

One of these is the so-called torsional moment M_x, which represents a moment with respect to the longitudinal axis of the beam x and is associated with a torsion of the beam. The topic of torsion is not dealt with in this book; this will be discussed in detail in Volume 2. On the other hand, the two so-called bending moments M_y and M_z occur, which are associated with a rotation of the cross-section about the $y-$axis and the $z-$axis respectively. In this chapter, we will look at how to determine these internal forces and moments in beams. The determination of internal forces and moments is a particularly important task for engineers in practice, because they allow a statement about the stresses of a structure, which is essential for the design and verification of a technical system.

We start the explanations in this chapter with beams or beam systems that are only loaded in one plane (Fig. 6.2). In the following, this is always the $xz-$plane. The circumstances under which a beam structure may be considered as acting only in one plane cannot be exhaustively explained in this book, this will be the topic of discussion of Volume 2. At this point, however, we want to assume that this assumption is valid. We will deal with arbitrarily loaded beams later on. If there is a beam that is only loaded in its $xz-$plane, then only the normal force N, the bending moment $M_y = M$ and the transverse shear force $Q_z = Q$ occur as relevant stress resultants, as shown in Fig. 6.2, where we think of the beam as a one-dimensional element represented by its centre of gravity line. The internal forces and moments are therefore forces and moments related to the centre of gravity S of the cross-section. The choice of the origin of the reference system x, y, z on the centre of gravity axis of the beam is arbitrary. In the system of Fig. 6.2, the coordinate system was placed arbitrarily in the left support.

Figure 6.2, bottom, shows the cut beam. The face from which the $x-$axis protrudes (or more precisely: whose normal vector $\underline{n}$ points in the positive $x-$direction) is referred to as the so-called positive face. Similarly, the other face is the so-called negative face. The internal forces and moments on these two faces must match due to the interaction principle. In order to maintain equilibrium, the stress resultants cut free in this way must be in equilibrium with the forces and moments on the two beam sections. Therefore, after performing a cut, the stress resultants of the beam can be determined by the already known equilibrium conditions.

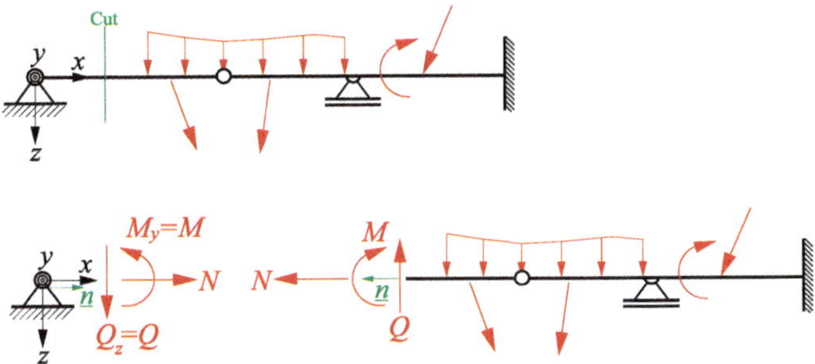

Fig. 6.2 Plane stress resultants

These are the sums of the forces in the horizontal $x-$direction and the vertical $z-$direction as well as the sum of the moments about the $y-$axis, which points out of the plane in the shown perspective. It should be noted at this point, however, that this procedure can only be carried out for statically determinate systems. For statically indeterminate beams and beam systems, further considerations are necessary, which we will discuss in detail in Volume 2. Accordingly, in this chapter we limit ourselves exclusively to the determination of internal forces and moments in statically determinate beams and beam systems.

The following sign convention applies to the stress resultants of a beam: At a positive face, positive forces and moments point in a positive coordinate direction. Similarly, at a negative face, positive stress resultants point in the negative coordinate directions. The forces and moments shown in Fig. 6.2 are therefore positive stres resultants: The normal force N points in the positive $x-$direction at the positive face, and the shear force Q also points in the positive $z-$direction. The bending moment M has a positive direction of rotation around the $y-$axis.

Often, only the horizontal $x-$axis is drawn in the planar view of beams and it is tacitly assumed that the $z-$axis points downwards, so that the $y-$axis points out of the plane in the sense of a right-handed coordinate system. However, we will also always draw the $z-$axis in all the following explanations in order to ensure clarity, but will refrain from labelling the $y-$axis. It always points out of the plane.

6.2 Stress Resultants in Straight Beams

6.2.1 Beams Under Point Forces

As an introductory elementary example for determining stress resultants we consider the beam in Fig. 6.3. Given a straight beam of length l, which has a two-valued support at its left end. At the right end there is a single-valued support. The beam is loaded in the centre of the span by a force F acting perpendicular to the beam axis. We want to determine the internal forces and moments N, Q, M for this beam.

In many cases, the calculation of internal forces and moments begins with the determination of the support reactions, which is elementarily simple in the present case: The horizontal support force A_H becomes zero under this load, and the two vertical support forces A_V and B_V are both half of the force F:

$$A_H = 0, \quad A_V = B_V = \frac{F}{2}. \tag{6.1}$$

We first consider cut 1, which is located at an arbitrary point x to the left of the applied force F. The corresponding free-body diagram is also shown in Fig. 6.3. The two beam segments have the lengths x and $l - x$ respectively, and we consider the left beam segment. The equilibrium conditions are as follows:

$$\begin{aligned}
\overset{\rightarrow}{\sum} H = 0: \quad & N + A_H = 0 \quad \rightarrow \quad N = 0, \\
\overset{\downarrow}{\sum} V = 0: \quad & Q - A_V = 0 \quad \rightarrow \quad Q = A_V = \frac{F}{2}, \\
\overset{\curvearrowleft}{\sum} M_{S_1} = 0: \quad & M - A_V x = 0 \quad \rightarrow \quad M = A_V x = \frac{F}{2} x.
\end{aligned} \tag{6.2}$$

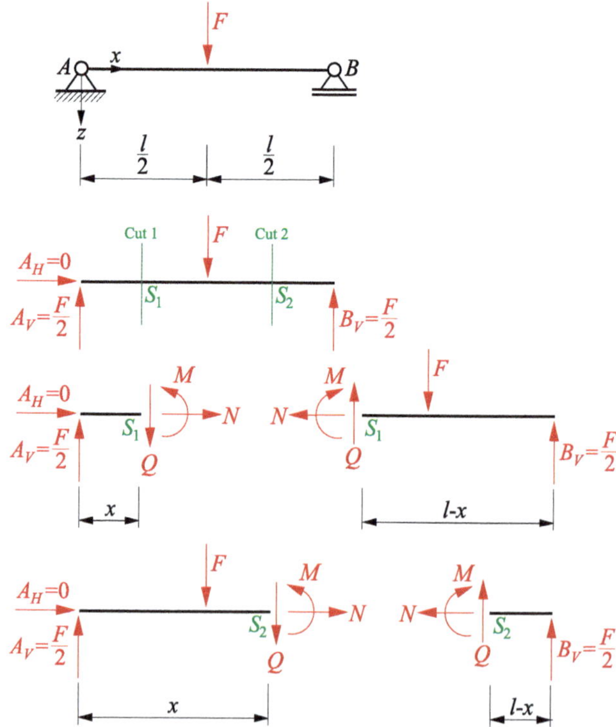

Fig. 6.3 Beam under vertical force F

It can be seen that the normal force N disappears in the left beam segment for $0 \le x \le \frac{l}{2}$, regardless of the position x of the section point S_1. It also shows that the shear force is constant with the value $Q = \frac{F}{2}$, which is also independent of the position x of S_1. For the bending moment M, on the other hand, there is a linear distribution $M(x) = \frac{F}{2}x$, whereby the bending moment becomes zero at the point $x = 0$ (support point A) and assumes the maximum value $M = \frac{Fl}{4}$ at the point of force application $x = \frac{l}{2}$.

It can be easily shown that we would obtain the same results if we looked at the right-hand beam segment. The equilibrium conditions are then

$$\begin{aligned}
&\overset{\leftarrow}{\sum} H = 0: \quad N = 0,\\
&\overset{\uparrow}{\sum} V = 0: \quad Q - F + B_V = 0 \quad \rightarrow \quad Q = F - B_V = \frac{F}{2},\\
&\overset{\curvearrowleft}{\sum} M_{S_1} = 0: \quad M + F\left(\frac{l}{2} - x\right) - B_V\,(l - x) = 0 \quad \rightarrow \quad M = \frac{F}{2}x.
\end{aligned} \tag{6.3}$$

It can be seen that the consideration of the right-hand beam segment is more complex than the treatment of the left-hand segment.

We now consider cut 2 and want to look at the resulting right-hand beam segment for $\frac{l}{2} \le x \le l$, as shown in Fig. 6.3. The equilibrium conditions are as follows:

$$\begin{aligned}
&\overset{\leftarrow}{\sum} H = 0: \quad N = 0,\\
&\overset{\uparrow}{\sum} V = 0: \quad Q + B_V = 0 \quad \rightarrow \quad Q = -B_V = -\frac{F}{2},\\
&\overset{\curvearrowleft}{\sum} M_{S_2} = 0: \quad M - B_V\,(l - x) = 0 \quad \rightarrow \quad M = \frac{F}{2}\,(l - x)\,.
\end{aligned} \tag{6.4}$$

It can be seen that the normal force N also becomes zero to the right of the acting force F, whereas the shear force Q is also constant here, but now occurs as a negative force with the value $Q = -\frac{F}{2}$. The bending moment is again a linear function of x, whereby it disappears in the right support B. Obviously we obtain the same results if we consider the left beam segment. This is not explained further here.

A graphical representation of the internal forces and moments is referred to as normal force diagram, shear force diagram and moment diagram for N, Q and M. The diagrams for this example are shown in Fig. 6.4. Usually, the internal forces are applied over the beam lengths in such a way that the normal force N and shear force Q are applied positively upwards and the bending moment M is applied positively downwards. Even if in this case the normal force becomes zero on both sides of the acting force, a separate diagram is provided for this and the value zero is entered there, as shown in Fig. 6.4. The essential ordinates are added to the diagrams, whereby these are always provided without a sign. The sign is usually entered inside the diagram.

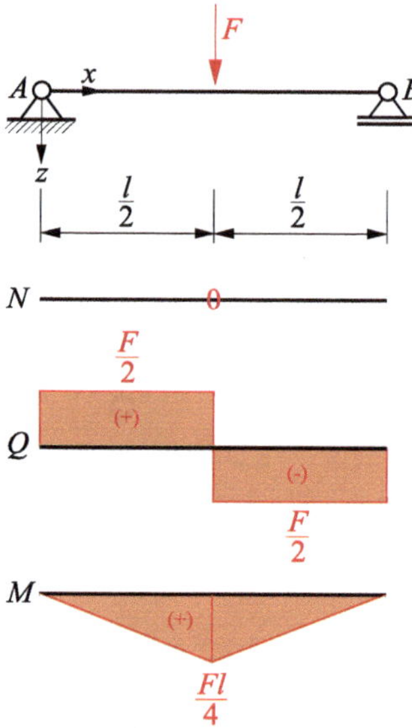

Fig. 6.4 Normal force diagram, shear force diagram and moment diagram

If the resulting diagrams are discussed in detail, it can be seen that the shear force Q is constant on both sides of the force F, but shows a discontinuity by exactly the value F precisely at the point where the force F is applied. The diagram for the bending moment M, on the other hand, is linear on both sides of the force F and exhibits a kink at the point where the force F is applied. The maximum bending moment occurs at the point of application of the force F. These are typical observations on beams under point forces, a more detailed explanation of these facts will follow later.

Example 6.1

Consider the beam shown in Fig. 6.5 with length l under a central point force $\sqrt{2}F$, which acts at an angle $\alpha = 45°$ to the beam axis. We want to determine the diagrams for N, Q, M.

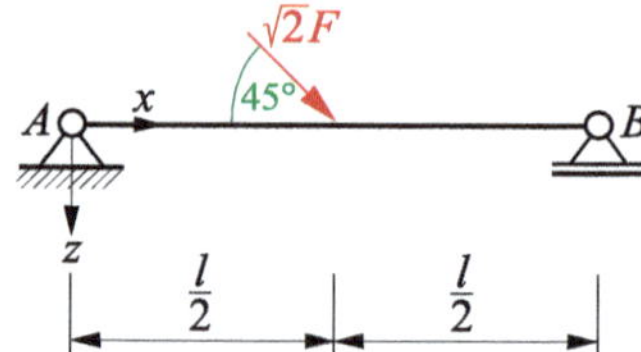

Fig. 6.5 Beam under point force F

Solution:

We cut the beam free and divide the applied force into its horizontal component F_H and its vertical component F_V. With $\sin 45° = \cos 45° = \frac{1}{\sqrt{2}}$ they result as $F_H = F_V = F$ (Fig. 6.6). Firstly, the support reactions are determined. The moment equilibrium with respect to support B results in the support force A_V as follows:

$$\overset{\curvearrowright}{\sum} M_B = 0: \quad A_V \cdot l - F_V \cdot \frac{l}{2} = 0 \quad \rightarrow \quad A_V = \frac{F}{2}. \tag{6.5}$$

The support force B_V then results from the sum of the vertical forces:

$$\overset{\uparrow}{\sum} V = 0: \quad B_V + A_V - F_V = 0 \quad \rightarrow \quad B_V = \frac{F}{2}. \tag{6.6}$$

Finally, the horizontal support force A_H can be determined from the sum of the horizontal forces:

$$\overset{\rightarrow}{\sum} V = 0: \quad A_H + F_H = 0 \quad \rightarrow \quad A_H = -F. \tag{6.7}$$

Accordingly, this support force actually points to the left.

To determine the internal forces and moments N, Q and M, we perform two sections as shown in Fig. 6.7 and first consider the free-body diagram for Section 1 ($0 \leq x \leq \frac{l}{2}$). Since the observations of both resulting free body images of the two beam segments are equivalent, we decide to look at the left side for the sake of simplicity. It should be noted that we have applied the support force A_H according to its true direction of action. Equilibrium of forces and moments yields:

$$\overset{\rightarrow}{\sum} H = 0: \quad N - A_H = 0 \quad \rightarrow \quad N = F,$$

$$\overset{\downarrow}{\sum} V = 0: \quad Q - A_V = 0 \quad \rightarrow \quad Q = A_V = \frac{F}{2},$$

$$\overset{\curvearrowleft}{\sum} M_{S_1} = 0: \quad M - A_V x = 0 \quad \rightarrow \quad M = A_V x = \frac{F}{2} x. \tag{6.8}$$

It can be readily shown that we obtain the same results if we consider at the right-hand beam segment.

We now consider Section 2 ($\frac{l}{2} \leq x \leq l$) and obtain for the right bar segment:

$$\overset{\leftarrow}{\sum} H = 0: \quad N = 0,$$

$$\overset{\uparrow}{\sum} V = 0: \quad Q + B_V = 0 \quad \rightarrow \quad Q = -B_V = -\frac{F}{2},$$

$$\overset{\curvearrowright}{\sum} M_{S_2} = 0: \quad M - B_V (l - x) = 0 \quad \rightarrow \quad M = \frac{F}{2} (l - x). \tag{6.9}$$

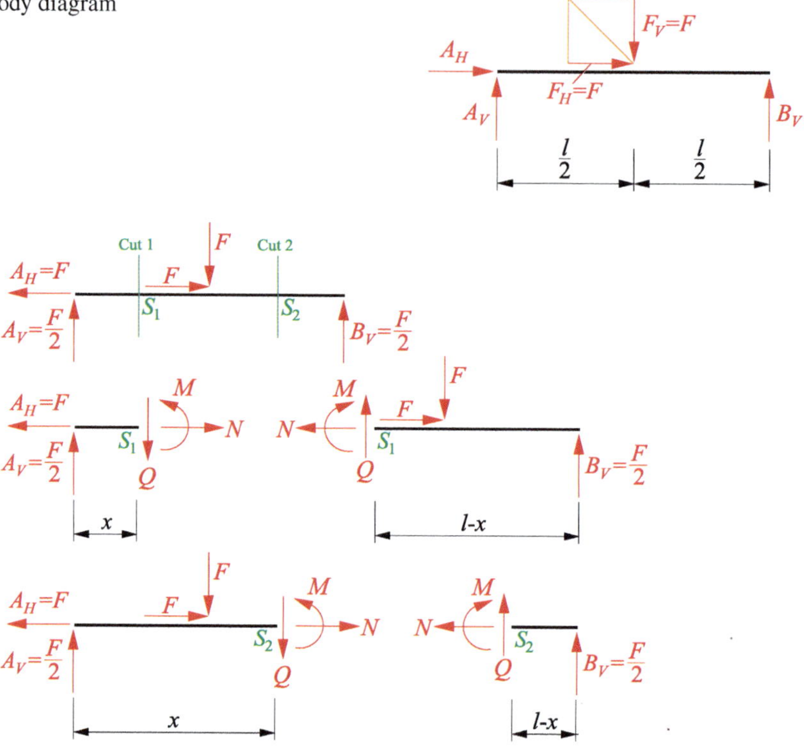

Fig. 6.6 Free-body diagram

Fig. 6.7 Free-body diagrams for Sections 1 and 2

The internal forces and moments determined in this way are shown in Fig. 6.8. Apparently the diagrams for Q and M correspond exactly to those of the introductory example in this section. In addition, it can be seen that the left half of the beam is under a normal tensile force, whereas the right half of the beam is free of normal forces. ◀

We now want to investigate how to determine stress resultants when a beam is considered under a number of point forces. As an introductory example, we consider the beam in Fig. 6.9, which is subjected to three point forces F_1, F_2, F_3, all of which act perpendicular to the beam axis. The forces F_1, F_2, F_3 have the distances l_1, l_2, l_3 to the left support point A.

We first determine the support reactions and consider the free-body diagram of Fig. 6.10. From the sums of moments with respect to A and B follows:

$$\overset{\curvearrowleft}{\sum} M_B = 0: \quad A_V l - F_1 (l - l_1) - F_2 (l - l_2) - F_3 (l - l_3) = 0 \quad \rightarrow \quad A_V = \frac{1}{l} \sum_{i=1}^{3} F_i (l - l_i),$$

$$\overset{\curvearrowright}{\sum} M_A = 0: \quad B_V l - F_1 l_1 - F_2 l_2 - F_3 l_3 = 0 \quad \rightarrow \quad B_V = \frac{1}{l} \sum_{i=1}^{3} F_i l_i. \tag{6.10}$$

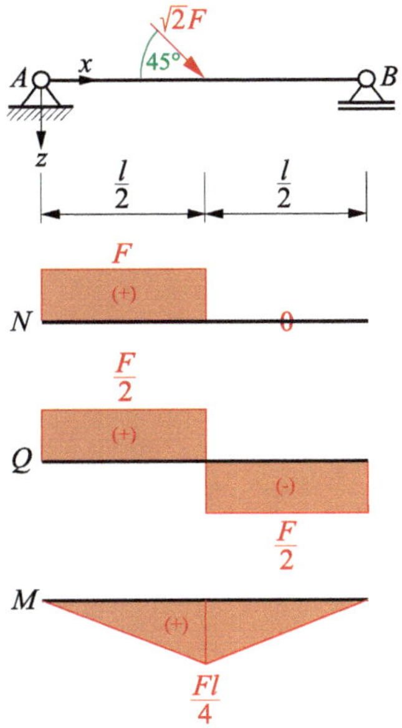

Fig. 6.8 Normal force diagram, shear force diagram and moment diagram

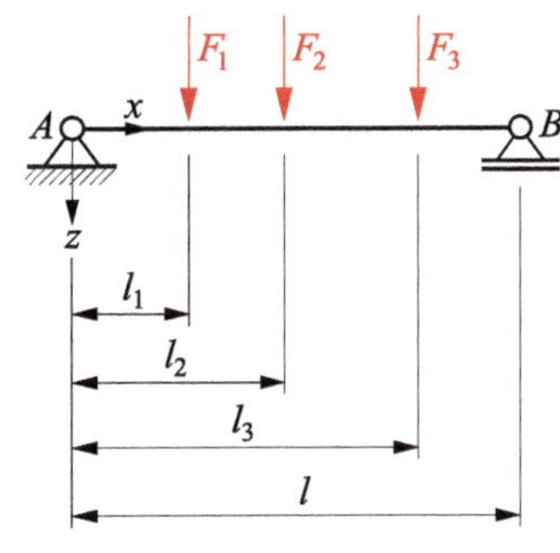

Fig. 6.9 Bar under three point forces F_1, F_2, F_3

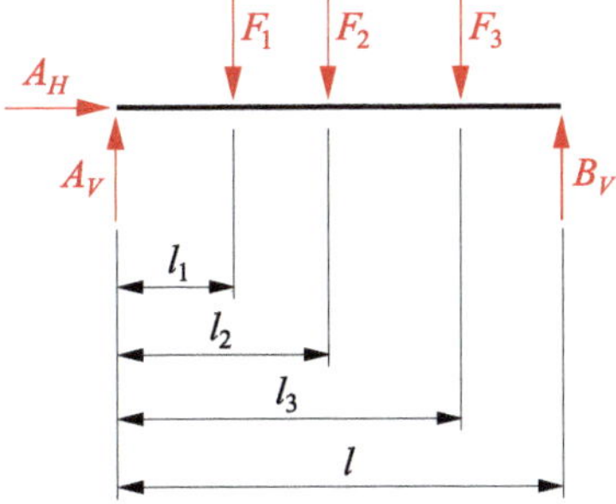

Fig. 6.10 Free-body diagram for the determination of the support forces

The horizontal support force A_H is identical to zero for this beam: $A_H = 0$.

The above result can be extended to the case of n vertically acting point forces, and the following applies:

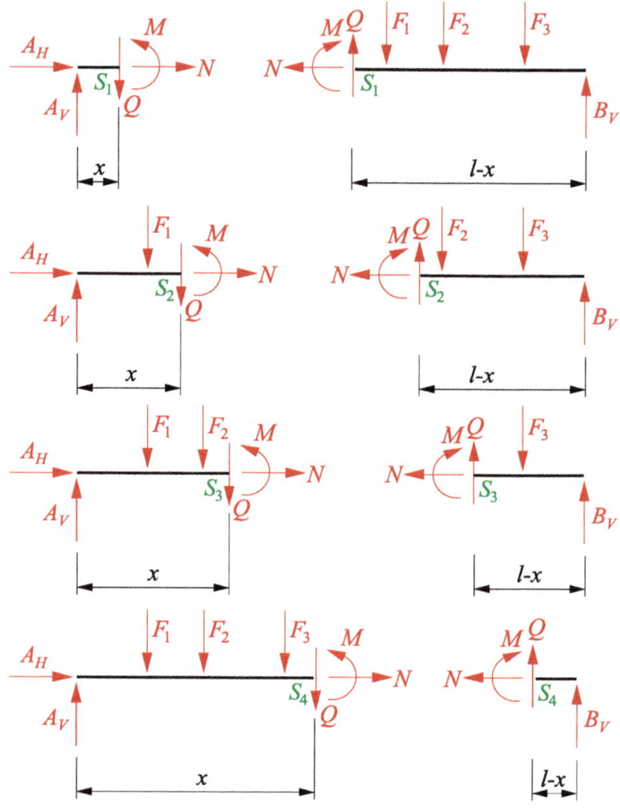

Fig. 6.11 Free-body diagrams for Sections 1–4

$$A_H = 0, \quad A_V = \frac{1}{l} \sum_{i=1}^{n} F_i \left(l - l_i \right), \quad B_V = \frac{1}{l} \sum_{i=1}^{n} F_i l_i. \tag{6.11}$$

We now want to determine the internal forces N, Q and moments M, whereby in this example the normal force N becomes zero at every point x, since none of the applied forces has a horizontal component and therefore the horizontal support force A_H also disappears. We now make a total of four cuts through the beam as follows:

- Section 1 between support A and force F_1 ($0 \leq x \leq l_1$)
- Section 2 between the forces F_1 and F_2 ($l_1 \leq x \leq l_2$)
- Section 3 between the forces F_2 and F_3 ($l_2 \leq x \leq l_3$)
- Section 4 between force F_3 and support B ($l_3 \leq x \leq l$).

The corresponding free-body diagrams are shown in Fig. 6.11.

We start the considerations with Section 1 ($0 \leq x \leq l_1$) and look at the left beam segment on the corresponding free-body diagram. The equilibrium conditions are as follows:

$$
\begin{aligned}
&\overset{\rightarrow}{\sum} H = 0: \quad N = 0, \\
&\overset{\downarrow}{\sum} V = 0: \quad Q - A_V = 0 \quad \rightarrow \quad Q = \frac{1}{l} \sum_{i=1}^{3} F_i \, (l - l_i) \, , \\
&\overset{\curvearrowleft}{\sum} M_{S_1} = 0: \quad M - A_V x = 0 \quad \rightarrow \quad M = \frac{x}{l} \sum_{i=1}^{3} F_i \, (l - l_i) \, .
\end{aligned}
\tag{6.12}
$$

For Section 2 with $l_1 \leq x \leq l_2$ considering the left beam segment we obtain:

$$
\begin{aligned}
&\overset{\rightarrow}{\sum} H = 0: \quad N = 0, \\
&\overset{\downarrow}{\sum} V = 0: \quad Q - A_V + F_1 = 0 \quad \rightarrow \quad Q = \frac{1}{l} \sum_{i=1}^{3} F_i \, (l - l_i) - F_1, \\
&\overset{\curvearrowleft}{\sum} M_{S_3} = 0: \quad M - A_V x + F_1 \, (x - l_1) = 0 \\
&\rightarrow \quad M = \frac{x}{l} \sum_{i=1}^{3} F_i \, (l - l_i) - F_1 \, (x - l_1) \, .
\end{aligned}
\tag{6.13}
$$

At Section 3 with $l_2 \leq x \leq l_3$ we have for the left beam segment:

$$
\begin{aligned}
&\overset{\rightarrow}{\sum} H = 0: \quad N = 0, \\
&\overset{\downarrow}{\sum} V = 0: \quad Q - A_V + F_1 + F_2 = 0 \quad \rightarrow \quad Q = \frac{1}{l} \sum_{i=1}^{3} F_i \, (l - l_i) - F_1 - F_2, \\
&\overset{\curvearrowleft}{\sum} M_{S_3} = 0: \quad M - A_V x + F_1 \, (x - l_1) + F_2 \, (x - l_2) = 0 \\
&\rightarrow \quad M = \frac{x}{l} \sum_{i=1}^{3} F_i \, (l - l_i) - F_1 \, (x - l_1) - F_2 \, (x - l_2) \, .
\end{aligned}
\tag{6.14}
$$

For Section 4 with $l_3 \leq x \leq l$ it finally follows using the left beam segment:

$$
\begin{aligned}
&\overset{\rightarrow}{\sum} H = 0: \quad N = 0, \\
&\overset{\downarrow}{\sum} V = 0: \quad Q - A_V + F_1 + F_2 + F_3 = 0 \quad \rightarrow \quad Q = \frac{1}{l} \sum_{i=1}^{3} F_i \, (l - l_i) - F_1 - F_2 - F_3, \\
&\overset{\curvearrowleft}{\sum} M_{S_4} = 0: \quad M - A_V x + F_1 \, (x - l_1) + F_2 \, (x - l_2) + F_3 \, (x - l_3) = 0 \\
&\rightarrow \quad M = \frac{x}{l} \sum_{i=1}^{3} F_i \, (l - l_i) - F_1 \, (x - l_1) - F_2 \, (x - l_2) - F_3 \, (x - l_2) \, .
\end{aligned}
\tag{6.15}
$$

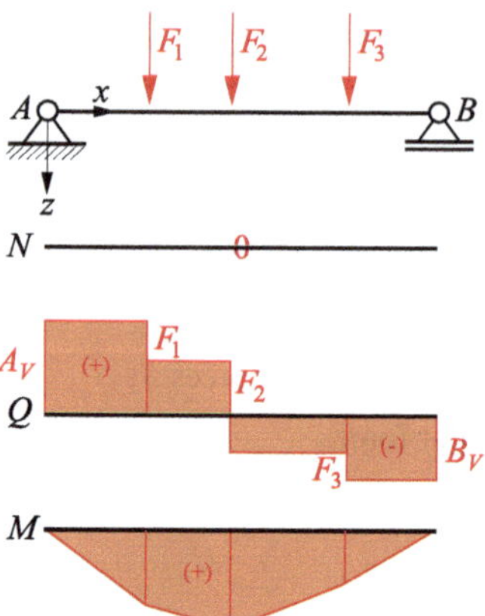

Fig. 6.12 Normal force diagram, shear force diagram and moment diagram

The stress resultants determined in this way are shown in the diagrams in Fig. 6.12. It can be seen that the shear force diagram represents a constant line for all sections, which has the value A_V in the interval $0 \leq x \leq l_1$ and the value $-B_V$ in the interval $l_3 \leq x$. At the points of application of the forces F_1, F_2 and F_3, discontinuities occur by exactly the magnitude of the applied force. The moment diagram, on the other hand, is composed of linear functions that exhibit kinks at the force application points. At the two support points A and B, the moment line assumes zero values.

The results obtained above for the internal forces and moments can be generalised as follows for the case of a beam of length l under n point forces F_i $(i = 1, 2, \ldots, n)$ acting perpendicular to the beam axis at the locations $x_i = l_i$. For segment i between the forces F_{i-1} and F_i at the point x, the following applies:

$$
\begin{aligned}
N &= 0, \\
Q &= \frac{1}{l} \sum_{i=1}^{n} F_i \, (l - l_i) - F_1 - F_2 - \cdots - F_{i-1}, \\
M &= \frac{x}{l} \sum_{i=1}^{n} F_i \, (l - l_i) - F_1 \, (x - l_1) - F_2 \, (x - l_2) - \cdots - F_{i-1} \, (x - l_{i-1}) \, .
\end{aligned} \tag{6.16}
$$

It should be noted that cutting a beam free always results in two segments, both of which can be used equally to determine the internal forces and moments. Which segment is used to set up the equilibrium conditions is always decided according to aspects of simplicity. One will therefore always endeavour to use the segment that is subject to fewer forces to make the analysis as simple as possible.

Example 6.2

Consider a cantilever beam of length l, which is loaded by several point forces (Fig. 6.13). We want to determine the diagrams for N, Q, M.

Solution:

We have already determined the support reactions in point A in Example 4.12. They are as follows:

$$A_H = \left(3 - \frac{1}{\sqrt{2}}\right) F,$$
$$A_V = \left(1 + \frac{1}{\sqrt{2}}\right) F,$$
$$M_A = -\frac{3Fl}{4\sqrt{2}}. \tag{6.17}$$

To determine the internal forces and moments, we perform a total of four sections on the beam as shown in Fig. 6.14. The individual sections therefore relate to the following intervals:

- Section 1: $0 \leq x \leq \frac{l}{4}$
- Section 2: $\frac{l}{4} \leq x \leq \frac{l}{2}$
- Section 3: $\frac{l}{2} \leq x \leq \frac{3l}{4}$
- Section 4: $\frac{3l}{4} \leq x \leq l$.

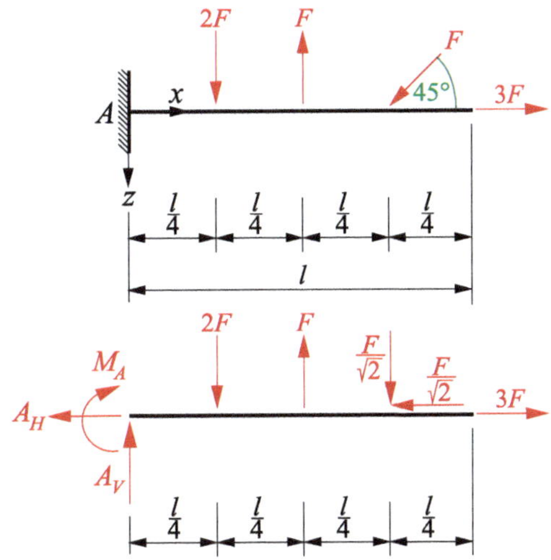

Fig. 6.13 Cantilever beam

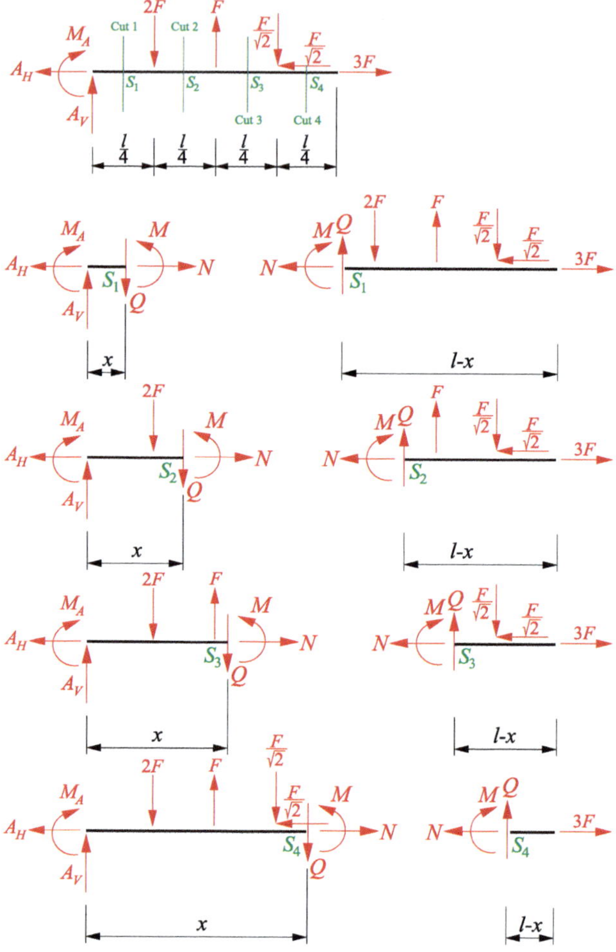

Fig. 6.14 Free-body diagrams

We start the considerations with Section 1 and obtain the following stress resultants from the equilibrium conditions on the left beam segment:

$$\overset{\rightarrow}{\sum} H = 0: \quad N - A_H = 0 \quad \rightarrow \quad N = \left(3 - \frac{1}{\sqrt{2}}\right) F,$$

$$\overset{\downarrow}{\sum} V = 0: \quad Q - A_V = 0 \quad \rightarrow \quad Q = \left(1 + \frac{1}{\sqrt{2}}\right) F,$$

$$\overset{\curvearrowleft}{\sum} M_{S_1} = 0: \quad M - M_A - A_V x = 0 \quad \rightarrow \quad M = \left[\left(1 + \frac{1}{\sqrt{2}}\right) x - \frac{3l}{4\sqrt{2}}\right] F. \tag{6.18}$$

While the normal force N and the shear force Q assume constant values in the interval $0 \leq x \leq \frac{l}{4}$, the bending moment M follows a linear distribution with the starting value $M\left(x=0\right) = -\frac{3Fl}{4\sqrt{2}}$ and the final value $M\left(x=\frac{l}{4}\right) = -\left(\sqrt{2}-1\right)\frac{Fl}{4}$.

Section 2 produces the following internal forces and moments using the left-hand beam segment:

$$\overset{\rightarrow}{\sum} H = 0: \quad N - A_H = 0 \quad \rightarrow \quad N = \left(3 - \frac{1}{\sqrt{2}}\right) F,$$

$$\overset{\downarrow}{\sum} V = 0: \quad Q - A_V + 2F = 0 \quad \rightarrow \quad Q = \left(\frac{1}{\sqrt{2}} - 1\right) F,$$

$$\overset{\curvearrowleft}{\sum} M_{S_2} = 0: \quad M - M_A - A_V x + 2F\left(x - \frac{l}{4}\right) = 0$$

$$\rightarrow \quad M = \left[\left(1 - \frac{3}{2\sqrt{2}}\right) l + \left(\frac{1}{\sqrt{2}} - 1\right) x\right] F. \tag{6.19}$$

The normal force and the shear force are also constant in the range $\frac{l}{4} \leq x \leq \frac{l}{2}$, while the bending moment is again linear with the starting value $M\left(x=\frac{l}{4}\right) = -\left(\sqrt{2}-1\right)\frac{Fl}{4}$ and the end value $M\left(x=\frac{l}{2}\right) = -\frac{Fl}{4\sqrt{2}}$.

For Section 3, the equilibrium conditions are as follows on the left beam segment:

$$\overset{\rightarrow}{\sum} H = 0: \quad N - A_H = 0 \quad \rightarrow \quad N = \left(3 - \frac{1}{\sqrt{2}}\right) F,$$

$$\overset{\downarrow}{\sum} V = 0: \quad Q - A_V + 2F - F = 0 \quad \rightarrow \quad Q = \frac{F}{\sqrt{2}},$$

$$\overset{\curvearrowleft}{\sum} M_{S_3} = 0: \quad M - M_A - A_V x + 2F\left(x - \frac{l}{4}\right) - F\left(x - \frac{l}{2}\right) = 0$$

$$\rightarrow \quad M = \frac{F}{\sqrt{2}}\left(x - \frac{3l}{4}\right). \tag{6.20}$$

This again results in constant values for normal force N and shear force Q as well as a linear function for the bending moment with the starting value $M\left(x=\frac{l}{2}\right) = -\frac{Fl}{4\sqrt{2}}$ and the end value $M\left(x=\frac{3l}{4}\right) = 0$.

Finally, we look at Section 4, for which using the left beam segment follows:

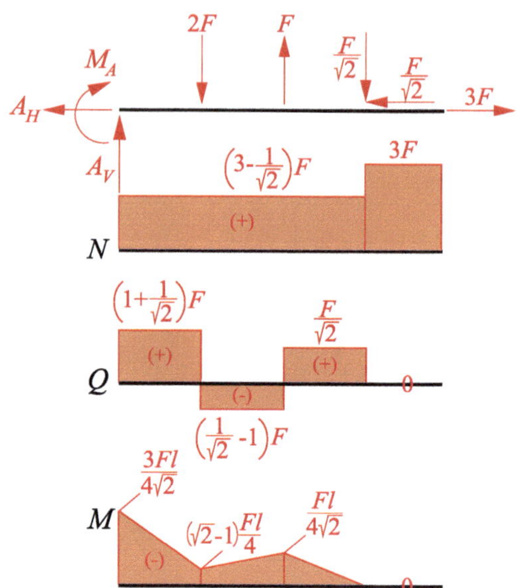

Fig. 6.15 Normal force diagram, shear force diagram and moment diagram

$$
\begin{aligned}
&\overset{\rightarrow}{\sum} H = 0: \quad N - A_H - \frac{F}{\sqrt{2}} = 0 \quad \rightarrow \quad N = 3F, \\
&\overset{\downarrow}{\sum} V = 0: \quad Q - A_V + 2F - F + \frac{F}{\sqrt{2}} = 0 \quad \rightarrow \quad Q = 0, \\
&\overset{\curvearrowleft}{\sum} M_{S_4} = 0: \quad M - M_A - A_V x + 2F\left(x - \frac{l}{4}\right) - F\left(x - \frac{l}{2}\right) + \frac{F}{\sqrt{2}}\left(x - \frac{3l}{4}\right) = 0 \\
&\rightarrow \quad M = 0.
\end{aligned}
\tag{6.21}
$$

This results in a constant normal force N, while both the shear force Q and the bending moment M for $\frac{3l}{4} \leq x \leq l$ are zero.

The determined stress resultants are shown in the diagrams given in Fig. 6.15. It can be seen that the normal force diagram N at the point of application of the force acting under 45° exhibits a discontinuity by exactly the horizontal force component $\frac{F}{\sqrt{2}}$. Similarly, the shear force diagram Q exhibits discontinuities at the force application points by exactly the amount of the vertical forces acting there and changes its sign twice over x, which is explained by the direction of action of the acting vertical forces. The beam section $\frac{3l}{4} \leq x \leq l$ is free of shear forces. The moment diagram is made up of linear functions in some areas and has kinks at the points of application of the vertical forces, it is negative throughout and becomes zero in the interval $\frac{3l}{4} \leq x \leq l$.

It should be noted here that the use of the right-hand beam segments in the free-body diagrams for Sections 3 and 4 would have been simpler than the use of the left-hand beam segments, but we have not done this here for illustrative purposes. It should also be noted that this is an example that could have been solved without first determining the support reactions. For this purpose, it would have been possible to determine the internal forces and moments 'from right to left', starting at the end of the cantilever beam. Of course, this approach would have produced the same results as in Fig. 6.15. ◄

6.2.2 Beams Under Point Forces and Moments

The procedure for the determination of stress resultants in beams is completely analogous to the explanations in Section 6.2.1 in the presence of applied moments. We consider the example of Fig. 6.16. Consider a beam on two supports of length l (partial lengths l_1 and l_2), which is loaded by a single moment M_0. We want to determine the force and moment diagrams N, Q and M for this example.

The support reactions were already determined in Example 4.3 as:

$$A_H = 0, \quad A_V = \frac{M_0}{l}, \quad B_V = -\frac{M_0}{l}. \tag{6.22}$$

With the known support reactions, the stress resultants in the beam can now be determined using the free-body diagram in Fig. 6.17. The internal forces and moments of the beam on the left side of the moment M_0 for Section 1 result as follows on the left beam segment:

$$\begin{aligned}
&\overset{\rightarrow}{\sum} H = 0: \quad N = 0, \\
&\overset{\downarrow}{\sum} V = 0: \quad Q - A_V = 0 \quad \rightarrow \quad Q = \frac{M_0}{l}, \\
&\overset{\curvearrowleft}{\sum} M_{S_1} = 0: \quad M - A_V x = 0 \quad \rightarrow \quad M = M_0 \frac{x}{l}.
\end{aligned} \tag{6.23}$$

While the normal force N is zero, the shear force Q is constant with the value $Q = \frac{M_0}{l}$. The bending moment, on the other hand, is linear over x and has the boundary values

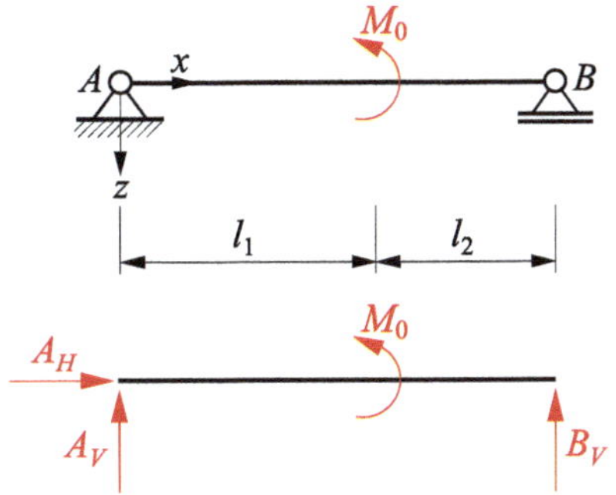

Fig. 6.16 Beam under single moment M_0

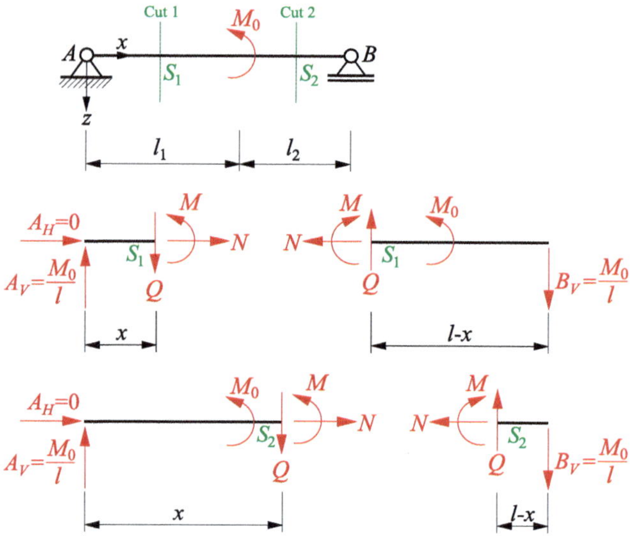

Fig. 6.17 Free-body diagrams

$M(x=0)=0$ at the hinged support at point A and $M\,(x=l_1)=M_0\frac{l_1}{l}$ to the left of the single moment M_0.

The following equilibrium conditions result for Section 2:

$$\overset{\rightarrow}{\sum} H=0: \quad N=0,$$

$$\overset{\downarrow}{\sum} V=0: \quad Q-A_V=0 \quad \rightarrow \quad Q=\frac{M_0}{l},$$

$$\overset{\curvearrowleft}{\sum} M_{S_1}=0: \quad M+M_0-A_V x=0 \quad \rightarrow \quad M=M_0\left(\frac{x}{l}-1\right). \tag{6.24}$$

This also results in a vanishing normal force N, a constant shear force $Q=\frac{M_0}{l}$ and a linear bending moment with the boundary values $M\,(x=l_1)=-M_0\frac{l_2}{l}$ and $M(x=l)=0$.

The force and moment diagrams for this example are shown in Fig. 6.18. It can be seen that the normal force N vanishes over the entire length of the beam, whereas the shear force Q is constant with the value $Q=\frac{M_0}{l}$. The bending moment is linear on both sides of the moment M_0 and has a discontinuity at the point of application of M_0 that corresponds exactly to the magnitude of M_0.

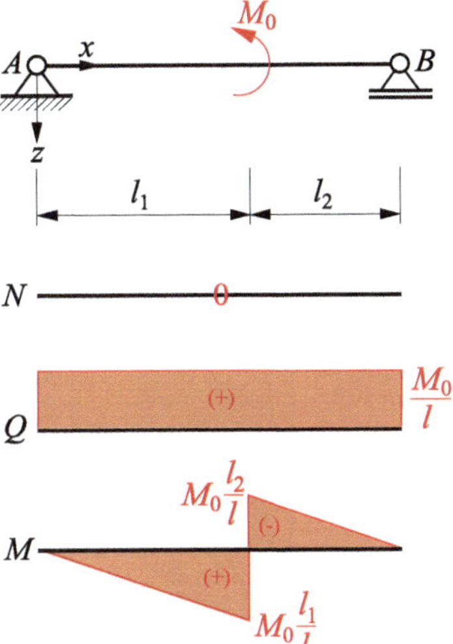

Fig. 6.18 Normal force diagram, shear force diagram and moment diagram

Example 6.3

Consider the cantilever beam shown in Fig. 6.19 with length $2l$, which is loaded by two vertically acting forces F, a horizontally acting force $2F$ and a moment $M_0 = 2Fl$. We want to determine the force and moment diagrams for this beam.

Solution:

This is an example in which it is not necessary to determine the support reactions in advance. We look at Sections 1 and 2 as shown in Fig. 6.19 and work our way from left to right. The following equilibrium conditions result at Section 1:

$$\overset{\rightarrow}{\sum} H = 0: \quad N + 2F = 0 \quad \rightarrow \quad N = -2F,$$

$$\overset{\downarrow}{\sum} V = 0: \quad Q + F = 0 \quad \rightarrow \quad Q = -F,$$

$$\overset{\curvearrowleft}{\sum} M_{S_1} = 0: \quad M + Fx = 0 \quad \rightarrow \quad M = -Fx. \tag{6.25}$$

The normal force N and the shear force Q both assume constant values, whereas the bending moment is a linear function with the boundary values $M(x = 0) = 0$ (free cantilever end) and $M(x = l) = -Fl$ to the left of the acting moment M_0.

Fig. 6.19 Cantilever beam

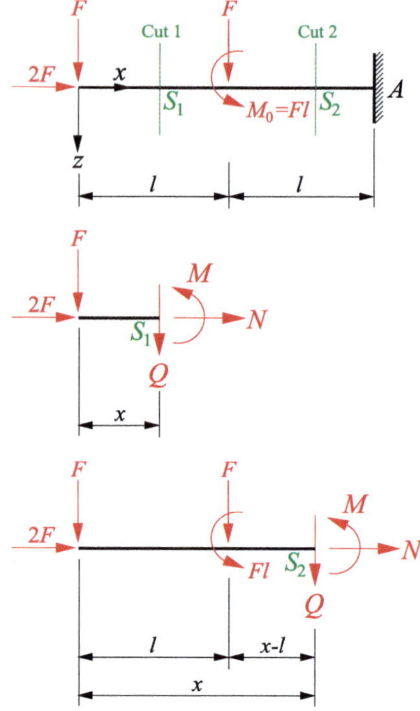

For Section 2 we obtain:

$$\overset{\rightarrow}{\sum} H = 0: \quad N + 2F = 0 \quad \rightarrow \quad N = -2F,$$

$$\overset{\downarrow}{\sum} V = 0: \quad Q + F + F = 0 \quad \rightarrow \quad Q = -2F,$$

$$\overset{\curvearrowleft}{\sum} M_{S_2} = 0: \quad M + Fx + F(x - l) + 2Fl = 0 \quad \rightarrow \quad M = -Fl\left(2\frac{x}{l} + 1\right). \quad (6.26)$$

Constant values for N and Q also occur here as well as a linear function for M with the boundary values $M(x = l) = -3Fl$ to the right of the applied moment M_0 and $M(x = 2l) = -5Fl$ at the clamped end.

The force and moment diagrams are shown in Fig. 6.20. It can be seen that the shear force diagram Q at the force application point $x = l$ has a discontinuity of exactly the magnitude of F. Similarly, the moment diagram M at the point of application of the moment $M_0 = 2Fl$ shows a discontinuity by exactly this value. ◀

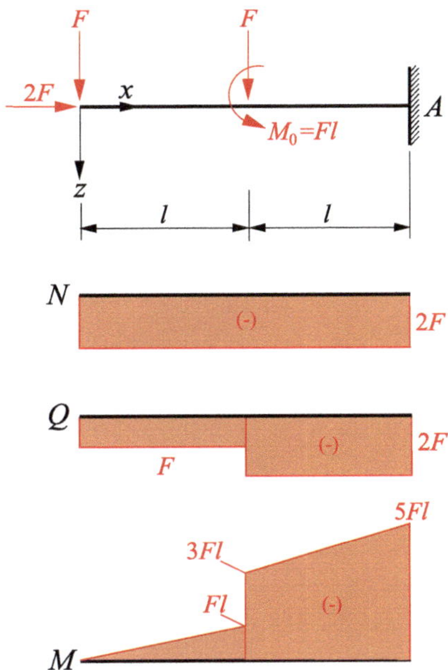

Fig. 6.20 Normal force diagram, shear force diagram and moment diagram

6.2.3 Beams Under Line Loads

We now want to extend the previous considerations to beams that are loaded by line loads. As an introduction, we consider the beam in Fig. 6.21 under the uniform line load q_0.

The support reactions can be determined as $A_V = B_V = \frac{q_0 l}{2}$ and $A_H = 0$ (see Example 4.4). To determine the stress resultants, we now cut the beam free at an arbitrary point x and look at the free-body diagram as shown in Fig. 6.21, centre. It should be noted that, in contrast to the determination of the support reactions, the line load q_0 must not be replaced by its total resultant $R = q_0 l$, but rather the resultant $R = q_0 x$ on the beam section of the length x must be considered. The equilibrium conditions are as follows:

$$\overset{\rightarrow}{\sum} H = 0: \quad N = 0,$$

$$\overset{\downarrow}{\sum} V = 0: \quad Q + q_0 x - \frac{q_0 l}{2} = 0 \quad \rightarrow \quad Q = \frac{q_0 l}{2}\left(1 - 2\frac{x}{l}\right),$$

$$\overset{\curvearrowleft}{\sum} M_{S_2} = 0: \quad M + q_0 x \cdot \frac{x}{2} - \frac{q_0 l}{2} \cdot x = 0 \quad \rightarrow \quad M = \frac{q_0 l^2}{2}\frac{x}{l}\left(1 - \frac{x}{l}\right). \quad (6.27)$$

The normal force N assumes the value zero over the entire length of the beam, whereas the shear force Q is a linear function over the length of the beam l with the boundary values $Q\,(x = 0) = \frac{q_0 l}{2}$ and $Q\,(x = l) = -\frac{q_0 l}{2}$. The bending moment is a quadratic function over x and disappears in the two supports A and B (hinged supports). The maximum value

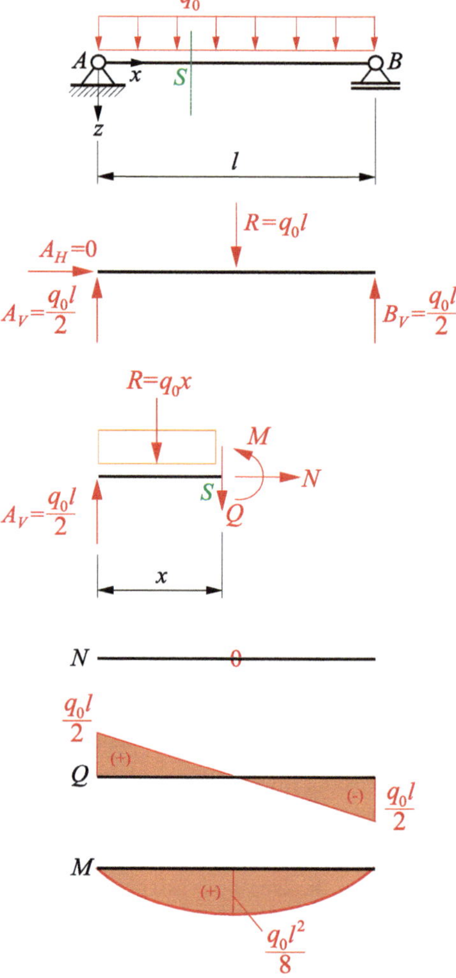

Fig. 6.21 Beam under constant line load

occurs in the centre of the beam at the point $x = \frac{l}{2}$, as can be concluded from considering the derivative $\frac{\mathrm{d}M}{\mathrm{d}x} = 0$. The maximum value is $M_{\max} = M\left(x = \frac{l}{2}\right) = \frac{q_0 l^2}{8}$. The force and moment diagrams are also shown in Fig. 6.21.

As a further example, consider the beam in Fig. 6.22 which is subjected to a linearly distributed line load $q(x) = q_0 \frac{x}{l}$. To determine the force and moment diagrams for this example, the support reactions are first determined, which result as $A_H = 0$, $A_V = \frac{q_0 l}{6}$ and $B_V = \frac{q_0 l}{3}$ (see Example 4.5). The internal forces and moments N, Q and M follow from the free-body diagram in Fig. 6.22. With the resultant of the line load $R = q_0 \frac{x}{l} \cdot \frac{1}{2} x = \frac{q_0 x^2}{2l}$ we obtain the following equilibrium conditions:

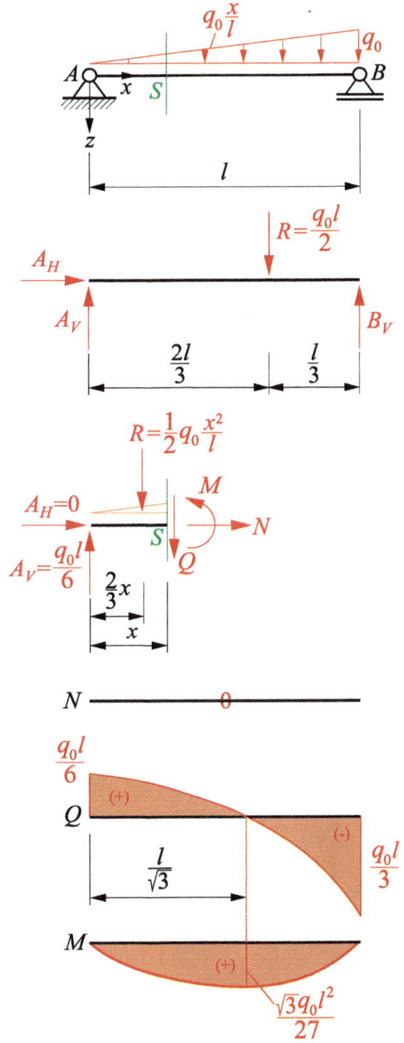

Fig. 6.22 Beam under linear line load

$$\overset{\rightarrow}{\sum} H = 0: \quad N = 0,$$

$$\overset{\downarrow}{\sum} V = 0: \quad Q + \frac{q_0 x^2}{2l} - \frac{q_0 l}{6} = 0 \quad \rightarrow \quad Q = \frac{q_0 l}{6}\left(1 - 3\frac{x^2}{l^2}\right),$$

$$\overset{\curvearrowleft}{\sum} M_{S_2} = 0: \quad M + \frac{q_0 x^2}{2l} \cdot \frac{x}{3} - \frac{q_0 l}{6} \cdot x = 0 \quad \rightarrow \quad M = \frac{q_0 l^2}{6}\frac{x}{l}\left(1 - \frac{x^2}{l^2}\right). \quad (6.28)$$

Accordingly, a linear line load results in a quadratic distribution of the shear force diagram Q, whereas the bending moment shows a cubic dependence on x. The force and moment diagrams for this example are shown in Fig. 6.22, bottom. The maximum value M_{max} for the bending moment is obtained by setting the first derivative $\frac{\text{d}M}{\text{d}x}$ to zero, which corresponds exactly to the shear force. Zeroing results in:

$$x = \frac{l}{\sqrt{3}}. \tag{6.29}$$

The value for $M = M_{\text{max}}$ at this point is $M_{\text{max}} = \frac{2q_0l^2}{9\sqrt{3}}$.

6.2.4 Relationship Between Load and Stress Resultants

There are relationships between the beam loads $n(x)$ and $q(x)$ on the one hand and the internal forces and moments $N(x)$, $Q(x)$ and $M(x)$ on the other hand, which we will consider below. To do this, we analyse the bar in Fig. 6.23, left, from which we separate an infinitesimally small section element (Fig. 6.23, right). We apply the stress resultants $N(x)$, $Q(x)$, $M(x)$ to this section element on the negative face. At the positive face at the point $x + \text{d}x$, these stress resultants occur with their infinitesimal increments, i.e. $N(x + \text{d}x) = N + \frac{\text{d}N}{\text{d}x}\text{d}x$, $Q(x + \text{d}x) = Q + \frac{\text{d}Q}{\text{d}x}\text{d}x$, $M(x + \text{d}x) = M + \frac{\text{d}M}{\text{d}x}\text{d}x$. Since this is an infinitesimally small section element, we can assume the two line loads $n(x)$ and $q(x)$ to be constant over the infinitesimal length $\text{d}x$. The equilibrium of the forces in the x-direction results in

$$\overset{\rightarrow}{\sum} H = 0: \quad N + \frac{\text{d}N}{\text{d}x}\text{d}x - N + n\text{d}x = 0, \tag{6.30}$$

or after reduction of N and $\text{d}x$:

$$\frac{\text{d}N}{\text{d}x} = N' = -n. \tag{6.31}$$

Accordingly, the first derivative of the normal force N corresponds to the negative applied line load n.

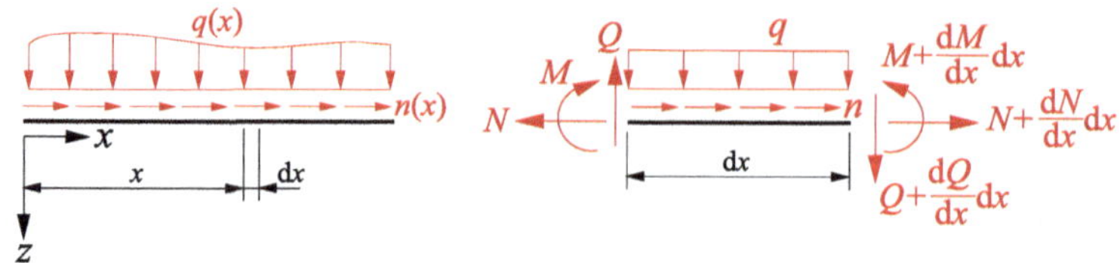

Fig. 6.23 Equilibrium for the infinitesimal beam element

The sum of the vertical forces results in:

$$\overset{\downarrow}{\sum} V = 0: \quad Q + \frac{\mathrm{d}Q}{\mathrm{d}x}\mathrm{d}x - Q + q\mathrm{d}x = 0, \tag{6.32}$$

which leads to:

$$\frac{\mathrm{d}Q}{\mathrm{d}x} = Q' = -q. \tag{6.33}$$

The first derivative of the shear force and thus its change corresponds accordingly to the negative applied line load q.

We now also consider the sum of the moments around the positive face and obtain

$$\overset{\curvearrowleft}{\sum} M = 0: \quad M + \frac{\mathrm{d}M}{\mathrm{d}x}\mathrm{d}x - M - Q\mathrm{d}x + q\mathrm{d}x \cdot \frac{\mathrm{d}x}{2} = 0. \tag{6.34}$$

This can be summarised as follows:

$$\frac{\mathrm{d}M}{\mathrm{d}x} - Q + q\frac{\mathrm{d}x}{2} = 0. \tag{6.35}$$

The third term in this equation is small compared to the other two terms, so it can be neglected. This leaves us with

$$\frac{\mathrm{d}M}{\mathrm{d}x} = M' = Q. \tag{6.36}$$

Obviously, the derivative of the bending moment M corresponds to the shear force Q. We derive this expression once and obtain

$$\frac{\mathrm{d}^2 M}{\mathrm{d}x^2} = M'' = Q'. \tag{6.37}$$

With $Q' = -q$ (see (6.33)) follows:

$$\frac{\mathrm{d}^2 M}{\mathrm{d}x^2} = M'' = -q. \tag{6.38}$$

From the relationships derived in this way, some important conclusions for the determination of internal forces and moments in beams can be derived:

- The line load $q(x)$ represents the gradient of the shear force diagram $Q(x)$ and the curvature of the moment diagram $M(x)$, as can be seen from (6.33) and (6.38). The diagram of the shear force $Q(x)$ represents the gradient of the bending moment diagram. For example, if there is a constant line load q_0, the shear force diagram is linear and the moment diagram is parabolic (see also Fig. 6.21). If, on the other hand, there is a linearly varying line load, the shear force diagram is quadratic and the moment diagram is cubic (Fig. 6.22). However, if there is no line load $q(x)$, the shear force curve is constant.

- Similarly, the distribution of the line load $n(x)$ describes the slope of the diagram of the normal force $N(x)$. If there is no line load $n(x)$, the normal force diagram is constant.
- If the shear force diagram $Q(x)$ has a zero point, then the bending moment $M(x)$ has an extreme value at this point (see e.g. Fig. 6.22).
- If a single force F acts on a beam in the z−direction, the shear force diagram $Q(x)$ at this point shows a discontinuity by exactly the magnitude of the acting force F. The moment diagram $M(x)$ shows a kink at such a point. Similarly, the normal force diagram $N(x)$ exhibits a discontinuity where a single force F acts in the x−direction, by exactly the magnitude of the applied force F. These rules also apply at locations where support forces occur.
- At a location x where a single moment acts, the bending moment diagram $M(x)$ exhibits a discontinuity by exactly the magnitude of the applied moment.

In the following, we will assume that there is no line load n. From Eq. (6.33) it follows that if there is a line load $q(x)$, the shear force follows by integrating the negative line load:

$$Q = -\int q \mathrm{d}x + C_1. \tag{6.39}$$

From Eqs. (6.36) and (6.38) it also follows that the bending moment $M(x)$ is obtained by double integration of the negative line load $q(x)$ or by single integration of the shear force $Q(x)$:

$$\begin{aligned} M &= -\int\int q \mathrm{d}x\mathrm{d}x + C_1 x + C_2 \\ &= \int Q \mathrm{d}x + C_2. \end{aligned} \tag{6.40}$$

The two integration constants C_1 and C_2 are to be determined from boundary conditions. A selection of typical support types is shown in Fig. 6.24. Note that conditions of the type $Q \neq 0$ or $M \neq 0$ cannot be used to determine the constants C_1 and C_2. If we want to determine the internal forces and moments of a beam by integrating the applied load, the support reactions do not have to be determined in advance. Instead, they are determined from the calculation as a result as we will show later.

As an example, let us consider the beam in Fig. 6.21 under the uniform line load q_0. For the shear force Q and the bending moment M it follows by integrating the line load q_0:

$$\begin{aligned} Q &= -q_0 x + C_1, \\ M &= -\frac{1}{2} q_0 x^2 + C_1 x + C_2. \end{aligned} \tag{6.41}$$

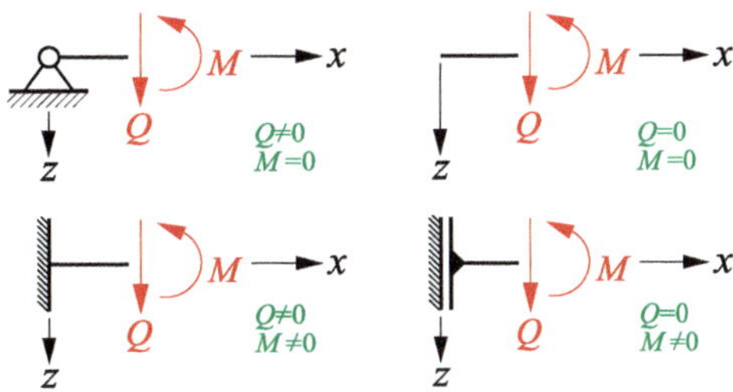

Fig. 6.24 Typical boundary conditions

The two constants C_1 and C_2 follow from the boundary conditions that the bending moment must become zero at the two support points $x = 0$ and $x = l$:

$$M(x = 0) = 0, \quad M(x = l) = 0. \tag{6.42}$$

If we use the expression for the bending moment M according to (6.41) we obtain:

$$\begin{aligned} M(x = 0) &= \quad \rightarrow \quad C_2 = 0, \\ M(x = l) &= 0 \quad \rightarrow \quad -\frac{1}{2}q_0 l^2 + C_1 l = 0 \quad \rightarrow \quad C_1 = \frac{1}{2}q_0 l. \end{aligned} \tag{6.43}$$

The shear force Q and the bending moment M can be specified accordingly as:

$$\begin{aligned} Q &= -q_0 x + \frac{1}{2}q_0 l = \frac{q_0 l}{2}\left(1 - 2\frac{x}{l}\right), \\ M &= -\frac{1}{2}q_0 x^2 + \frac{1}{2}q_0 l x = \frac{1}{2}q_0 l^2 \frac{x}{l}\left(1 - \frac{x}{l}\right). \end{aligned} \tag{6.44}$$

This corresponds to the results already provided with (6.27).

The support forces A_V and B_V (Fig. 6.21) follow directly from the shear force diagram (the support force A_H is identical to zero). The following applies:

$$A_V = Q(x = 0) = \frac{q_0 l}{2}, \quad B_V = -Q(x = l) = \frac{q_0 l}{2}. \tag{6.45}$$

Example 6.4

Consider the beam in Fig. 6.22 under a linearly varying line load $q(x) = q_0\frac{x}{l}$. We want to determine the shear force $Q(x)$ and the bending moment $M(x)$ by integration.

Solution:

The line load is given as $q(x) = q_0\frac{x}{l}$. This results in the shear force $Q(x)$ by integration as:

$$Q = -\int q \mathrm{d}x + C_1 = -q_0 \int \frac{x}{l} \mathrm{d}x + C_1 = -\frac{q_0 x^2}{2l} + C_1. \tag{6.46}$$

For the bending moment $M(x)$ then follows:

$$M = \int Q \mathrm{d}x + C_2 = -\int \left(\frac{q_0 x^2}{2l} + C_1 \right) \mathrm{d}x + C_2 = -\frac{q_0 x^3}{6l} + C_1 x + C_2. \tag{6.47}$$

The two integration constants C_1 and C_2 follow from the consideration of the boundary conditions:

$$\begin{aligned} M(x = 0) = \quad &\rightarrow \quad C_2 = 0, \\ M(x = l) = 0 \quad &\rightarrow \quad -\frac{q_0 l^3}{6l} + C_1 l = 0 \quad \rightarrow \quad C_1 = \frac{q_0 l}{6}. \end{aligned} \tag{6.48}$$

From this, the shear force and bending moment can be determined as:

$$\begin{aligned} Q &= \frac{q_0 l}{6} \left(1 - 3\frac{x^2}{l^2} \right), \\ M &= \frac{q_0 l^2}{6} \frac{x}{l} \left(1 - \frac{x^2}{l^2} \right). \end{aligned} \tag{6.49}$$

This result agrees with the result already determined from equilibrium considerations.

The support forces A_V and B_V follow from the shear force as:

$$A_V = Q(x = 0) = \frac{q_0 l}{6}, \quad B_V = -Q(x = l) = \frac{q_0 l}{3}. \tag{6.50}$$

◀

Example 6.5

For the cantilever beam shown in Fig. 6.25 the diagrams for N, Q, M are to be determined.

Solution:

The line load $q(x)$ can be represented as $q(x) = q_0 \left(1 - \frac{x}{l}\right)$. The shear force Q and bending moment M then follow as (the normal force N is identical to zero for this example):

$$\begin{aligned} Q &= -\int q \mathrm{d}x + C_1 = -q_0 \int \left(1 - \frac{x}{l} \right) \mathrm{d}x + C_1 = -\frac{q_0 x}{2} \left(2 - \frac{x}{l} \right) + C_1, \\ M &= \int Q \mathrm{d}x + C_2 = \int \left[-\frac{q_0 x}{2} \left(2 - \frac{x}{l} \right) + C_1 \right] \mathrm{d}x = -\frac{q_0 x^2}{6} \left(3 - \frac{x}{l} \right) + C_1 x + C_2. \end{aligned} \tag{6.51}$$

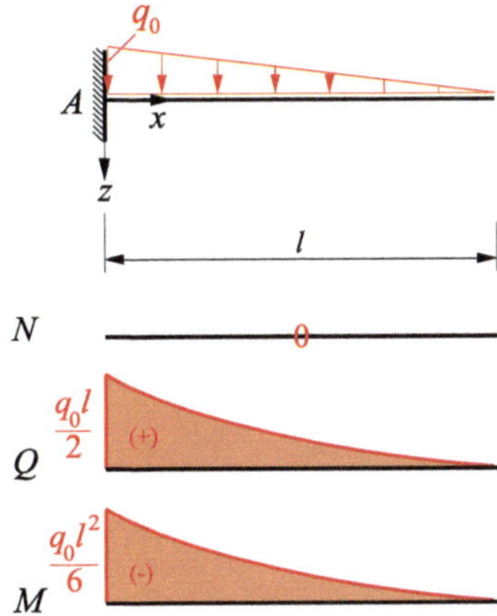

Fig. 6.25 Cantilever beam under linearly varying line load

The boundary conditions are as follows:

$$Q(x = l) = 0, \quad M(x = l) = 0. \tag{6.52}$$

Evaluating the boundary conditions results in the following integration constants C_1 and C_2:

$$C_1 = \frac{q_0 l}{2}, \quad C_2 = -\frac{q_0 l^2}{6}. \tag{6.53}$$

This allows the shear force Q and the bending moment M to be represented as:

$$\begin{aligned} Q &= \frac{q_0 l}{2}\left(1 - 2\frac{x}{l} + \frac{x^2}{l^2}\right), \\ M &= -\frac{q_0 l^2}{6}\left(1 - 3\frac{x}{l} + 3\frac{x^2}{l^2} - \frac{x^3}{l^3}\right). \end{aligned} \tag{6.54}$$

The diagrams described by these two functions are shown in Fig. 6.25. ◀

Example 6.6

For the cantilever beam shown in Fig. 6.26 the diagrams for N, Q, M are to be determined.

Solution:

Shear force Q and bending moment M follow here as (the normal force N is zero for this cantilever beam):

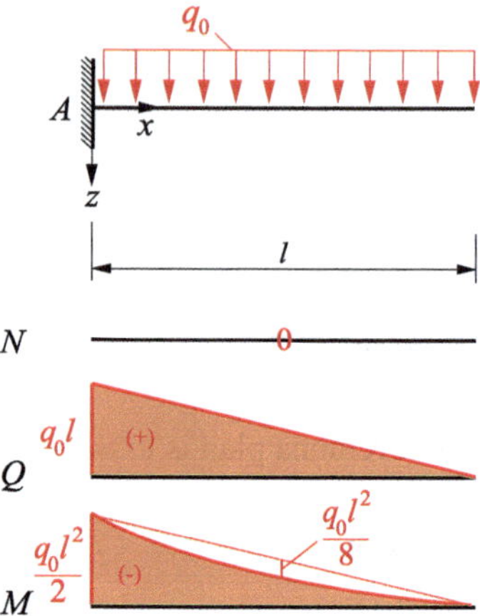

Fig. 6.26 Cantilever beam under constant line load

$$\begin{aligned} Q &= -\int q \mathrm{d}x + C_1 = -q_0 x + C_1, \\ M &= \int Q \mathrm{d}x + C_2 = \int (-q_0 x + C_1)\, \mathrm{d}x = -\frac{q_0 x^2}{2} + C_1 x + C_2. \end{aligned} \tag{6.55}$$

The boundary conditions for this example are as follows:

$$Q(x = l) = 0, \quad M(x = l) = 0. \tag{6.56}$$

The integration constants C_1 and C_2 follow from this as:

$$C_1 = q_0 l, \quad C_2 = -\frac{q_0 l^2}{2}. \tag{6.57}$$

Shear force Q and bending moment M can then be specified as:

$$\begin{aligned} Q &= q_0 l \left(1 - \frac{x}{l}\right), \\ M &= -\frac{q_0 l^2}{2} \left(1 - 2\frac{x}{l} + \frac{x^2}{l^2}\right). \end{aligned} \tag{6.58}$$

The force and moment diagrams are shown in Fig. 6.26. ◀

6.3 Superposition Principle

The superposition principle applies to the determination of the stress resultants for the linear problems considered in this chapter. As an illustration, we consider the cantilever beam of Fig. 6.27 with the length $2l$, which is loaded by two point forces F and by the line load q. The moment diagram M is to be determined. In many cases, it proves to be advantageous not to calculate the internal forces and moments for all simultaneously acting loads, but in many cases it can greatly simplify the considerations if each applied load is first considered for itself and the resulting internal forces and moments are superimposed at the end, i.e. added together. Using the example of Fig. 6.27, this means that the moment lines due to the two point forces F are determined first and then the moment line due to the line load q. The final moment line M results from the addition of the three partial moment lines.

It will also prove useful for some applications (see Volume 2) to break down the moment diagrams into elementary forms as far as possible. Using the example of Fig. 6.27, this means for the triangular moment diagram due to the force F at the point $x = 2l$ that it can be divided into a rectangle and two triangles with the base lengths l and the magnitude Fl each. Similarly, the moment diagram due to the line load q can be broken down into

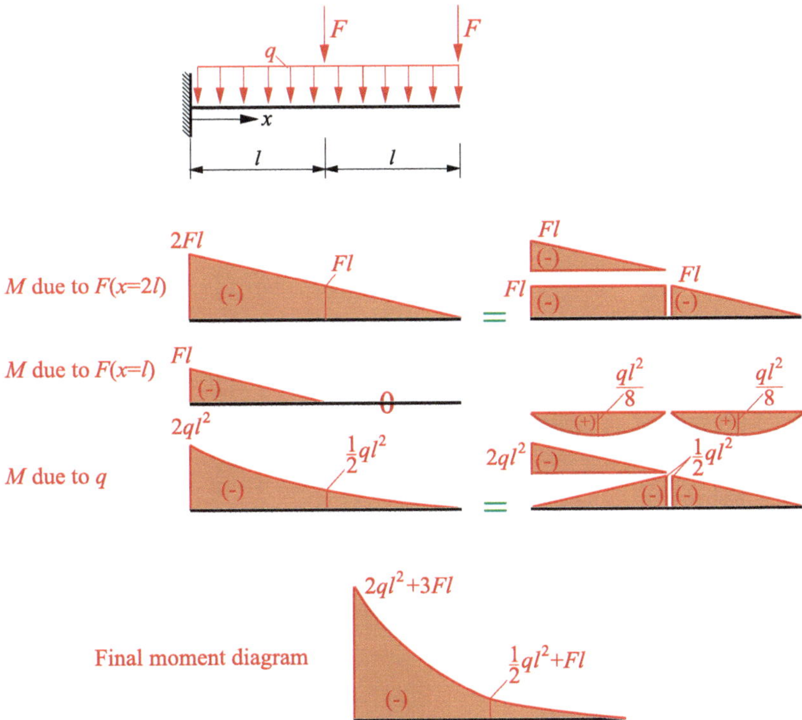

Fig. 6.27 Beam under line load q and point loads F (top), moment diagrams (bottom)

rectangles, triangles and parabolic areas as shown in Fig. 6.27, bottom.

Example 6.7

For the cantilever beam shown in Fig. 6.28 the force and moment diagrams N, Q, M are to be determined.

Solution:

The line load $q(x)$ can be specified as $q(x) = q_0 \left(3 - 2\frac{x}{l}\right)$. Then the shear force and the bending moment follow as:

$$\begin{aligned} Q &= -\int q \mathrm{d}x + C_1 = -q_0 x \left(3 - \frac{x}{l}\right) + C_1, \\ M &= \int Q \mathrm{d}x + C_2 = -\frac{3}{2} q_0 x^2 \left(1 - \frac{2}{9}\frac{x}{l}\right) + C_1 x + C_2. \end{aligned} \tag{6.59}$$

The boundary conditions are as follows for the given situation:

$$Q(x = l) = 0, \quad M(x = l) = 0. \tag{6.60}$$

The integration constants C_1 and C_2 can be determined from this as:

$$C_1 = 2q_0 l, \quad C_2 = -\frac{5q_0 l^2}{6}. \tag{6.61}$$

The shear force Q and the bending moment M can then be written as:

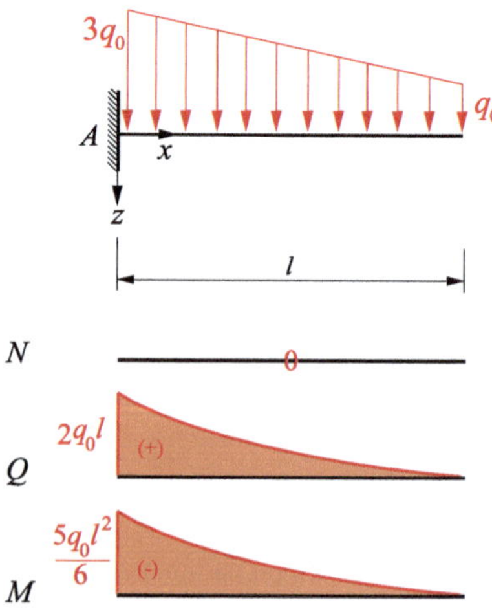

Fig. 6.28 Cantilever beam under linearly varying line load

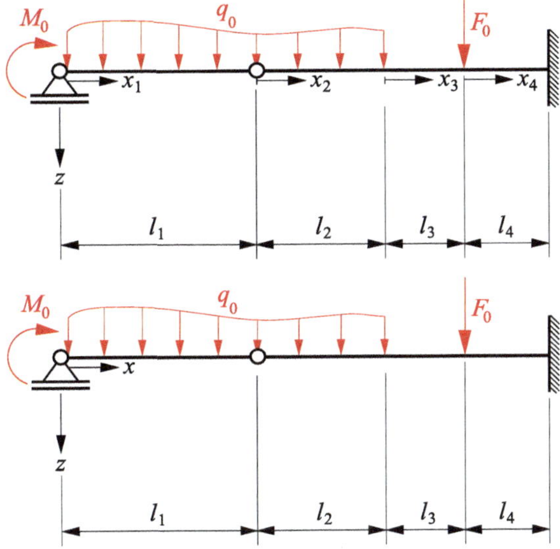

Fig. 6.29 Beam under point force F_0, moment M_0 and line load q_0

$$
\begin{aligned}
Q &= q_0 l \left(2 - 3\frac{x}{l} + \frac{x^2}{l^2}\right), \\
M &= -\frac{5}{6} q_0 l^2 \left(1 - \frac{12}{5}\frac{x}{l} + \frac{9}{5}\frac{x^2}{l^2} - \frac{2}{5}\frac{x^3}{l^3}\right).
\end{aligned}
\tag{6.62}
$$

The force and moment diagrams are shown in Fig. 6.29.

It can easily be shown that the shear force Q and the bending moment M can be obtained from the superposition of the results of Examples 6.5 and 6.6. It follows:

$$
\begin{aligned}
Q &= q_0 l \left(1 - \frac{x}{l}\right) + 2\frac{q_0 l}{2}\left(1 - 2\frac{x}{l} + \frac{x^2}{l^2}\right) = q_0 l \left(2 - 3\frac{x}{l} + \frac{x^2}{l^2}\right), \\
M &= -\frac{q_0 l^2}{2}\left(1 - 2\frac{x}{l} + \frac{x^2}{l^2}\right) - 2\frac{q_0 l^2}{6}\left(1 - 3\frac{x}{l} + 3\frac{x^2}{l^2} - \frac{x^3}{l^3}\right) \\
&= -\frac{5}{6} q_0 l^2 \left(1 - \frac{12}{5}\frac{x}{l} + \frac{9}{5}\frac{x^2}{l^2} - \frac{2}{5}\frac{x^3}{l^3}\right).
\end{aligned}
\tag{6.63}
$$

◀

6.4 Multi-section Problems

In many cases, neither the load q of the beam nor the resulting force and moment diagrams are continuous functions over the entire length of the beam. Rather, individual forces and moments as well as hinges and supports, but also discontinuous changes in the applied line load q cause discontinuities in the diagrams Q and M. This is also referred to as a so-called multi-field problem, and to determine the force and moment diagrams Q and M, the beam under consideration must then be divided into sub-fields within which the forces and moments are continuous functions and the relationships (6.39) and (6.40) can be used to determine Q and M by integration. This is illustrated by the beam in Fig. 6.29, which is loaded by a point force F_0, a moment M_0 and a line load q_0. The beam can be divided into four sections or fields of lengths $l_1, \ldots, l_4$ and we introduce the local axes $x_1, \ldots, x_4$ as indicated (Fig. 6.29, top). The integration of the load is then carried out section by section as follows. For $0 \leq x_1 \leq l_1$ we obtain:

$$\begin{aligned} Q_1 &= -\int q \mathrm{d}x_1 + C_{1,1}, \\ M_1 &= \int Q_1 \mathrm{d}x_1 + C_{2,1} = \int \left(-\int q \mathrm{d}x_1 + C_{1,1} \right) \mathrm{d}x_1 + C_{2,1} \\ &= -\int \int q \mathrm{d}x_1 \mathrm{d}x_1 + C_{1,1} x_1 + C_{2,1}. \end{aligned} \tag{6.64}$$

In the range $0 \leq x_2 \leq l_2$ results:

$$\begin{aligned} Q_2 &= -\int q \mathrm{d}x_2 + C_{1,2}, \\ M_2 &= \int Q_2 \mathrm{d}x_2 + C_{2,2} = \int \left(-\int q \mathrm{d}x_2 + C_{1,2} \right) \mathrm{d}x_2 + C_{2,2} \\ &= -\int \int q \mathrm{d}x_2 \mathrm{d}x_2 + C_{1,2} x_2 + C_{2,2}. \end{aligned} \tag{6.65}$$

For $0 \leq x_3 \leq l_3$ we have:

$$\begin{aligned} Q_3 &= -\int q \mathrm{d}x_3 + C_{1,3} = C_{1,3}, \\ M_3 &= \int Q_3 \mathrm{d}x_3 + C_{2,3} = C_{1,3} x_3 + C_{2,3}. \end{aligned} \tag{6.66}$$

Finally, in the range $0 \leq x_4 \leq l_4$ we have:

$$\begin{aligned} Q_4 &= -\int q \mathrm{d}x_4 + C_{1,4} = C_{1,4}, \\ M_4 &= \int Q_4 \mathrm{d}x_4 + C_{2,4} = C_{1,4} x_4 + C_{2,4}. \end{aligned} \tag{6.67}$$

This means that a total of eight integration constants must be determined, for which suitable conditions must be found for their calculation. In addition to the boundary conditions, the so-called continuity conditions must be formulated, which establish relationships between the forces and moments of adjacent subsections. Using the specific example of Fig. 6.29, the applicable boundary and continuity conditions are as follows.

At the left support $x_1 = 0$, the bending moment M_1 corresponds to the applied moment M_0:

$$M_1(x_1 = 0) = M_0. \tag{6.68}$$

At the hinge point $x_1 = l_1$ or $x_2 = 0$, the shear force is continuous so that Q_1 and Q_2 are identical at this point:

$$Q_1(x_1 = l_1) = Q_2(x_2 = 0). \tag{6.69}$$

At the hinge point $x_1 = l_1$ or $x_2 = 0$ the two bending moments M_1 and M_2 must become zero:

$$M_1(x_1 = l_1) = 0, \quad M_2(x_2 = 0) = 0. \tag{6.70}$$

At the end point of the line load $q(x)$, both the shear forces and the bending moments of the two Sections 2 and 3 match:

$$Q_2(x_2 = l_2) = Q_3(x_3 = 0), \quad M_2(x_2 = l_2) = M_3(x_3 = 0). \tag{6.71}$$

At the point of application of the force F_0 at $x_3 = l_3$ or $x_4 = 0$, the shear force diagram exhibits a discontinuity of exactly the value F_0:

$$Q_3(x_3 = l_3) = Q_4(x_4 = 0) + F_0. \tag{6.72}$$

In addition, the bending moment is identical on both sides of the force F_0:

$$M_3(x_3 = l_3) = M_4(x_4 = 0). \tag{6.73}$$

This provides a total of eight boundary and continuity conditions from which the eight integration constants can be determined. In principle, 2 boundary conditions and $2(n-1)$ continuity conditions are available for a beam that is to be divided into n subsections.

It should also be noted that it is not absolutely necessary to introduce local axes $x_1, \ldots, x_4$ as shown in Fig. 6.29. Similarly, the internal forces and moments can be determined by introducing a single reference axis x as shown in Fig. 6.29, bottom. The boundary and continuity conditions for the example in Fig. 6.29, bottom, are then as follows:

$$\begin{aligned}
M_1(x=0) &= M_0, \\
Q_1(x=l_1) &= Q_2(x=l_1), \\
M_1(x=l_1) &= 0, \\
M_2(x=l_1) &= 0, \\
Q_2(x=l_1+l_2) &= Q_3(x=l_1+l_2), \\
M_2(x=l_1+l_2) &= M_3(x=l_1+l_2), \\
Q_3(x=l_1+l_2+l_3) &= Q_4(x=l_1+l_2+l_3)+F_0, \\
M_3(x=l_1+l_2+l_3) &= M_4(x=l_1+l_2+l_3).
\end{aligned} \tag{6.74}$$

Figure 6.30 shows some typical transitions between two sub-sections i and $i+1$ in which the line loads q_i and q_{i+1} occur.

If there are two segments i and $i+1$ of a beam that are separated by a support with a vertically acting support force (Fig. 6.30, top left), then the two bending moments M_i and M_{i+1} are identical on both sides of the support:

$$M_i = M_{i+1}. \tag{6.75}$$

The shear force diagram will show a discontinuity by exactly the magnitude of the support force due to the support applied at this point. However, since the support force is unknown

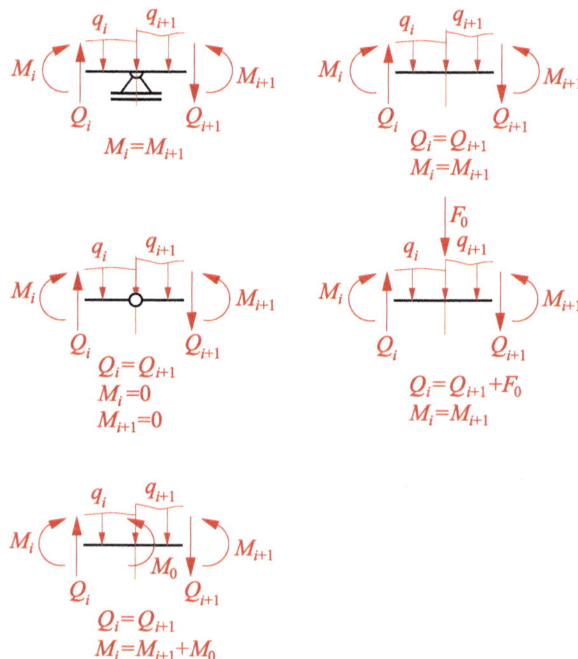

Fig. 6.30 Continuity conditions on the multi-section beam

and will rather be a result of the calculations, this condition cannot be used as a continuity condition. The moment line has a kink at this point, i.e. its gradient is discontinuous.

At the transition of the Fig. 6.30, top right, both the transverse forces and the bending moments are identical on both sides of the transition point:

$$Q_i = Q_{i+1}, \quad M_i = M_{i+1}. \tag{6.76}$$

The shear force line has a kink at this point due to the different line loads q_i and q_{i+1}, whereas the moment line is continuous.

If there is a hinge at the transition between the areas i and $i+1$ (Fig. 6.30, centre left), then the two bending moments on both sides of the joint are identical to zero, whereas the shear force on both sides of the joint is identical:

$$Q_i = Q_{i+1}, \quad M_i = 0, \quad M_{i+1} = 0. \tag{6.77}$$

If a point force F_0 acts on the beam (Fig. 6.30, centre right), the shear force diagram shows a discontinuity by exactly the magnitude of the force. The bending moment, on the other hand, is identical on both sides of the force application point, but the moment line has a kink:

$$Q_i = Q_{i+1} + F_0, \quad M_i = M_{i+1}. \tag{6.78}$$

In the event that a single moment M_0 acts on the beam (Fig. 6.30, bottom), the moment diagram shows a discontinuity by exactly the magnitude of the applied moment. The shear force diagram is continuous at this point:

$$Q_i = Q_{i+1}, \quad M_i = M_{i+1} + M_0. \tag{6.79}$$

It should be noted that the calculation of stress resultants using integration quickly becomes very involved and time-consuming if there are beams with many segments. This method is therefore only suitable for very simple beam situations. We will get to know a much more practically relevant method later on.

Example 6.8

For the beam in Fig. 6.31 we want to formulate all boundary and continuity conditions.

Solution:

We divide the beam into a total of four sections and introduce the local reference axes $x_1, \ldots, x_4$ as shown. We can formulate a total of eight boundary and continuity conditions for this beam.

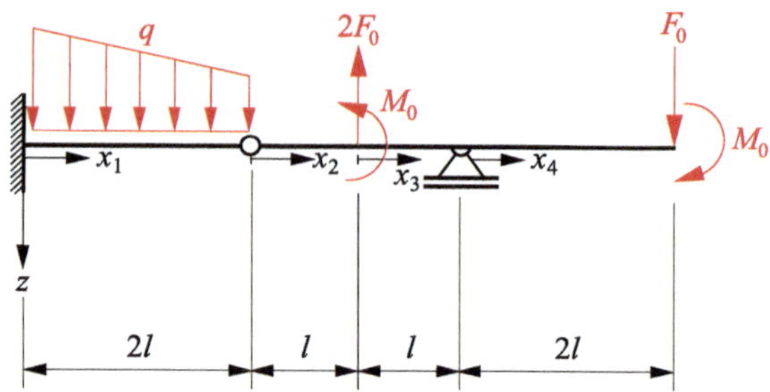

Fig. 6.31 Given beam

In the hinge at the transition between segments 1 and 2, the bending moments on both sides of the hinge must become zero. The shear force, on the other hand, is continuous at this point:

$$\begin{aligned} M_1(x_1 = 2l) &= 0, \\ M_2(x_2 = 0) &= 0, \\ Q_1(x_1 = 2l) &= Q_2(x_2 = 0). \end{aligned} \tag{6.80}$$

At the transition between segments 2 and 3, the shear force and the bending moment have discontinuities by exactly the magnitudes of the applied force $2F_0$ and the applied moment M_0 respectively:

$$\begin{aligned} Q_2(x_2 = l) &= Q_3(x_3 = 0) - 2F_0, \\ M_2(x_2 = l) &= M_3(x_3 = 0) + M_0. \end{aligned} \tag{6.81}$$

At the support point between segments 3 and 4, the moment diagram is continuous, but will have a kink:

$$M_3(x_3 = l) = M_4(x_4 = 0). \tag{6.82}$$

At this point, the shear force diagram will show a discontinuity by exactly the magnitude of the support force. However, as the support force is unknown and will result from the calculation, this cannot be used as a continuity condition.

At the free end of the beam, the shear force Q corresponds to the point force F_0 acting there. The bending moment there assumes the value of the negative single moment M_0. The following applies:

$$\begin{aligned} Q_4(x_4 = 2l) &= F_0, \\ M_4(x_4 = 2l) &= -M_0. \end{aligned} \tag{6.83}$$

◀

Example 6.9

For the beam on two supports shown in Fig. 6.32 under point force F, the internal forces and moments are to determined.

Solution:

We divide the beam into two segments with the reference axes x_1 and x_2 and determine the transverse forces Q_1 and Q_2 as well as the bending moments M_1 and M_2 by integration (the normal force N is identical to zero under this load over the entire length of the beam):

$$
\begin{aligned}
Q_1 &= -\int q \mathrm{d}x_1 + C_{1,1} = C_{1,1}, \\
M_1 &= \int Q_1 \mathrm{d}x_1 + C_{2,1} = C_{1,1}x_1 + C_{2,1}, \\
Q_2 &= -\int q \mathrm{d}x_2 + C_{1,2} = C_{1,2}, \\
M_2 &= \int Q_2 \mathrm{d}x_2 + C_{2,2} = C_{1,2}x_2 + C_{2,2}.
\end{aligned}
\tag{6.84}
$$

To determine the integration constants, we use the boundary and continuity conditions as follows:

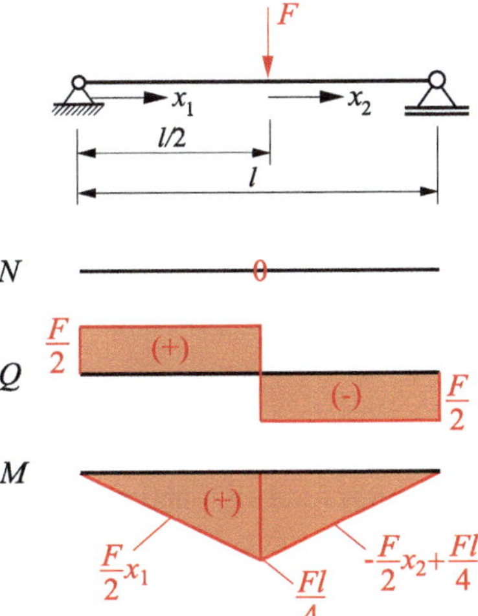

Fig. 6.32 Beam on two supports under point force F

$$
\begin{aligned}
M_1(x_1 = 0) &= 0 \quad \rightarrow \quad C_{2,1} = 0, \\
Q_1\left(x_1 = \frac{l}{2}\right) &= Q_2\left(x_2 = 0\right) + F_0 \quad \rightarrow \quad C_{1,1} = C_{1,2} + F_0, \\
M_1\left(x_1 = \frac{l}{2}\right) &= M_2\left(x_2 = 0\right) \quad \rightarrow \quad C_{1,1}\frac{l}{2} = C_{2,2}, \\
M_2\left(x_2 = \frac{l}{2}\right) &= 0 \quad \rightarrow \quad C_{1,2}\frac{l}{2} + C_{2,2} = 0.
\end{aligned} \tag{6.85}
$$

While the integration constant $C_{2,1}$ results in zero from the first condition, the remaining three conditions represent a linear system of equations for the three constants $C_{1,1}$, $C_{1,2}$, $C_{2,2}$, which can be easily solved. It follows:

$$
C_{1,1} = \frac{F_0}{2}, \quad C_{1,2} = -\frac{F_0}{2}, \quad C_{2,2} = \frac{F_0 l}{4}. \tag{6.86}
$$

This allows the stress resultants Q and M to be specified as follows:

$$
\begin{aligned}
Q_1 &= \frac{F_0}{2}, \\
M_1 &= \frac{F_0}{2}x_1, \\
Q_2 &= -\frac{F_0}{2}, \\
M_2 &= \frac{F_0 l}{4}\left(1 - 2\frac{x_2}{l}\right).
\end{aligned} \tag{6.87}
$$

The stress resultants are shown in Fig. 6.32, bottom.

◀

Example 6.10

For the beam in Fig. 6.33 the internal forces and moments are to be determined.

Solution:

The beam is divided into two sections with the reference axes x_1 and x_2. We determine the transverse forces Q_1 and Q_2 and the bending moments M_1 and M_2 by means of integration. The normal force N is zero under the given load. It follows:

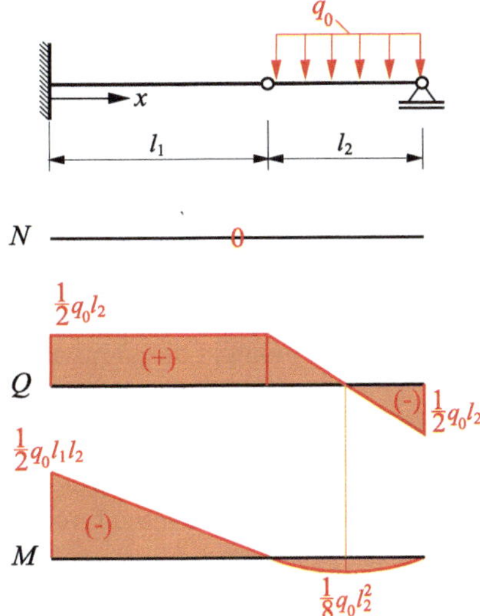

Fig. 6.33 Given beam

$$
\begin{aligned}
Q_1 &= -\int q \mathrm{d}x_1 + C_{1,1} = C_{1,1}, \\
M_1 &= \int Q_1 \mathrm{d}x_1 + C_{2,1} = C_{1,1}x_1 + C_{2,1}, \\
Q_2 &= -\int q \mathrm{d}x_2 + C_{1,2} = -q_0x_2 + C_{1,2}, \\
M_2 &= \int Q_2 \mathrm{d}x_2 + C_{2,2} = -\frac{1}{2}q_0x_2^2 + C_{1,2}x_2 + C_{2,2}.
\end{aligned} \tag{6.88}
$$

We can use the following boundary and continuity conditions:

$$
\begin{aligned}
&M_1(x_1 = l_1) = 0: \quad \rightarrow \quad C_{1,1}l_1 + C_{2,1} = 0, \\
&M_2(x_2 = 0) = 0: \quad \rightarrow \quad C_{2,2} = 0, \\
&M_2(x_2 = l_2) = 0: \quad \rightarrow \quad C_{1,2} = \frac{1}{2}q_0l_2, \\
&Q_1(x_1 = l_1) = Q_2x_2 = 0: \quad \rightarrow \quad C_{1,1} = \frac{1}{2}q_0l_2.
\end{aligned} \tag{6.89}
$$

Finally, $C_{2,1} = -\frac{1}{2}q_0l_1l_2$ can be derived from the first equation in (6.89). The stress resultants can be specified as:

$$
\begin{aligned}
Q_1 &= \frac{1}{2} q_0 l_2, \\
M_1 &= \frac{1}{2} q_0 l_1 l_2 \left(\frac{x_1}{l_1} - 1 \right), \\
Q_2 &= -q_0 x_2 + \frac{1}{2} q_0 l_2, \\
M_2 &= \frac{1}{2} q_0 l_2 x_2 \left(1 - \frac{x_2}{l_2} \right).
\end{aligned} \tag{6.90}
$$

The force and moment diagrams are shown in Fig. 6.33. ◀

Example 6.11

For the beam in Fig. 6.34, the force and moment diagrams are to be determined.

Solution:

The beam is divided into a total of two sections. At the centre support point, the shear force and the slope of the bending moment diagram will be discontinuous, so that a new reference axis is introduced at this point. On the other hand, no new division is necessary for this beam at the hinge point, since neither individual forces nor moments act here and, in addition, the applied line load q_0 runs continuously across the hinge. The following applies to the stress resultants in the two segments 1 and 2:

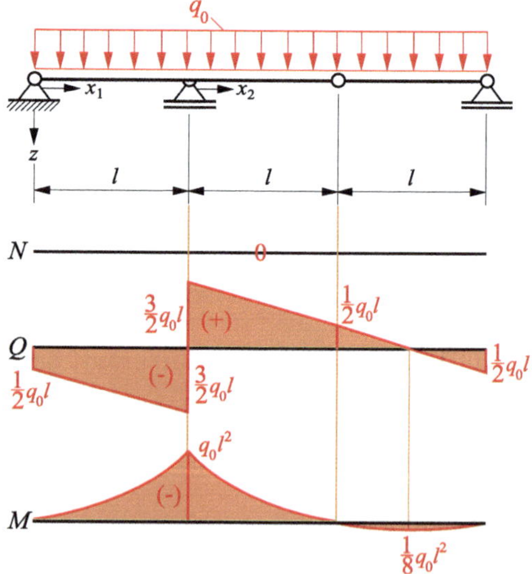

Fig. 6.34 Given beam

$$
\begin{aligned}
Q_1 &= -\int q\mathrm{d}x_1 + C_{1,1} = -q_0 x_1 + C_{1,1}, \\
M_1 &= \int Q_1 \mathrm{d}x_1 + C_{2,1} = -\frac{1}{2} q_0 x_1^2 + C_{1,1} x_1 + C_{2,1}, \\
Q_2 &= -\int q\mathrm{d}x_2 + C_{1,2} = -q_0 x_2 + C_{1,2}, \\
M_2 &= \int Q_2 \mathrm{d}x_2 + C_{2,2} = -\frac{1}{2} q_0 x_2^2 + C_{1,2} x_2 + C_{2,2}.
\end{aligned} \tag{6.91}
$$

The boundary and continuity conditions to be applied are as follows:

$$
\begin{aligned}
M_1(x_1 = 0) &= 0, \\
M_2(x_2 = 2l) &= 0, \\
M_1(x_1 = l) &= M_2(x_2 = 0), \\
M_2(x_2 = l) &= 0.
\end{aligned} \tag{6.92}
$$

From this, the integration constants can be obtained as follows:

$$
C_{1,1} = -\frac{1}{2} q_0 l, \quad C_{2,1} = 0, \quad C_{1,2} = \frac{3}{2} q_0 l, \quad C_{2,2} = -q_0 l^2. \tag{6.93}
$$

The stress resultants can then be specified as follows (the normal force is identical to zero for this beam):

$$
\begin{aligned}
Q_1 &= -\frac{1}{2} q_0 l \left(2\frac{x_1}{l} + 1\right), \\
M_1 &= -\frac{1}{2} q_0 l x_1 \left(\frac{x_1}{l} + 1\right), \\
Q_2 &= \frac{3}{2} q_0 l \left(1 - \frac{2}{3}\frac{x_2}{l}\right), \\
M_2 &= q_0 l^2 \left(-\frac{1}{2}\frac{x_2^2}{l^2} + \frac{3}{2}\frac{x_2}{l} - 1\right).
\end{aligned} \tag{6.94}
$$

The force and moment diagrams are shown in Fig. 6.34.

◀

6.5 Internal Forces in Bars

So far, we have dealt with beams, i.e. structures that are primarily subjected to bending. However, there may also be load-bearing structures for which there is a purely normal force effect. In this case, we speak of so-called bars. A bar effect always exists if the axial load in the form of individual forces or line loads acts centrally, i.e. its line of action is identical to the centre of gravity line of the member. The infinitesimal equilibrium has already been specified with (6.31) as $\frac{\mathrm{d}N}{\mathrm{d}x} = N' = -n$. The first derivative of the normal force of the bar therefore results in the negative applied axial line load n. The normal force N can therefore be determined by integrating the load as follows:

$$N = -\int n\mathrm{d}x + C. \tag{6.95}$$

The integration constant C follows from a suitable boundary condition. We consider the example of Fig. 6.35. We consider a straight bar of length l, which is loaded by the constant line load n and by a point force F at its free end. The integration (6.95) is as follows:

$$N = -\int n\mathrm{d}x + C = -nx + C. \tag{6.96}$$

The boundary condition to be applied here is:

$$N(x = l) = F. \tag{6.97}$$

The constant of integration C follows from this as:

$$C = F + nl. \tag{6.98}$$

The normal force $N(x)$ can thus be specified as:

$$N(x) = F + n(l - x). \tag{6.99}$$

The normal force diagram is shown in Fig. 6.35. Due to the applied line load n, the normal force is a linear function over x.

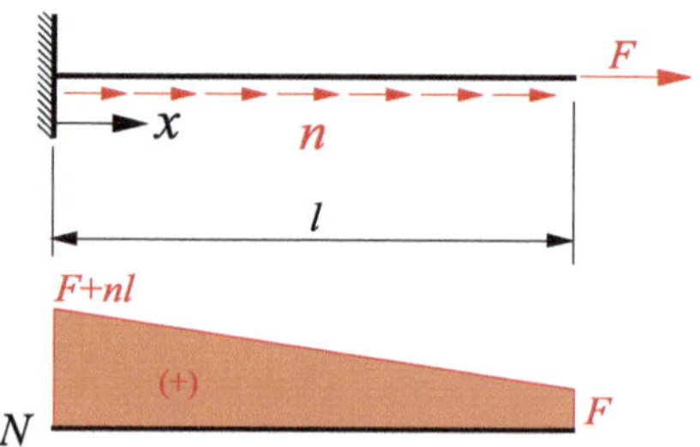

Fig. 6.35 Determination of the normal force N in a straight bar clamped on one side under line load and point force

6.6 Practical Determination of Stress Resultants

In many practically relevant cases, there is no interest in describing the internal forces N, Q and M by mathematical functions. Rather, in many cases one will determine the internal forces and moments at selected points of a structure and use general rules for the distribution between those points, which we have already examined in detail in this chapter. This approach proves particularly useful when a large number of discontinuities occur in a structure and the previously used procedures reach their limits.

To illustrate this, let us consider a beam on two supports under point force F, as shown in Fig. 6.3. We have already determined the support forces as $A_H = 0$, $A_V = \frac{F}{2}$, $B_V = \frac{F}{2}$. Since only a point force F acts on this beam, we know without any calculation that the normal force N becomes zero over the entire length of the beam. Since no line load q acts, we can immediately conclude that the shear force diagram will be constant in certain areas, whereby we have to divide the beam into two sub-fields, since the shear force will show a discontinuity at the point of force application, whereas the moment line will show a kink at this point. In addition, we know without any calculation that the moment diagram will have zero values at the support points. Consequently, it is sufficient to consider two sections and to determine the internal forces and moments in these two sections, whereby one section is taken immediately to the left at infinitesimal distance next to the point force, whereas the second section is to be taken immediately to the right at infinitesimal distance next to the point force (Fig. 6.36). We form the equilibrium conditions for the section S_1 on the left side of the single force and consider the resulting free-body diagram of the left side:

$$\overset{\rightarrow}{\sum} H = 0: \quad N = 0,$$
$$\overset{\downarrow}{\sum} V = 0: \quad Q - A_V = 0 \quad \rightarrow \quad Q = \frac{F}{2},$$
$$\overset{\curvearrowleft}{\sum} M_{S_1} = 0: \quad M - A_V \cdot \frac{l}{2} = 0 \quad \rightarrow \quad M = \frac{Fl}{4}. \tag{6.100}$$

In addition, we consider section S_2 to the right of the point force and obtain the following equilibrium conditions for the resulting right free-body diagram:

$$\overset{\leftarrow}{\sum} H = 0: \quad N = 0,$$
$$\overset{\uparrow}{\sum} V = 0: \quad Q + B_V = 0 \quad \rightarrow \quad Q = -\frac{F}{2},$$
$$\overset{\curvearrowright}{\sum} M_{S_2} = 0: \quad M - B_V \cdot \frac{l}{2} = 0 \quad \rightarrow \quad M = \frac{Fl}{4}. \tag{6.101}$$

With the known values for the shear force Q to the left and right of the point force, we can conclude that in the left half of the beam for $0 \leq x \leq \frac{l}{2}$ the shear force diagram is constant

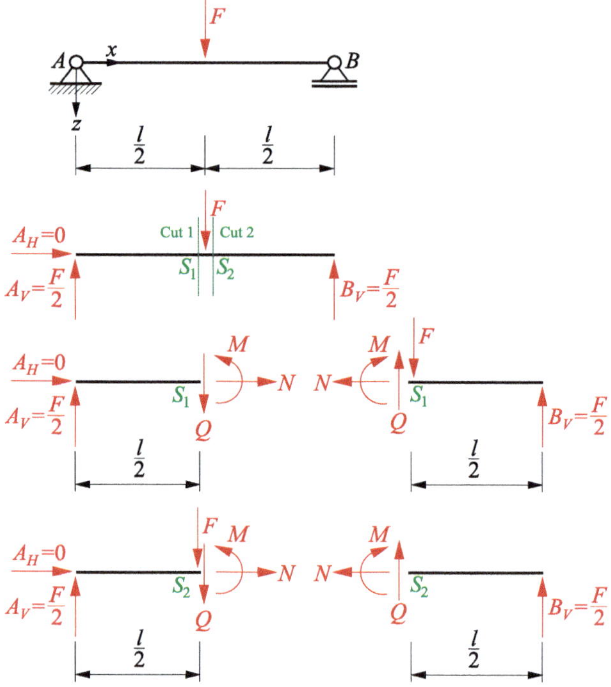

Fig. 6.36 Beam on two supports under point force

with the value $Q = \frac{F}{2}$, whereas a constant value $Q = -\frac{F}{2}$ is found in the right half of the beam for $\frac{l}{2} \leq x \leq l$. This means that the resulting discontinuity in the shear force diagram by exactly the magnitude of the point force F is also clearly defined. The bending moment, on the other hand, is linear in the two segments of the beam and assumes the value $M = \frac{Fl}{4}$ exactly at the point of force application, while zero values occur in the hinged supports. The force and moment diagrams can thus be drawn and result as shown in Fig. 6.4.

Example 6.12

Consider the beam in Fig. 6.27. We want to determine the stress resultants N, Q, M at all relevant points as well as the corresponding force and moment diagrams.

Solution:

Due to the applied load, the shear force will be a linear function, whereas the moment diagram will show a parabolic distribution. In addition, it is immediately apparent that the shear force diagram at the point of application of the point force in the centre of the beam will show a discontinuity by the magnitude of the point force, and the bending moment will show a kink at this point. The normal force is identical to zero at every point on the beam.

The beam must therefore be divided into two segments due to the applied load, and we want to determine the stress resultants at relevant points.

We first determine the support reactions using the free-body diagram in Fig. 6.37, top, and obtain:

$$A_H = 0, \quad A_V = 2F + 2ql, \quad M_A = 3Fl + 2ql^2. \tag{6.102}$$

We now perform a total of four cuts on the beam, namely at $x = 0$, at $x = l$ to the left of the point force, at $x = l$ to the right of the point force and at $x = 2l$. The resulting free-body diagrams are also shown in Fig. 6.37. We obtain the following internal forces and moments in these sections:

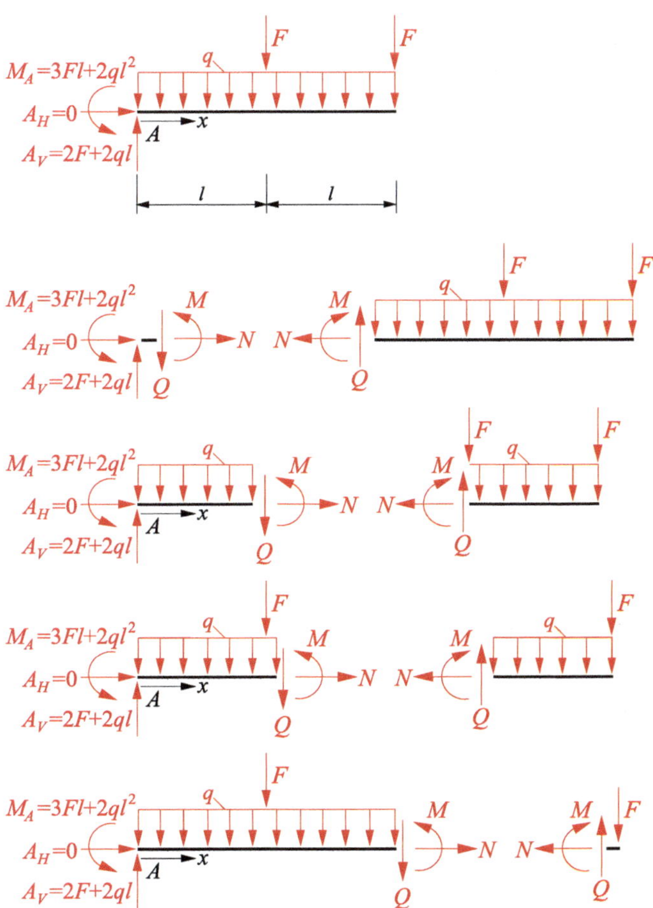

Fig. 6.37 Determination of the support reactions and stress resultants

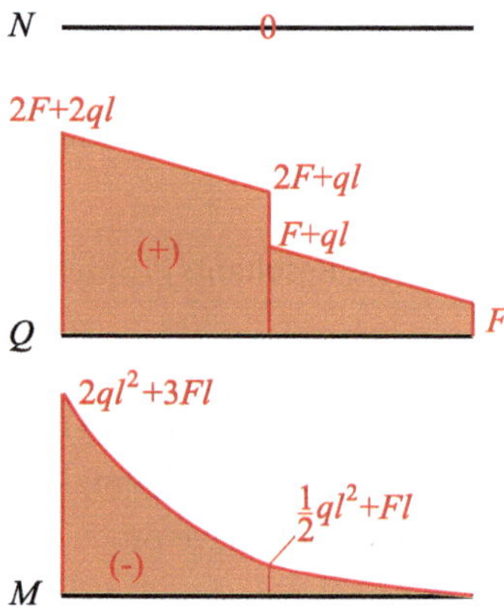

Fig. 6.38 Force and moment diagrams

$$\begin{aligned} Q(x=0) &= 2F+2ql, \\ Q(x=l) &= \begin{cases} 2F+ql & \text{to the left of } F \\ F+ql & \text{to the right of } F \end{cases}, \\ Q(x=2l) &= F, \\ M(x=0) &= -3Fl-2ql^2, \\ M(x=l) &= -Fl-\frac{1}{2}ql^2, \\ M(x=2l) &= 0. \end{aligned} \tag{6.103}$$

The force and moment diagrams are shown in Fig. 6.38.

◀

Example 6.13

For the beam in Fig. 6.39 the internal forces and moments are to be determined.

Solution:

The support and hinge reactions for this beam have already been determined in Example 4.8 as follows:

$$\begin{aligned} A_H &= 0 \quad A_V = \frac{4}{3}F_0, \quad M_A = -\frac{1}{3}F_0 l, \\ G_H &= 0, \quad G_V = -\frac{2}{3}F_0, \quad B_V = \frac{19}{6}F_0. \end{aligned} \tag{6.104}$$

We determine the stress resultants at relevant points of this beam, whereby we want to consider the points $x=0$, $x=l$, $x=2l$, $x=3l$ and $x=4l$. At the point $x=0$, the stress resultants result from the support reactions, and we have:

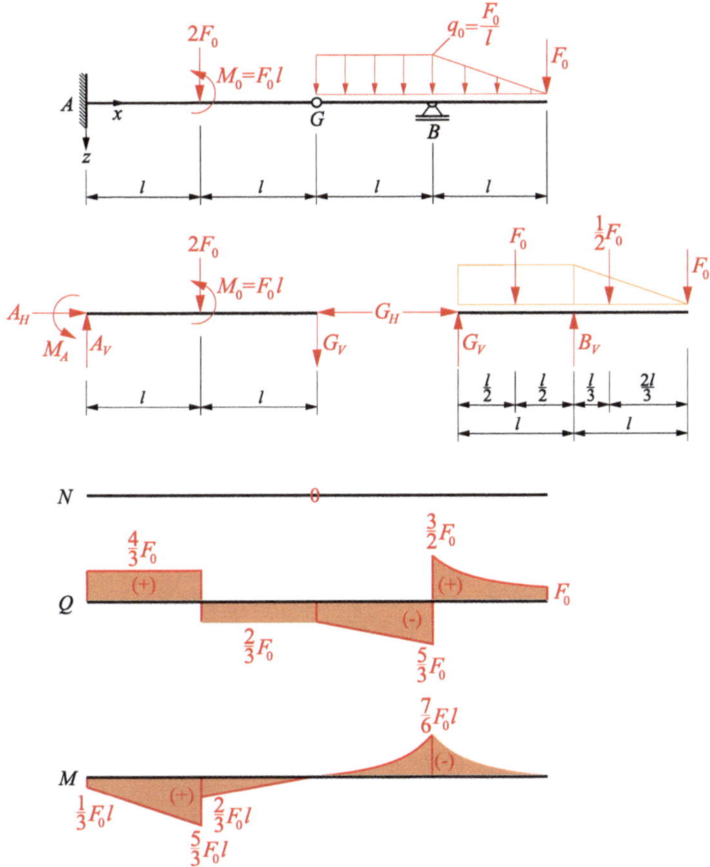

Fig. 6.39 Beam (top), determination of support and hinge reactions (centre), force and moment diagrams (bottom)

$$
\begin{aligned}
Q(x=0) &= A_V = \frac{4}{3}F_0, \\
M(x=0) &= -M_A = \frac{1}{3}F_0 l.
\end{aligned}
\tag{6.105}
$$

Since there is no applied load in the range $0 \leq x \leq l$, the shear force Q will be constant. Similarly, the bending moment M is linear in this range.

At the point $x = l$, the shear force diagram Q will show a discontinuity by exactly the applied force $2F_0$, whereas the moment diagram M will show a discontinuity by the applied single moment $M_0 = F_0 l$. We must therefore determine both Q and M directly on both sides of this point. We obtain:

$$Q(x=l) = \begin{cases} \frac{4}{3}F_0 & \text{to the left of } 2F_0 \\ -\frac{2}{3}F_0 & \text{to the right of } 2F_0 \end{cases},$$
$$M(x=l) = \begin{cases} \frac{5}{3}F_0 l & \text{to the left of } M_0 \\ \frac{2}{3}F_0 l & \text{to the right of } M_0 \end{cases}. \tag{6.106}$$

As no load is applied in the range $l \leq x \leq 2l$, the shear force diagram Q is constant in this range with the value $Q = -\frac{2}{3}F_0$. The moment diagram is linear and has a zero value at the joint $x = 2l$. The following applies:

$$Q(x=2l) = -\frac{2}{3}F_0,$$
$$M(x=2l) = 0. \tag{6.107}$$

We now consider the range $2l \leq x \leq 3l$, in which the constant line load $q_0 = \frac{F_0}{l}$ is applied. Due to this load, the shear force diagram Q is linear, while the moment diagram M is parabolic. The support force B_V also acts at the support point B, so that the shear force diagram at this point has a discontinuity by exactly the support force B_V. We obtain

$$Q(x=3l) = \begin{cases} -\frac{5}{3}F_0 & \text{to the left of } B \\ \frac{3}{2}F_0 & \text{to the right of } B \end{cases},$$
$$M(x=3l) = -\frac{7}{6}F_0 l. \tag{6.108}$$

Finally, a linear line load is applied in the range $3l \leq x \leq 4l$, so that the shear force diagram Q is parabolic in this range. The moment diagram, on the other hand, is cubic. At the free end of the beam, the bending moment M is identical to zero. The shear force Q is identical to the applied force F_0 at this point:

$$Q(x=4l) = F_0,$$
$$M(x=4l) = 0. \tag{6.109}$$

The force and moment diagrams for this beam are shown in Fig. 6.39, bottom. ◀

6.7 Angled Beams and Frames

So far we have dealt with the determination of stress resultants in straight beams and beam systems in this chapter. In many technical applications, however, we are dealing with angled beams or so-called frames, which we will analyse in this section. As in beam systems, stress resultants in angled beams and frames are determined section by section, whereby a separate coordinate system is usually introduced for each area in which the internal forces and moments are continuous. Figure 6.40 shows an example for an angled beam. We consider a cantilever beam of the length l, to which a vertical segment of the length h is connected at its right end by means of a rigid connection. A horizontally acting point force F acts at the free end of the vertical segment.

Local reference axes x_1, z and x_2, z are introduced as shown in Fig. 6.40, left, and the stress resultants in the two beam segments are determined based on this. We do not use the integration method here and determine the stress resultants by cutting the beam at relevant points. This angled beam is an example in which we do not have to determine the support reactions in advance and first consider the section $0 \leq x_2 \leq h$. The corresponding free-body diagram created by a section at an arbitrary point x_2 is shown in Fig. 6.40, top right. The equilibrium conditions are as follows:

$$\overset{\leftarrow}{\sum} H = 0: \quad Q_2 - F = 0 \quad \rightarrow \quad Q_2 = F,$$

$$\overset{\downarrow}{\sum} V = 0: \quad N_2 = 0,$$

$$\overset{\curvearrowright}{\sum} M_S = 0: \quad M_2 + F(h - x_2) = 0 \quad \rightarrow \quad M_2 = F(x_2 - h). \tag{6.110}$$

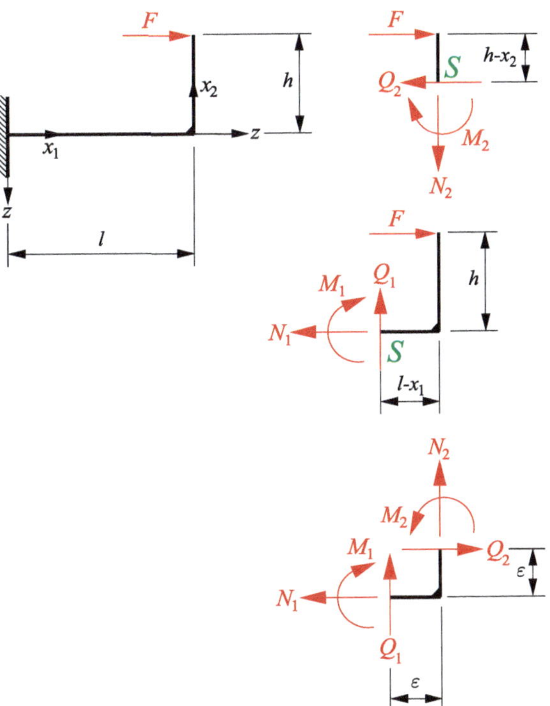

Fig. 6.40 Angled beam

Obviously, the normal force N_2 in this beam segment is identical to zero, and the shear force Q_2 is constant with the value $Q_2 = F$. The bending moment M_2 runs linearly over x_2, disappears at the free end $x_2 = h$ and assumes the value $M_2(x_2 = 0) = -Fh$ at the corner point.

We also cut free at an arbitrary point x_1 in the horizontal segment of the beam (Fig. 6.40, centre right) and obtain

$$\overset{\leftarrow}{\sum} H = 0: \quad N_1 - F = 0 \quad \rightarrow \quad N_1 = F,$$

$$\overset{\uparrow}{\sum} V = 0: \quad Q_1 = 0,$$

$$\overset{\curvearrowright}{\sum} M_S = 0: \quad M_1 + Fh = 0 \quad \rightarrow \quad M_1 = -Fh. \tag{6.111}$$

Accordingly, the normal force N_1 is constant with the value $N_1 = F$, and the shear force Q_1 disappears. The bending moment M_1 is constant with the value $M_1 = -Fh$. The force and moment diagrams are shown in Fig. 6.41.

Particular attention must be paid to the conditions at a corner point where the internal forces and moments are transferred from one area to the next. From the equilibrium conditions on the free-body diagram with the segment lengths $\varepsilon \rightarrow 0$ of Fig. 6.40, bottom right, we obtain:

$$N_2 = -Q_1, \quad Q_2 = N_1, \quad M_2 = M_1. \tag{6.112}$$

As a general conclusion, it can be stated that the bending moment at a right-angled rigid corner is transferred unchanged from one area to the other (in the moment diagram M of Fig. 6.41 indicated by a dashed arc), whereas the normal force of one area is transferred to the shear force of the other area.

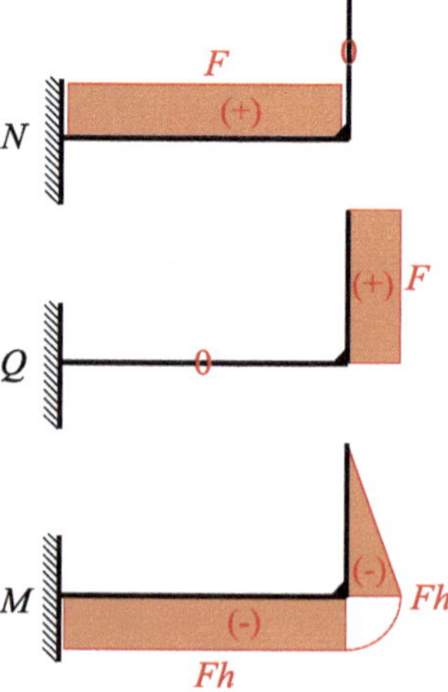

Fig. 6.41 Force and moment diagrams

In the following, we will limit ourselves exclusively to determining the internal forces and moments at relevant points and inferring the distributions between these selected points by means of the general rules that have already been established in this chapter.

Example 6.14

Consider the frame of Fig. 6.42, which consists of two uprights of height h and a horizontal beam of length l and is loaded by a uniform line load q. The horizontal beam is interrupted exactly in its centre at point G by a moment hinge. We want to determine the diagrams for N, Q, M.

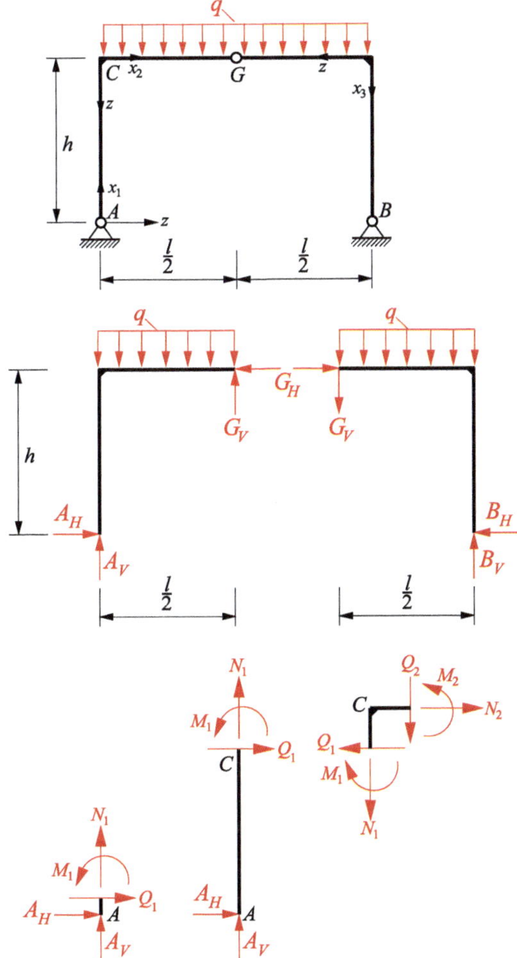

Fig. 6.42 Frame (top), determination of the support and hinge reactions (centre), free-body diagrams to determine the stress resultants (bottom)

Solution:

To determine the internal forces and moments, the local reference axes x_1, z, x_2, z and x_3, z are introduced as shown in Fig. 6.42. The support and hinge reactions for this frame were already determined in Example 4.9 and are given again here for the sake of clarity. They are as follows:

$$A_V = B_V = \frac{ql}{2}, \quad A_H = B_H = G_H = \frac{ql^2}{8h}, \quad G_V = 0. \tag{6.113}$$

We first consider segment 1 with $0 \le x_1 \le h$. At the support point A, the moment diagram $M_1(x_1 = 0)$ has a zero value (hinged support), whereas the support force $A_V = \frac{ql}{2}$ causes a compressive normal force $N_1(x_1 = 0) = -\frac{ql}{2}$. The horizontal support force A_H results in a negative shear force $Q_1(x_1 = 0)$ with the value $Q_1(x_1 = 0) = -\frac{ql^2}{8h}$:

$$N_1(x_1 = 0) = -\frac{ql}{2}, \quad Q_1(x_1 = 0) = -\frac{ql^2}{8h}, \quad M_1(x_1 = 0) = 0. \tag{6.114}$$

Since there is no applied line load in area 1, both the normal force $N(x_1)$ and the shear force $Q(x_1)$ are constant with the values determined above. The bending moment $M(x_1)$, on the other hand, is linear and assumes the value $M_1(x_1 = h) = -\frac{ql^2}{8}$ at the upper end of segment 1 at $x_1 = h$, as can be seen from the moment sum with regard to the point C at $x_1 = h$ directly below the rigid corner:

$$\begin{aligned} N_1(x_1 = h) &= -\frac{ql}{2}, \\ Q_1(x_1 = h) &= -\frac{ql^2}{8h}, \\ M_1(x_1 = h) &= -\frac{ql^2}{8}. \end{aligned} \tag{6.115}$$

In segment 2, i.e. the horizontal beam with $0 \le x_2 \le l$, the uniform line load q is applied. Accordingly, the shear force $Q_2(x_2)$ will be a linear function and the bending moment will show a parabolic distribution. The normal force, on the other hand, shows a constant value $N_2(x_2)$. From the free-body diagram of the rigid corner at point C according to Fig. 6.42, bottom right, the following values of the internal forces and moments N_2, Q_2, M_2 for $x_2 = 0$ result:

$$\begin{aligned} N_2(x_2 = 0) &= Q_1(x_1 = h) = -\frac{ql^2}{8h}, \\ Q_2(x_2 = 0) &= -N_1(x_1 = h) = \frac{ql}{2}, \\ M_2(x_2 = 0) &= M_1(x_1 = h) = -\frac{ql^2}{8}. \end{aligned} \tag{6.116}$$

At the hinge point G, both the transverse force $Q_2\left(x_2 = \frac{l}{2}\right)$ (the vertical hinge force G_V is identical to zero) and the bending moment $M_2\left(x_2 = \frac{l}{2}\right)$ (no bending moment can occur in a moment hinge) disappear. The normal force is constant over the entire length of the horizontal beam with the value $N_2\left(x_2 = \frac{l}{2}\right) = -\frac{ql^2}{8h}$:

$$
\begin{aligned}
N_2\left(x_2 = \frac{l}{2}\right) &= -\frac{ql^2}{8h},\\
Q_2\left(x_2 = \frac{l}{2}\right) &= 0,\\
M_2\left(x_2 = \frac{l}{2}\right) &= 0.
\end{aligned}
\tag{6.117}
$$

All further internal forces and moments can be deduced analogously, whereby it can be seen that the normal force line N and the moment line M are symmetrical with respect to the hinge point, whereas the shear force line is antimetric. The force and moment diagrams are shown in Fig. 6.43.

◄

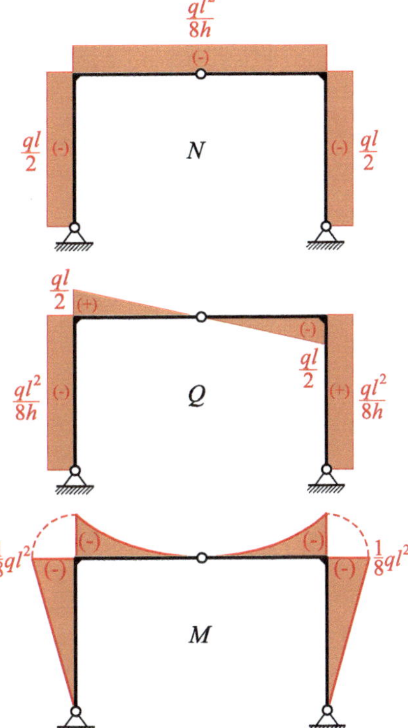

Fig. 6.43 Force and moment diagrams

Example 6.15

For the clamped angled beam of Fig. 6.44 under the point force F_0 the force and moment diagrams N, Q, M are to be determined.

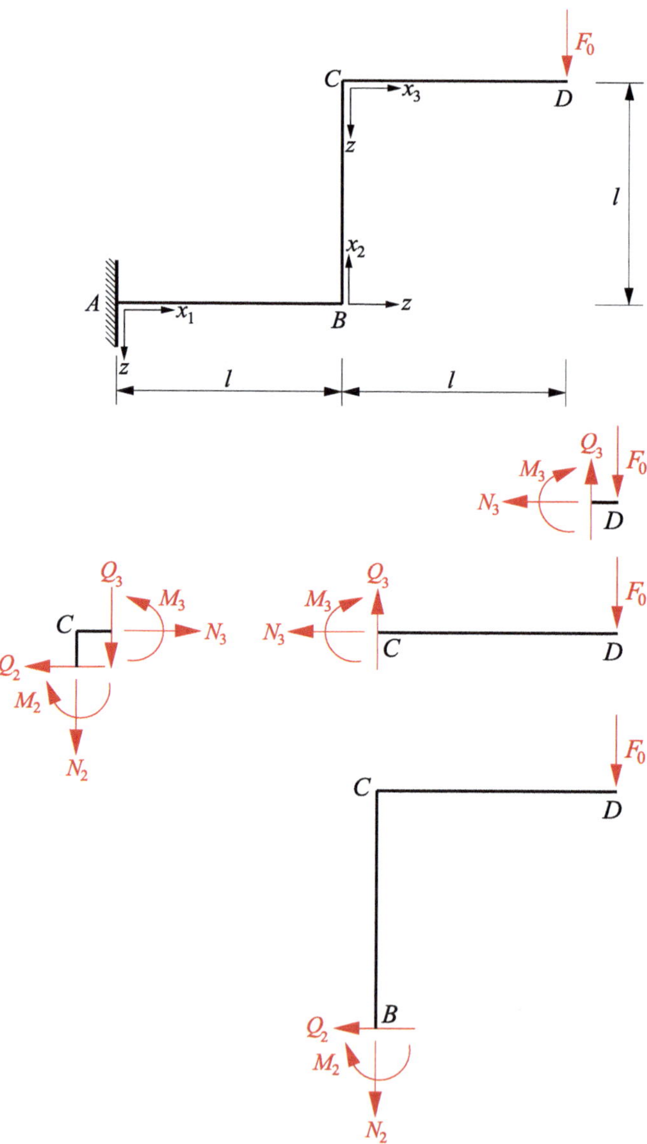

Fig. 6.44 Angled beam (top), free-body diagrams to determine the stress resultants (bottom)

Solution:

The internal forces and moments can be determined for this beam without determining the support reactions in point A in advance. At the point $x_3 = l$ at point D, both the normal force $N_3(x = l)$ and the bending moment $M_3(x_3 = l)$ disappear. The shear force $Q_3(x_3 = l)$ corresponds to the applied point force F_0:

$$
\begin{aligned}
N_3\,(x_3 = l) &= \;0,\\
Q_3\,(x_3 = l) &= \;F_0,\\
M_3\,(x_3 = l) &= \;0.
\end{aligned}
\tag{6.118}
$$

For point C at the location $x_3 = 0$, the corresponding free-body diagram in Fig. 6.44, bottom, leads the following values:

$$
\begin{aligned}
N_3\,(x_3 = 0) &= \;0,\\
Q_3\,(x_3 = 0) &= \;F_0,\\
M_3\,(x_3 = 0) &= \;-F_0 l.
\end{aligned}
\tag{6.119}
$$

The shear force Q_3 is constant, whereas M_3 is distributed linearly over x_3.

The stress resultants in segment 2 for $x_2 = l$ can be determined on the free body diagram of the rigid corner C. The following applies:

$$
\begin{aligned}
N_2(x_2 = l) &= \;-Q_3\,(x_3 = 0) = -F_0,\\
Q_2(x_2 = l) &= \;N_3\,(x_3 = 0) = 0,\\
M_2\,(x_2 = l) &= \;M_3\,(x_3 = 0) = -F_0 l.
\end{aligned}
\tag{6.120}
$$

The stress resultants at the corner B at $x_2 = 0$ follow in a similar way.

Accordingly, all stress resultants in segment 2 are constantly distributed over x_2.

The internal forces and moments in segment 1 can also be determined point by point in the same way, which is not shown here. The force and moment diagrams for the angled beam are shown in Fig. 6.45.

◄

Example 6.16

For the angled beam of Fig. 6.46 (see Example 4.6), the moment diagram is to be determined.

Solution:

We want to show that the moment diagram M for this example can be determined in a very simple way without prior calculation of the support forces. To do this, we start the considerations at the free end of the beam at $x_2 = l$ at point D. There is no external moment

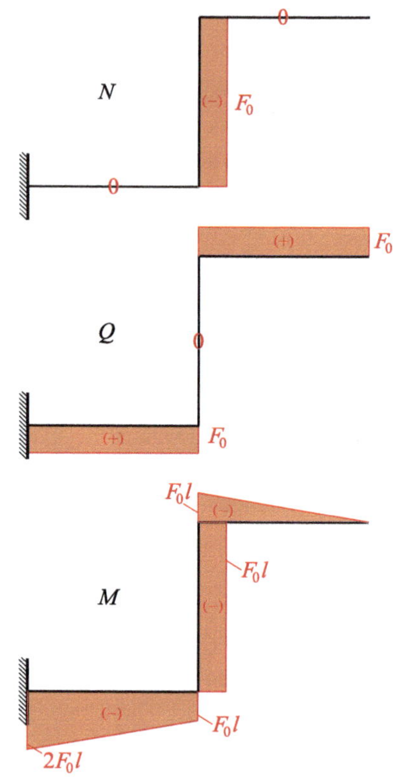

Fig. 6.45 Force and moment diagrams

applied at this end, so the bending moment must be zero:

$$M_2\,(x_2 = l) = 0. \tag{6.121}$$

As there is no applied external load in the form of a line load in the range $0 \leq x_2 \leq l$, the moment diagram will be linear in this range. To determine the bending moment diagram M, it is therefore sufficient to determine the moment at the point C at $x_2 = 0$ and to connect the two values determined in this way in this area with a straight line. We obtain the value for $M(x_2 = 0)$ from the free-body diagram in Fig. 6.46, centre left. It follows:

$$M_2\,(x_2 = 0) = -q_0 l^2. \tag{6.122}$$

At the rigid corner C, the bending moment is transferred unchanged between the two segments 1 and 2, so that we can directly conclude that $M_1(x_1 = 2l) = M_2\,(x_2 = 0)$ applies:

$$M_1\,(x_2 = 2l) = -q_0 l^2. \tag{6.123}$$

Based on the free-body diagram in Fig. 6.46, centre right, we can conclude that the force $F_0 = q_0 l$ has a constant lever arm l for all values $l \leq x_2 \leq 2l$, so that the bending moment

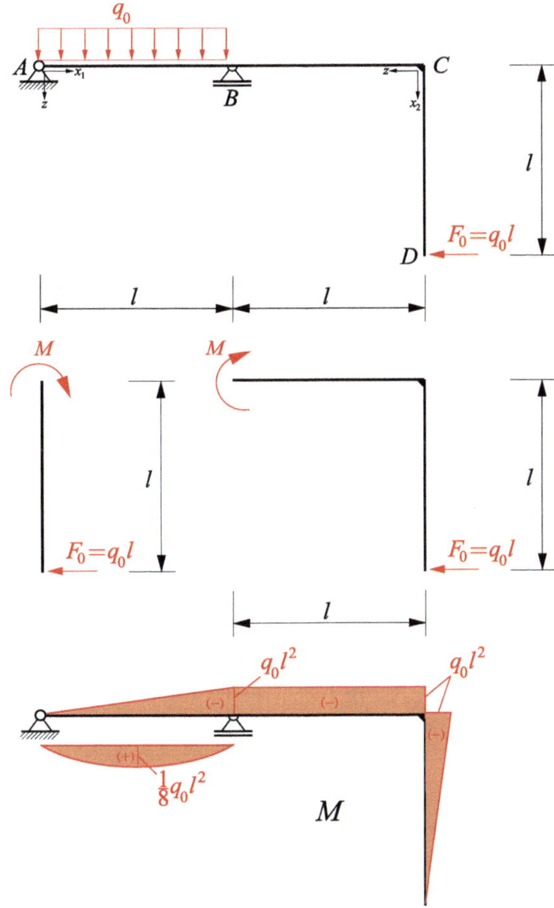

Fig. 6.46 Angled beam (top), free-body diagrams (centre), moment diagram (bottom)

in this area is constant with the value $M = -q_0 l^2$. It therefore applies in particular at the support point B:

$$M_1 (x_2 = l) = -q_0 l^2. \tag{6.124}$$

Finally, the bending moment M must disappear in the support A so that the following applies:

$$M_1 (x_2 = 0) = 0. \tag{6.125}$$

Between the supports A and B, the moment line M will show a parabolic curve due to the applied uniform line load q_0.

With the values for the bending moment determined in this way and the conclusions drawn, the moment diagram can be drawn as shown in Fig. 6.46, bottom. In the area $0 \leq$

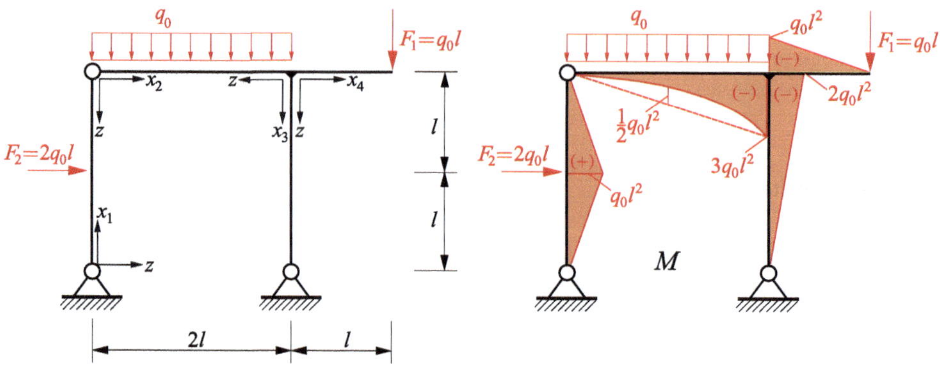

Fig. 6.47 Frame (left), moment diagram (right)

$x_1 \leq l$, we have made use of the fact that the moment diagram can be divided into a triangle with the value $-q_0 l^2$ and a parabola with the value $\frac{q_0 l^2}{8}$.

◀

Example 6.17

For the frame of Fig. 6.47 (see Example 4.10), the moment diagram is to be determined.

Solution:

The moment diagram can be constructed in a similar way to the previous examples by determining individual values point by point with the aid of the support and joint reactions already determined in Example 4.10, whereby it must be noted here that a linear function for M results in the areas without line load, whereas a parabolic diagram results in the area $0 \leq x_2 \leq 2l$. The calculation steps are not reproduced here. The moment line M is shown in Fig. 6.47, right.

◀

6.8 Curved Beams

Curved beams are beams that are eiter curved in one plane or curved in space. At this point, we want to limit our considerations to those curved beams whose curvature is limited to one plane. The considered curved beam (see Fig. 6.48) is loaded by the line load q acting in the normal direction (characterised by the normal vector $\underline{n}$). In addition, a tangential line load n may also be permitted. Furthermore, point forces F_i and moments M_j can also act. As

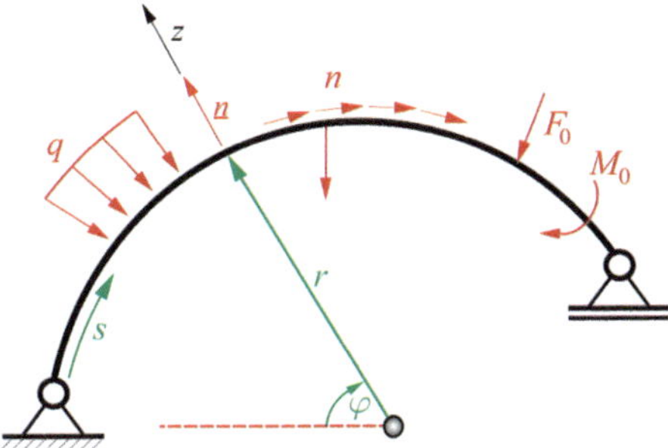

Fig. 6.48 Curved beam

a reference axis, we introduce the circumferential coordinate s as indicated. Let the radius of curvature of the curved beam be r, whereby r can vary locally, i.e. $r = r(s)$. For later purposes, it will be helpful to introduce another reference coordinate in the normal direction, namely the z-axis, which has its origin in the centre line of the beam.

Due to the curvature of the beam, a combination of beam and bar action results. Consequently, in addition to the shear force Q and the bending moment M, the normal force N will also occur in general. We now consider the local equilibrium at an infinitesimal element of a curved beam, where the corresponding opening angle is $\mathrm{d}\varphi$. The length of this element is $\mathrm{d}s$. The infinitesimal element is shown in addition to the internal forces and moments in Fig. 6.49, where a division into three subfigures was made here for reasons of clarity. While the stress resultants N, Q and M occur on the negative face, the stress resultants N, Q and M can also be found on the positive face, along with their incremental changes $\mathrm{d}N$, $\mathrm{d}Q$ and $\mathrm{d}M$. For the following equilibrium considerations, we want to use the angle φ as a reference instead of the circumferential axis s. The relationship between s and φ is $\mathrm{d}s = r\mathrm{d}\varphi$. The force sum in the normal direction then results in

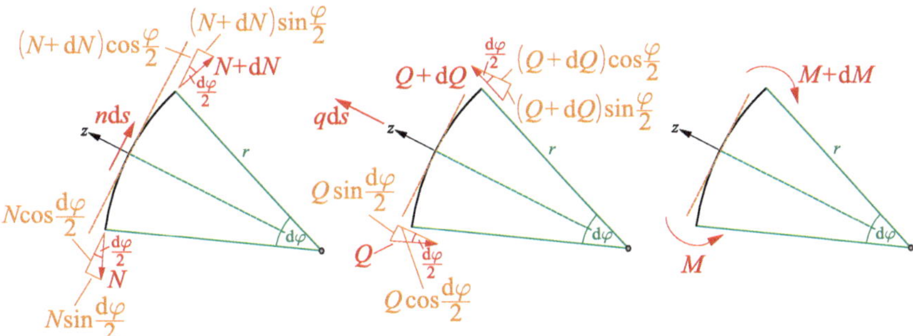

Fig. 6.49 Free-body diagrams of an infinitesimal beam element

$$-Q\cos\frac{\mathrm{d}\varphi}{2}+(Q+\mathrm{d}Q)\cos\frac{\mathrm{d}\varphi}{2}-N\sin\frac{\mathrm{d}\varphi}{2}-(N+\mathrm{d}N)\sin\frac{\mathrm{d}\varphi}{2}+q\mathrm{d}s=0, \quad (6.126)$$

or when neglecting terms of higher order and the assumption that $\cos\frac{\mathrm{d}\varphi}{2}\simeq 1$ and $\sin\frac{\mathrm{d}\varphi}{2}\simeq\frac{\mathrm{d}\varphi}{2}$:

$$\mathrm{d}Q-N\mathrm{d}\varphi+q\mathrm{d}s=0. \quad (6.127)$$

Similarly, we obtain the force sum in the tangential direction:

$$(N+\mathrm{d}N)\cos\frac{\mathrm{d}\varphi}{2}-N\cos\frac{\mathrm{d}\varphi}{2}+(Q+\mathrm{d}Q)\sin\frac{\mathrm{d}\varphi}{2}+Q\sin\frac{\mathrm{d}\varphi}{2}+n\mathrm{d}s=0, \quad (6.128)$$

or

$$\mathrm{d}N+Q\mathrm{d}\varphi+n\mathrm{d}s=0. \quad (6.129)$$

With $\mathrm{d}\varphi=\frac{\mathrm{d}s}{r}$, the two equations (6.127) and (6.129) can also be written as:

$$\begin{aligned}\frac{\mathrm{d}Q}{\mathrm{d}s}-\frac{N}{r}+q&=0,\\ \frac{\mathrm{d}N}{\mathrm{d}s}+\frac{Q}{r}+n&=0.\end{aligned} \quad (6.130)$$

The sum of moments around the origin of the $z-$axis shown in Fig. 6.49 yields:

$$Q\frac{\mathrm{d}s}{2}+(Q+\mathrm{d}Q)\frac{\mathrm{d}s}{2}+M-(M+\mathrm{d}M)=0, \quad (6.131)$$

which can be brought into the following form:

$$\frac{\mathrm{d}M}{\mathrm{d}s}=Q. \quad (6.132)$$

Equations (6.130) and (6.132) represent three coupled differential equations from which the internal forces and moments N, Q and M can be calculated for a given load q and n.

The Eqs. (6.130) and (6.132) contain the straight beam as a special case. This is achieved by setting $r\to\infty$ and $\mathrm{d}s=\mathrm{d}x$. Then:

$$\frac{\mathrm{d}Q}{\mathrm{d}x}+q=0,\quad \frac{\mathrm{d}N}{\mathrm{d}x}+n=0,\quad \frac{\mathrm{d}M}{\mathrm{d}x}=Q. \quad (6.133)$$

An interesting special case is the beam with a constant radius R, where the two line loads q and n do not occur. The Eqs. (6.130) and (6.132) then read:

$$\begin{aligned}\frac{\mathrm{d}Q}{\mathrm{d}s}-\frac{N}{R}&=0,\\ \frac{\mathrm{d}N}{\mathrm{d}s}+\frac{Q}{R}&=0,\\ \frac{\mathrm{d}M}{\mathrm{d}s}&=Q.\end{aligned} \tag{6.134}$$

If we set $\mathrm{d}s = R\mathrm{d}\varphi$ here, we obtain:

$$\begin{aligned}\frac{\mathrm{d}Q}{\mathrm{d}\varphi}-N&=0,\\ \frac{\mathrm{d}N}{\mathrm{d}\varphi}+Q&=0,\\ \frac{\mathrm{d}M}{\mathrm{d}\varphi}&=QR.\end{aligned} \tag{6.135}$$

Deriving the first equation in (6.135) with respect to φ and then eliminating N from the second equation results in:

$$\frac{\mathrm{d}^2Q}{\mathrm{d}\varphi^2}+Q=0. \tag{6.136}$$

This is a linear ordinary homogeneous second-order differential equation with constant coefficients for the transverse force Q. Its solution reads:

$$Q = C_1\cos\varphi + C_2\sin\varphi. \tag{6.137}$$

From the first equation in (6.135), the normal force N can then be determined as:

$$N=\frac{\mathrm{d}Q}{\mathrm{d}\varphi}=-C_1\sin\varphi+C_2\cos\varphi. \tag{6.138}$$

From the third equation in (6.135), the bending moment follows as:

$$M=\int_\varphi QR\mathrm{d}\varphi = C_1R\sin\varphi - C_2R\cos\varphi + C_3. \tag{6.139}$$

The constants C_1, C_2 and C_3 are adapted to given boundary conditions.

We illustrate the procedure using the two beams shown in Fig. 6.50 (constant radius R) with the opening angle 90°. In the first case a moment M_0 is applied as shown. The corresponding boundary conditions are thus as follows:

$$N\left(\varphi=\frac{\pi}{2}\right)=0,\quad Q\left(\varphi=\frac{\pi}{2}\right)=0,\quad M\left(\varphi=\frac{\pi}{2}\right)=-M_0. \tag{6.140}$$

Evaluating the solutions

$$
\begin{aligned}
N &= -C_1 \sin\varphi + C_2 \cos\varphi, \\
Q &= C_1 \cos\varphi + C_2 \sin\varphi, \\
M &= C_1 R \sin\varphi - C_2 R \cos\varphi + C_3
\end{aligned}
\tag{6.141}
$$

yields:

$$
C_1 = 0, \quad C_2 = 0, \quad C_3 = -M_0, \tag{6.142}
$$

so that:

$$
N = 0, \quad Q = 0, \quad M = -M_0. \tag{6.143}
$$

The curved beam under single moment at its free end is therefore free of normal and transverse forces, and the acting bending moment is constant over the entire beam.

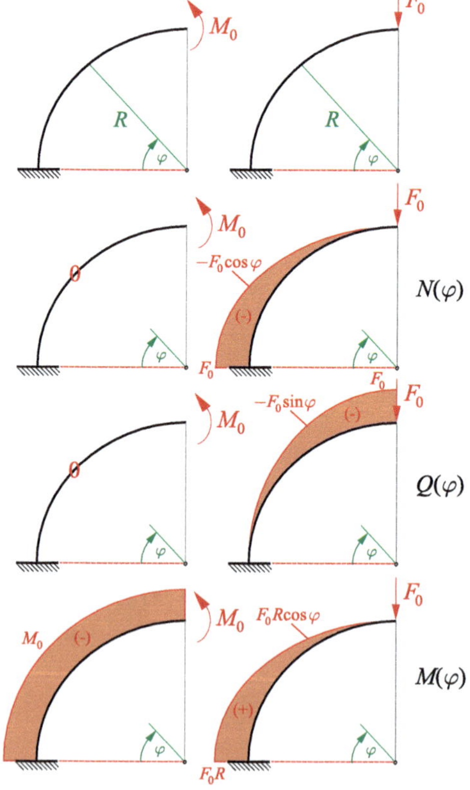

Fig. 6.50 Curved beam with constant radius clamped on one side under moment M_0 (top left) and force F_0 (top right); resulting internal forces and moments (bottom)

The second case consists of an identical curved beam with constant radius, whereby the load now consists of a single vertical force F_0 at the free end of the beam. Evaluating the boundary conditions

$$N\left(\varphi=\frac{\pi}{2}\right)=0, \quad Q\left(\varphi=\frac{\pi}{2}\right)=F_0, \quad M\left(\varphi=\frac{\pi}{2}\right)=0 \tag{6.144}$$

gives the constants C_1, C_2 and C_3 as:

$$C_1=0, \quad C_2=F_0, \quad C_3=0, \tag{6.145}$$

so that:

$$\begin{aligned} N &= -F_0\cos\varphi, \\ Q &= -F_0\sin\varphi, \\ M &= F_0 R\cos\varphi. \end{aligned} \tag{6.146}$$

It can be shown that the internal forces and moments of the curved beam can also be obtained by using elementary equilibrium considerations, as already shown for the straight beam. Let us consider the example in Fig. 6.50, right, again and first determine the support reactions at the clamped end, as shown in Fig. 6.51, left. We obtain:

$$A_H=0, \quad A_V=F_0, \quad M_A=F_0 R. \tag{6.147}$$

We now cut the beam free at an arbitrary point φ and determine the internal forces and moments which results in the following equilibrium conditions. The sum of all forces in the direction of the normal force N results in

$$N+A_V\cos\varphi=0 \quad \rightarrow \quad N=-F_0\cos\varphi. \tag{6.148}$$

For the force equilibrium in the direction of the shear force Q follows:

$$Q+A_V\sin\varphi=0 \quad \rightarrow \quad Q=-F_0\sin\varphi. \tag{6.149}$$

Finally, the moment equilibrium with respect to the point S results in:

$$M-M_a+A_V R\,(1-\cos\varphi)=0 \quad \rightarrow \quad M=F_0 R\cos\varphi. \tag{6.150}$$

It can be seen that the stress resultants determined in this way are consistent with the results of Fig. 6.50, right.

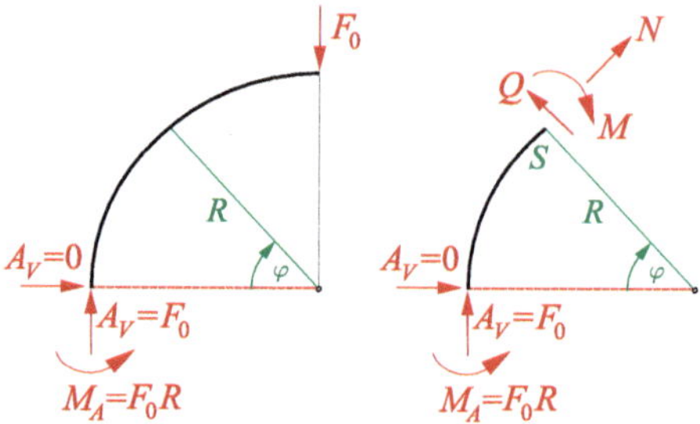

Fig. 6.51 Circular beam clamped on one side under force F_0

Example 6.18

For the semicircular curved beam on two supports with radius R under the force F_0 (Fig. 6.52), the internal forces and moments are to be determined.

Solution:

We cut the beam free at an arbitrary point φ and obtain the free-form diagram as shown in Fig. 6.52, right. The vertical support reaction A_V is identical to zero under the given load case. The equilibrium conditions can be written as follows. The sum of all forces in the direction of the normal force N results in

$$N - F_0 \sin\varphi = 0 \quad \rightarrow \quad N = F_0 \sin\varphi. \tag{6.151}$$

For the force equilibrium in the direction of the shear force Q follows:

$$Q + F_0 \cos\varphi = 0 \quad \rightarrow \quad Q = -F_0 \cos\varphi. \tag{6.152}$$

The moment equilibrium with respect to the point S results in:

$$M + F_0 R \sin\varphi = 0 \quad \rightarrow \quad M = -F_0 R \sin\varphi. \tag{6.153}$$

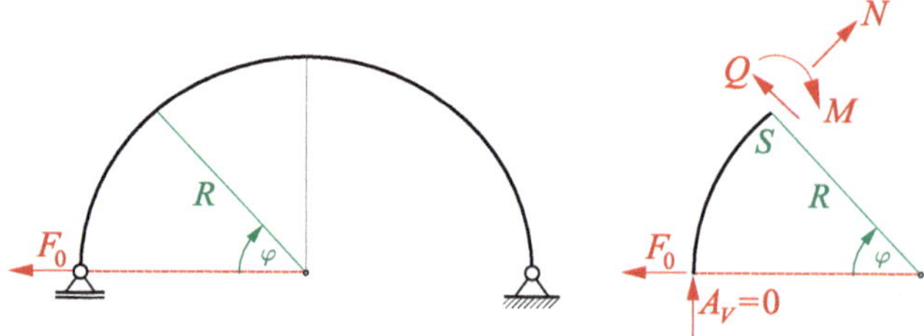

Fig. 6.52 Curved beam on two supports under force F_0

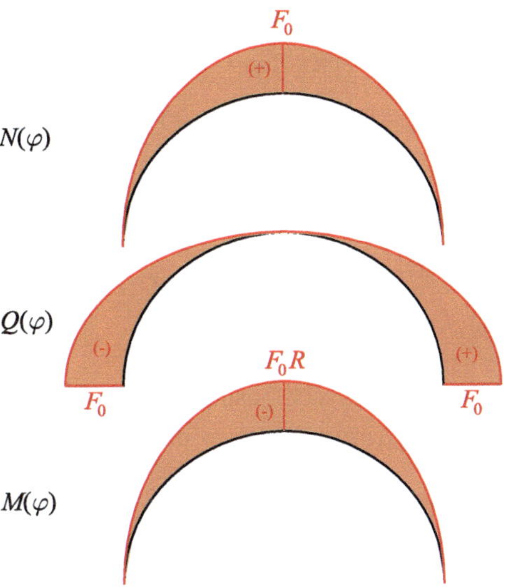

Fig. 6.53 Force and moment diagrams

The resultant force and moment diagrams are shown in Fig. 6.53. ◀

Let us now consider the case where the given curved beam is under a line load q. The load n parallel to the axis is excluded here. The equilibrium conditions (6.130) and (6.132) then read as follows:

$$\begin{aligned}\frac{\mathrm{d}Q}{\mathrm{d}s} &= \frac{N}{r} - q,\\ \frac{\mathrm{d}N}{\mathrm{d}s} &= -\frac{Q}{r},\\ \frac{\mathrm{d}M}{\mathrm{d}s} &= Q.\end{aligned} \tag{6.154}$$

If we derive the third equation once, we obtain

$$\frac{\mathrm{d}^2 M}{\mathrm{d}s^2} = \frac{\mathrm{d}Q}{\mathrm{d}s}, \tag{6.155}$$

which can be represented using the first equation in (6.154) as:

$$\frac{\mathrm{d}^2 M}{\mathrm{d}s^2} = \frac{N}{r} - q. \tag{6.156}$$

From the first equation in (6.154) we also obtain the following:

$$N = r\frac{\mathrm{d}Q}{\mathrm{d}s} + qr. \tag{6.157}$$

Differentiating by s then gives (note that the product rule of differential calculus is to be applied here since $r = r(s)$):

$$\frac{\mathrm{d}N}{\mathrm{d}s} = \frac{\mathrm{d}}{\mathrm{d}s}\left(r\frac{\mathrm{d}Q}{\mathrm{d}s} + qr\right) = \frac{\mathrm{d}r}{\mathrm{d}s}\left(\frac{\mathrm{d}Q}{\mathrm{d}s} + q\right) + r\left(\frac{\mathrm{d}^2Q}{\mathrm{d}s^2} + \frac{\mathrm{d}q}{\mathrm{d}s}\right). \tag{6.158}$$

Inserting into the second equation in (6.154) yields:

$$\frac{\mathrm{d}r}{\mathrm{d}s}\left(\frac{\mathrm{d}Q}{\mathrm{d}s} + q\right) + r\left(\frac{\mathrm{d}^2Q}{\mathrm{d}s^2} + \frac{\mathrm{d}q}{\mathrm{d}s}\right) = -\frac{Q}{r}. \tag{6.159}$$

If we use the relationships $\frac{\mathrm{d}^2M}{\mathrm{d}s^2} = \frac{\mathrm{d}Q}{\mathrm{d}s}$ and $\frac{\mathrm{d}M}{\mathrm{d}s} = Q$ and subsequently take into account circularly curved beams with R =const. so that $\frac{\mathrm{d}R}{\mathrm{d}s} = 0$, then we obtain:

$$R\frac{\mathrm{d}^3M}{\mathrm{d}s^3} + R\frac{\mathrm{d}q}{\mathrm{d}s} = -\frac{1}{R}\frac{\mathrm{d}M}{\mathrm{d}s}. \tag{6.160}$$

Integrating over s results in:

$$\frac{\mathrm{d}^2M}{\mathrm{d}s^2} + \frac{M}{R^2} = -q. \tag{6.161}$$

This is a linear inhomogeneous second-order differential equation for determining the bending moment $M(s)$. It can be converted into the following form with $\mathrm{d}s = R\mathrm{d}\varphi$:

$$\frac{\mathrm{d}^2M}{\mathrm{d}\varphi^2} + M = -R^2q. \tag{6.162}$$

Once the bending moment has been determined, the normal force and shear force of the curved beam can be determined as:

$$Q = \frac{\mathrm{d}M}{\mathrm{d}s}, \quad N = -\frac{M}{R}. \tag{6.163}$$

It can be shown that curved beam with a certain shape under a specific load transfers its loads exclusively via normal forces and that no transverse forces or bending moments occur. It is therefore a relevant task for engineers to choose the shape of the curved beam for a given load in such a way that this type of load-bearing effect is achieved exclusively via normal forces and thus without bending moments. We want to pursue this question in this subsection and assume a Cartesian coordinate system x, y when determining this specific shape of the curved beam. As a special case, we only consider a vertically acting line load $q(x)$. The shape of the beam is described by the function $y(x)$. An important prerequisite for moment-free load transfer is a corresponding support in which the support forces are introduced exactly tangentially into the beam (Fig. 6.54, left). If, for example, transverse forces are introduced (Fig. 6.54, right), moment-free load transfer cannot be guaranteed.

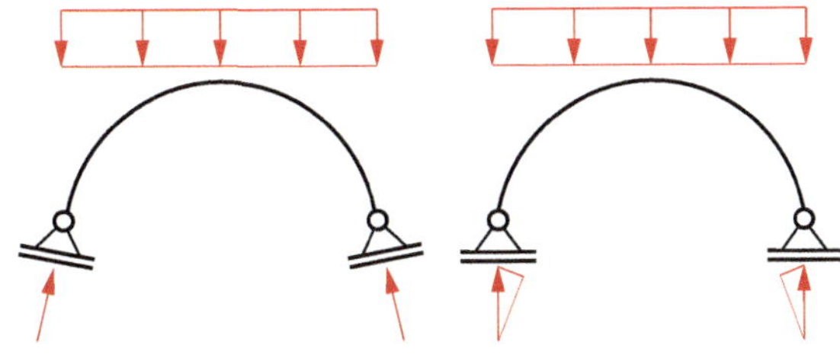

Fig. 6.54 Suitable supports for moment-free load transfer (left), unsuitable supports (right)

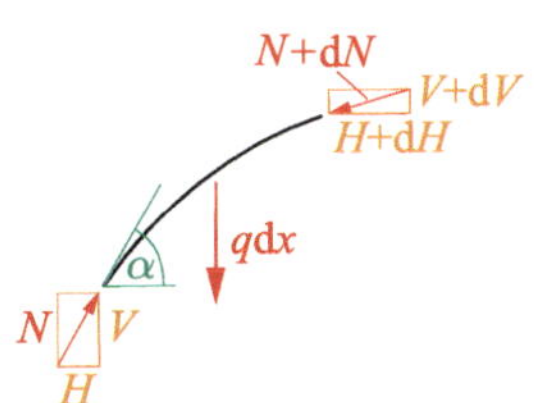

Fig. 6.55 Infinitesimal element of the curved beam with moment-free load transfer

We now analyse an infinitesimal section element of length $\mathrm{d}x$ (Fig. 6.55) and consider the normal force N, which we want to assume to be positive when it is a compressive force, as the only occurring internal force. It has the horizontal component H and the vertical component V. The resultant $q\mathrm{d}x$ of the line load $q(x)$ is also shown here. The sum of the horizontal forces results in

$$\mathrm{d}H = 0, \tag{6.164}$$

from which it can be concluded that the horizontal component of the normal compressive force N is constant. The vertical force sum provides:

$$\mathrm{d}V + q\mathrm{d}x = 0, \tag{6.165}$$

which can be transformed into:

$$\frac{\mathrm{d}V}{\mathrm{d}x} = -q. \tag{6.166}$$

Furthermore, from Fig. 6.55 it can be concluded that

$$\frac{V}{H} = \frac{\mathrm{d}y}{\mathrm{d}x} = \tan\alpha, \tag{6.167}$$

which, after derivation to x with the equilibrium conditions, leads to the following expression:

$$\frac{\mathrm{d}^2 y}{\mathrm{d}x^2} = \frac{\mathrm{d}}{\mathrm{d}x}\left(\frac{V}{H}\right) = \frac{1}{H}\frac{\mathrm{d}V}{\mathrm{d}x} = -\frac{q}{H}. \tag{6.168}$$

This is a differential equation for determining the shape of the moment-free curved beam. It can be solved by double integration, in which case two integration constants must be taken into account. It should also be noted here that the horizontal force component H is still unknown. These unknown quantities can be determined from two suitable boundary conditions and a further condition, e.g. via a statement on the geometric shape of the curved beam.

Once both the form $y(x)$ of the curved beam and the horizontal force component H have been determined, then the normal compressive force N can be determined as:

$$N = \sqrt{H^2 + V^2}. \tag{6.169}$$

Using $V = H\frac{\mathrm{d}y}{\mathrm{d}x}$ the result is

$$N = H\sqrt{1 + \left(\frac{\mathrm{d}y}{\mathrm{d}x}\right)^2}. \tag{6.170}$$

The procedure is illustrated using the example of a constant line load $q(x) = q_0$. From (6.168) we obtain the form of the curved beam after double integration with respect to x as:

$$\begin{aligned}
\frac{\mathrm{d}^2 y}{\mathrm{d}x^2} &= -\frac{q_0}{H}, \\
\frac{\mathrm{d}y}{\mathrm{d}x} &= -\frac{q_0 x}{H} + C_1, \\
y &= -\frac{q_0 x^2}{2H} + C_1 x + C_2.
\end{aligned} \tag{6.171}$$

The form of the beam is therefore a parabola in this particular case.

We will now assume a curved beam in which the support is suitable for moment-free load transfer and in which the two supports are at the same height (see Fig. 6.54, left). The distance between the two supports is l. If the origin of the coordinate system x, y is then placed in the left support, the boundary conditions for the beam shape are as follows:

$$y(x = 0) = 0, \quad y(x = l) = 0, \tag{6.172}$$

which leads to:

$$C_1 = \frac{q_0 l}{2H}, \quad C_2 = 0. \tag{6.173}$$

The shape of the curved beam then results as:

$$y(x) = \frac{q_0 l^2}{2H}\left[\frac{x}{l} - \left(\frac{x}{l}\right)^2\right]. \tag{6.174}$$

However, the horizontal force H is still unknown at this point. We arrive at an expression for H if, for example, we specify the vertex height $y\left(x = \frac{l}{2}\right)$ of the parabola at the point $x = \frac{l}{2}$ with the value f. Then the following results:

$$H = \frac{q_0 l^2}{8f}. \tag{6.175}$$

This also determines the horizontal compressive force H and the normal compressive force N using (6.170). It is identical to the horizontal compressive force H at the apex of the curved beam and reaches the maximum value in the two supports with the value $N = \frac{q_0 l^2}{8f}\sqrt{1 + \frac{16f^2}{l^2}}$.

6.9 Spatial Beams

Finally, we consider the determination of the stress resultants in spatial beam. As we have already shown with Fig. 6.1, three internal forces and three moments must be determined for a spatial beam:

- The normal force N in the $x-$ direction, i.e. in the longitudinal direction of the beam,
- the two transverse shear forces Q_y and Q_z in $y-$ and $z-$direction, respectively,
- the torsional moment M_x about the longitudinal axis x of the beam,
- the two bending moments M_y and M_z about the $y-$ and $z-$axes, respectively.

In general, spatial support reactions must also be determined in the same way. For the internal forces and moments $N, Q_y, Q_z, M_x, M_y, M_z$, the sign convention already introduced applies that a positive force or moment on a positive face points in the positive coordinate direction. Accordingly, the stress resultants of Fig. 6.1 are all positive.

The stress resultants in spatial beam structures can be determined exactly as already shown for plane beams by equilibrium considerations. It is also possible to derive a general relationship between load and internal forces and moments which, however, will not be shown here.

Example 6.19

For the spatial angled frame of Fig. 6.56 (cf. Example 4.11), the force and moment diagrams are to be determined.

Solution:

The support reactions have already been determined in Example 4.11, they are shown in Fig. 6.56 and read as follows:

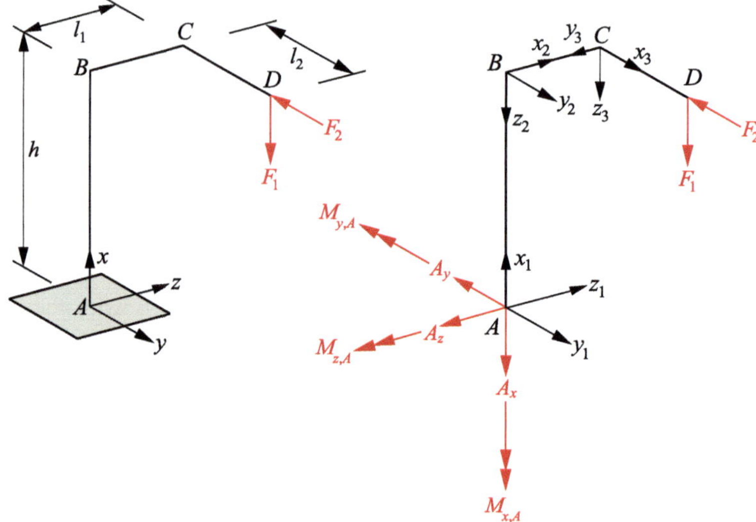

Fig. 6.56 Angled spatial frame (left), free-body diagram and local axis systems (right)

$$\begin{aligned} &A_x = -F_1, \quad A_y = -F_2, \quad A_z = 0, \\ &M_{x,A} = F_2 l_1, \quad M_{y,A} = -F_1 l_1, \quad M_{z,A} = F_1 l_2 - F_2 h. \end{aligned} \tag{6.176}$$

We also provide each section of the frame with its own coordinate system as shown in Fig. 6.56, right.

In the first step, we cut the frame in area 1 at an arbitrary point x_1 and look at the resulting free-body diagram of Fig. 6.57, left. The force and moment equilibrium conditions

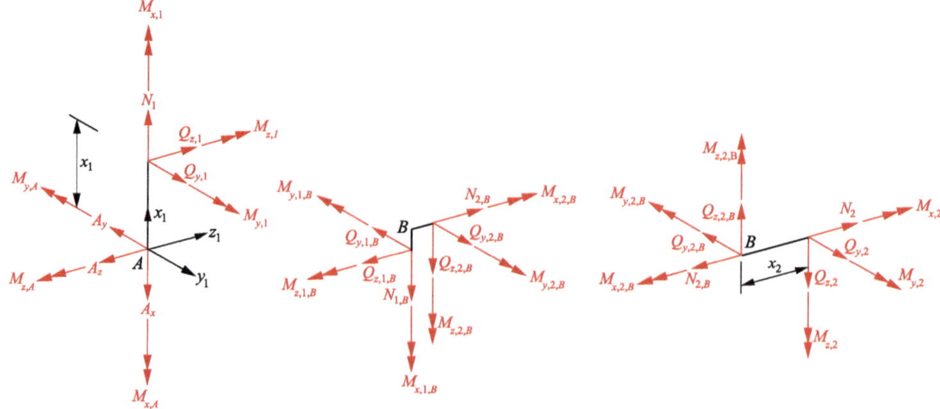

Fig. 6.57 Free-body diagrams

then result in the following stress resultants:

$$\begin{aligned}
N_1 &= -F_1, \\
Q_{y,1} &= -F_2, \\
Q_{z,1} &= 0, \\
M_{x,1} &= F_2 l_1, \\
M_{y,1} &= -F_1 l_1, \\
M_{z,1} &= F_1 l_2 + F_2 (x_1 - h) .
\end{aligned} \tag{6.177}$$

For the transition between areas 1 and 2 at the rigid corner B, we consider the free-body diagram of Fig. 6.57, centre. The following stress resultants result in segment 2 in the immediate vicinity of the rigid corner B:

$$\begin{aligned}
N_{2,B} &= Q_{z,1,B} = 0, \\
Q_{y,2,B} &= Q_{y,1,B} = -F_2, \\
Q_{z,2,B} &= -N_{1,B} = F_1, \\
M_{x,2,B} &= M_{z,1,B} = F_1 l_2, \\
M_{y,2,B} &= M_{y,1,B} = -F_1 l_1, \\
M_{z,2,B} &= -M_{x,1,B} = -F_2 l_1 .
\end{aligned} \tag{6.178}$$

To determine the stress resultants in segment 2, we use the free-body diagram of Fig. 6.57, right. It follows:

$$\begin{aligned}
N_2 &= 0, \\
Q_{y,2} &= -F_2, \\
Q_{z,2} &= F_1, \\
M_{x,2} &= F_1 l_2, \\
M_{y,2} &= F_1 (x_1 - l_1) , \\
M_{z,2} &= F_2 (x_1 - l_1) .
\end{aligned} \tag{6.179}$$

We can continue in the same way for segment 3, whereby we do not depict a free-body diagram at this point. The following stress resultants can be determined:

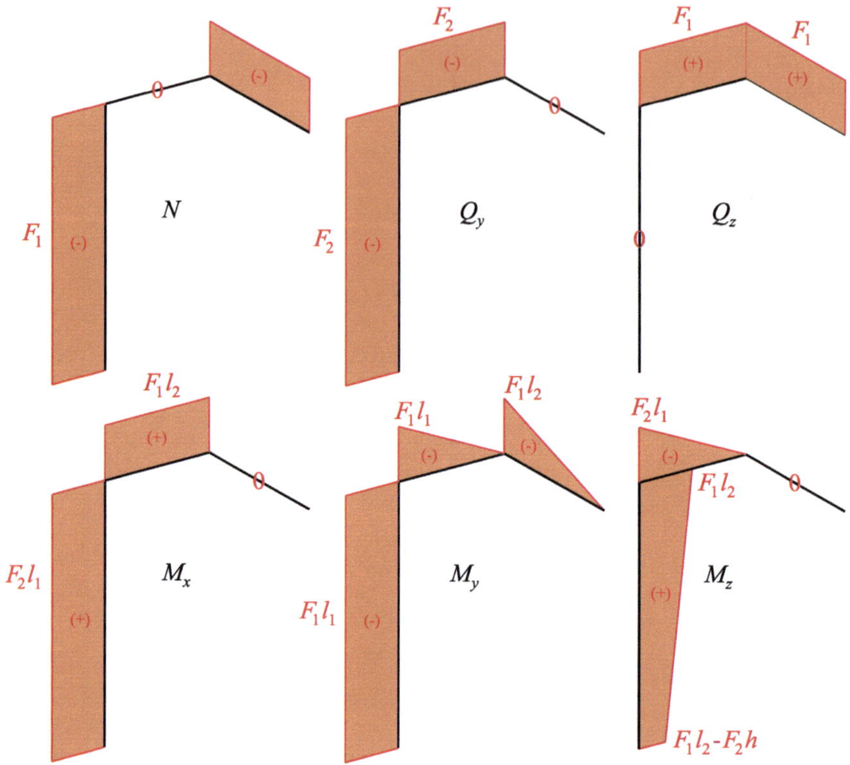

Fig. 6.58 Stress resultants

$$
\begin{aligned}
N_3 &= -F_2, \\
Q_{y,3} &= 0, \\
Q_{z,3} &= F_1, \\
M_{x,3} &= 0, \\
M_{y,3} &= F_1 \left(x_3 - l_2\right), \\
M_{z,3} &= 0.
\end{aligned}
\tag{6.180}
$$

The diagrams of the stress resultants are shown in Fig. 6.58. ◀

Work

7

This chapter introduces the concept of work and, based on this, presents the principle of virtual displacements, which is of fundamental importance for applied mechanics and has a number of interesting applications. The concepts of potential and potential forces are then introduced and we also analyse non-rigid systems. The chapter concludes with a consideration of the concept of stability, and we will analyse how the stability of equilibrium positions can be classified. In addition, we will conclude with the determination of so-called influence lines for forces and moments on static systems.

7.1 Work

Figure 7.1 shows a particle that is under a force F and is displaced by this force by the distance u in the direction of action of the force.

The work W performed by the force F is calculated as follows:

$$W = Fu. \tag{7.1}$$

This simple result arises from the fact that we have assumed that the force F does not change over the distance u travelled and that the displacement u also occurs exactly in the positive direction of action of the force F. We will now extend the considerations to the general case as shown in Fig. 7.2 and which can no longer be treated with a simple definition as given in (7.1). Given a particle under the force

C. Mittelstedt, *Engineering Mechanics 1: Statics*,
https://doi.org/10.1007/978-3-662-71852-0_7

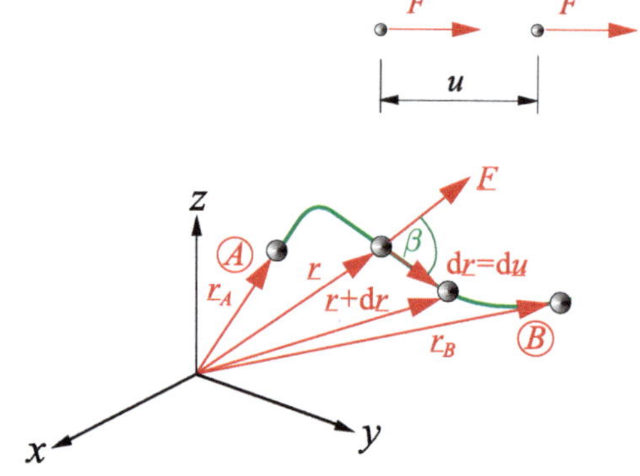

Fig. 7.1 Displacement of a particle under the force F by the distance u in the direction of action of the force

Fig. 7.2 Particle under the force $\underline{F}$ on a spatial trajectory between the points A and B

$$\underline{F} = \begin{pmatrix} F_x \\ F_y \\ F_z \end{pmatrix}, \tag{7.2}$$

which is located on an arbitrary spatial trajectory with the position vector

$$\underline{r} = \begin{pmatrix} r_x \\ r_y \\ r_z \end{pmatrix} \tag{7.3}$$

and which moves from point A to point B. Let the position vectors of points A and B be $\underline{r}_A$ and $\underline{r}_B$, respectively. The force $\underline{F}$ is location-dependent, i.e. $\underline{F}$ can have different magnitudes and also directions of action at different points on the trajectory. The work increment $\mathrm{d}W$, which is determined by $\underline{F}$ along a displacement increment

$$\mathrm{d}\underline{u} = \begin{pmatrix} \mathrm{d}u \\ \mathrm{d}v \\ \mathrm{d}w \end{pmatrix} \tag{7.4}$$

can be written as:

$$\mathrm{d}W = \underline{F}\mathrm{d}\underline{u} = F_x \mathrm{d}u + F_y \mathrm{d}v + F_z \mathrm{d}w. \tag{7.5}$$

The expression $\mathrm{d}W$ is an instantaneous value, and it is clear that the total work performed between A and B is the sum of all work increments. The quantity $\mathrm{d}W$ is the scalar product of $\underline{F}$ and $\mathrm{d}\underline{u}$, and the following applies:

$$\underline{F}\mathrm{d}\underline{u} = \left|\underline{F}\right| \left|\mathrm{d}\underline{u}\right| \cos \sphericalangle \left(\underline{F}\mathrm{d}\underline{u}\right) = F \mathrm{d}u \cos\beta = F\cos\beta \mathrm{d}u. \tag{7.6}$$

Accordingly, this is the product of the force component $F\cos\beta$ that is tangential to the path curve multiplied by the displacement increment $\mathrm{d}u$. As the work increment results from

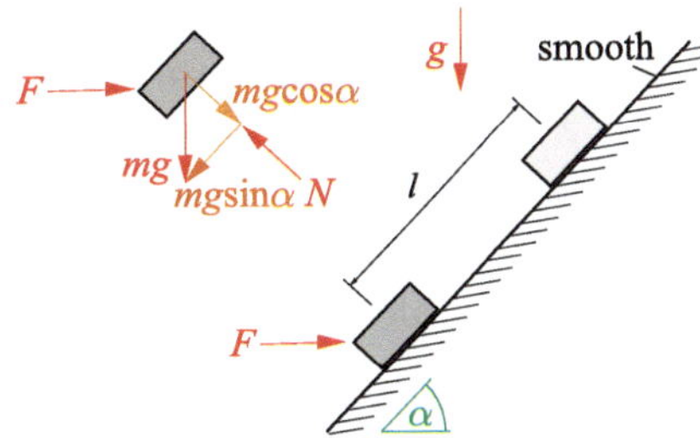

Fig. 7.3 Mass m on a smooth inclined plane under force F

the scalar product of $\underline{F}$ and $\mathrm{d}\underline{u}$, it is a scalar quantity. The work increment is therefore a quantity with a magnitude that is not directed.

The total work W performed by $\underline{F}$ on the trajectory between point A and point B is then the sum of all work increments or the integral over the scalar product $\underline{F}\mathrm{d}\underline{u}$:

$$W = \int_A^B \underline{F}\mathrm{d}\underline{u}. \tag{7.7}$$

Accordingly, work is a scalar quantity. It is expressed in Newton metres [Nm] or in Joule [J],[1] where the relationship 1Nm = 1J applies. The joule is the SI unit for work and energy:

$$1\mathrm{J} = 1\frac{\mathrm{kg} \cdot \mathrm{m}^2}{\mathrm{s}^2}. \tag{7.8}$$

Accordingly, a work of 10J is performed when an object weighing 1kg is lifted by 1 m in the Earth's gravitational field, whereby we have assumed for the sake of simplicity that the acceleration due to gravity is $10\frac{\mathrm{m}}{\mathrm{s}^2}$.

The work performed by a force can be positive, negative or zero. As an example, consider a mass m that is displaced upwards by a constant force F on a smooth inclined plane (angle of inclination α) without friction by the amount l (Fig. 7.3). The free-body diagram of the mass is shown in Fig. 7.3, left. The movement causes the constant force F to shift upwards and to the right. While no work is performed by the upward part of the movement, the force F performs work of $W_F = Fl\cos\alpha$ along the horizontal displacement. Of the weight force mg of the mass, only the component $mg\sin\alpha$ parallel to the path performs work, which in this case is negative (direction of action of the force component against the displacement): $W_G = -mgl\sin\alpha$. The component $mg\cos\alpha$ of the weight force perpendicular to the path, on the other hand, does not perform work; it is perpendicular to the direction of movement. The contact force N, which acts between the mass and the plane, also does no work; it acts perpendicular to the direction of movement.

We can summarise the above observations a little more generally as follows (Fig. 7.2). If the angle β lies in the interval $0° \leq \beta < 90°$ or in the interval $270° < \beta \leq 360°$, then

[1] James Prescott Joule, 1818–1889, English physicist.

the work increment $\mathrm{d}W$ is positive: $\mathrm{d}W > 0$. If, on the other hand, the angle β has the values $\beta = 90°$ or $\beta = 270°$, then the work increment $\mathrm{d}W$ is identical to zero: $\mathrm{d}W = 0$. For $90° < \beta < 270°$, on the other hand, the work increment $\mathrm{d}W$ is negative, i.e. $\mathrm{d}W < 0$. We can visualise this using the example in Fig. 7.3 to see that this generalisation applies here. In particular, it can be stated that weight forces acting perpendicular to the direction of motion do not perform any work. The same applies to contact forces in such a guided motion. If the motion of the Fig. 7.3 were also subject to friction, then the resultant friction force would counteract the motion (the friction force would have to be overcome to enable the motion). Therefore, the work of friction forces is always negative.

In the case of a moment $\underline{M} = (M_x, M_y, M_z)^T$, which is subject to the rotation $\underline{\varphi} = (\varphi_x, \varphi_y, \varphi_z)^T$, we can proceed analogously and specify a work increment $\mathrm{d}W$ as follows:

$$\mathrm{d}W = \underline{M}\mathrm{d}\underline{\varphi} = M_x \mathrm{d}\varphi_x + M_y \mathrm{d}\varphi_y + M_z \mathrm{d}\varphi_z. \tag{7.9}$$

If we consider the total work performed between an initial and a final rotation φ_A and φ_B, respectively, we obtain the following:

$$W = \int_{\varphi_A}^{\varphi_B} \underline{M}\mathrm{d}\underline{\varphi}. \tag{7.10}$$

7.2 The Principle of Virtual Displacements

In the previous considerations we assumed that a force shifts by a certain distance. However, there are a number of applications where we do not assume an actual displacement, but rather virtual displacements. The associated work principle is the so-called principle of virtual work or the principle of virtual displacements, which is based on imaginary, i.e. virtual displacements (these can be both displacements and rotations), which are subject to the following requirements:

- Virtual displacements are assumed to be infinitesimally small.
- They are imaginary and do not exist in reality.
- It is required that they are consistent with the given geometric boundary conditions of the considered static system.

Such displacements, which can represent both displacements and rotations, are labelled with the $\delta-$symbol of the calculus of variations. Accordingly, virtual displacements will be denoted as δu, for example, and virtual rotations as $\delta\varphi$. If a force $\underline{F}$ is subjected to a virtual displacement $\delta\underline{u}$, then it performs a virtual work δW as follows:

$$\delta W = \underline{F}\delta\underline{u}. \tag{7.11}$$

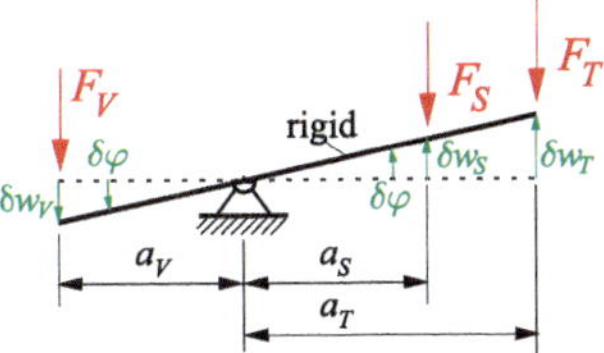

Fig. 7.4 Kinematics of the beam

If, on the other hand, a moment $\underline{M}$ is subjected to a virtual rotation $\delta\underline{\varphi}$, then the associated virtual work is $\underline{M}$:

$$\delta W = \underline{M}\delta\underline{\varphi}. \tag{7.12}$$

The principle of virtual displacements can be verbalised as follows:

A body is in equilibrium exactly when the entire virtual work of the applied forces and moments disappears at any permissible virtual displacement from the equilibrium position.

This can be formulated as follows:

$$\delta W = \sum_{i=1}^{n} \underline{F}_i \delta \underline{u}_i + \sum_{j=1}^{m} \underline{M}_j \delta \underline{\varphi}_j = 0. \tag{7.13}$$

This is a work statement under virtual displacements, which is why the requirement $\delta W = 0$ is often simply referred to as the principle of work. As we will show in more detail later, the principle of virtual displacements always leads to an equilibrium statement. It has some very interesting and essential applications for our purposes, which we will discuss in the following. In component notation, (7.13) is in the $xy-$plane:

$$\delta W = \sum_{i=1}^{n} \begin{pmatrix} F_{xi} \\ F_{yi} \end{pmatrix} \begin{pmatrix} \delta u_i \\ \delta v_i \end{pmatrix} + \sum_{j=1}^{m} M_j \delta\varphi_j = \sum_{i=1}^{n} \left(F_{xi}\delta u_i + F_{yi}\delta v_i \right) + \sum_{j=1}^{m} M_j \delta\varphi_j. \tag{7.14}$$

A simple example is shown in Fig. 7.4 (cf. Example 2.14). Let a rigid beam be given, which is loaded by three forces F_V (distance a_V from the support point), F_S (distance a_S) and F_T (distance a_T). The beam is assumed to be ideally rigid so that no deformations occur due to the applied load. The forces F_V, F_S and F_T then perform virtual work with a virtual rotation $\delta\varphi$ around the support point.

The virtual rotation $\delta\varphi$ is now assumed to take place. The force application points then pass through the virtual displacements δw_V, δw_S and δw_T. The virtual work δ is then:

$$\delta W = \delta W_a = F_V \delta w_V - F_S \delta w_S - F_T \delta w_T. \tag{7.15}$$

The components relating to F_S and F_T therefore have negative signs, as these two forces are deflected in the opposite direction to their direction of action. It should be noted here

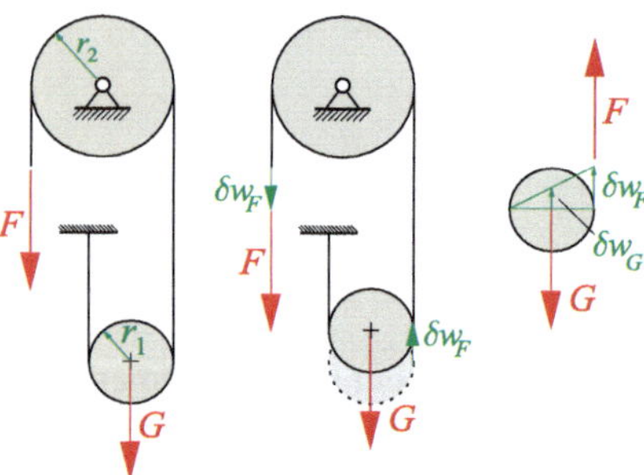

Fig. 7.5 Pulley (left), virtual displacements (centre), kinematics (right)

that the vertical support force that also occurs at point A does not perform any virtual work, the support point remains stationary and is not displaced in any way.

The kinematics, i.e. the relationships between the virtual rotation $\delta\varphi$ and the virtual displacements δw_V, δw_S and δw_T, follows from geometric considerations, taking into account that the virtual displacements are infinitesimally small. From Fig. 7.4 we obtain:

$$\delta w_V = \delta\varphi a_V, \quad \delta w_S = \delta\varphi a_S, \quad \delta w_T = \delta\varphi a_T. \tag{7.16}$$

The performed virtual work δW follows as:

$$\delta W = (F_V a_V - F_S a_S - F_T a_T)\, \delta\varphi. \tag{7.17}$$

If we now demand at this point that the virtual work δW disappears, then the following applies:

$$\delta W = (F_V a_V - F_S a_S - F_T a_T)\, \delta\varphi = 0. \tag{7.18}$$

There are now two ways to solve this equation. On the one hand, it can be demanded that the virtual rotation $\delta\varphi$ becomes zero: $\delta\varphi = 0$. This would mean that no virtual rotation has taken place, so that this solution is also referred to as the so-called trivial solution. The aim here is therefore to set the bracket expression to zero, i.e.:

$$F_V a_V - F_S a_S - F_T a_T = 0. \tag{7.19}$$

Obviously this expression corresponds exactly to the moment equilibrium with respect to the support point. The principle of virtual displacements therefore always leads to a statement of equilibrium.

Another simple example is the pulley shown in Fig. 7.5, to which a weight with the weight force G is attached. We want to determine the necessary force F at the left end of the rope to keep the given system in equilibrium.

The system is deflected by the virtual displacement δw_F (Fig. 7.5, centre). From the kinematic relationships, as shown in Fig. 7.5, right, we can see that the weight force G runs through the virtual displacement δw_G, which is only half as large as δw_F. The virtual work performed is then

$$\delta W = F\delta w_F - G \cdot \frac{\delta w_F}{2} = 0. \tag{7.20}$$

We obtain the following non-trivial solution:

$$F = \frac{G}{2}. \tag{7.21}$$

Overall, it can therefore be stated that an equilibrium statement can always be determined from the principle of virtual displacements. This principle is therefore equivalent to the equilibrium conditions of a system; the equilibrium conditions follow from the validity of the principle of virtual displacements.

7.3 Determination of Forces and Moments at Statically Determinate Systems

The first elementary application of the principle of virtual displacements consists is the determination of forces and moments (e.g. support reactions or internal forces and moments) for statically determinate beam structures. For this purpose, the quantity that needs to be determined is cut free and thus gives the system the possibility of motion, which it would not have without this cut. The quantity cut free in this way now performs a virtual work, which is included in the principle of virtual displacements, just like for the applied forces and moments. The requirement $\delta W = 0$ then leads to the quantity to be determined.

For motivation, let us consider the beam shown in Fig. 7.6, top left, on two supports. The beam is loaded by the uniform line load q_0, the force P_0 (which acts exactly in the centre of the beam) and the moment M_0. The beam has the length l. We want to determine the support reaction at support B and the bending moment at point $x = l/2$. To determine the support force at point B, we cut away the support and thus release the support force B (Fig. 7.6, top right). This makes the beam, which is actually statically determinate, simply kinematically displaceable (i.e. there is a clearly defined degree of freedom from which a unique displacement figure can be constructed) and performs a virtual rotation $\delta\varphi$ around the support A. It should be noted that this is a pure rigid body rotation, so that the beam itself remains undeformed and straight. The force P_0 and the resultant of the line load q_0 with the magnitude $q_0 l$ both act in the centre of the beam and have the virtual displacement δw_p. Since the virtual displacements are assumed to be infinitesimally small, $\delta w_p = \delta\varphi\frac{l}{2}$ can be written as a kinematic relationship. The support force B undergoes the virtual displacement $\delta w_B = \delta\varphi l$. The moment M_0 performs a virtual work along the virtual rotation $\delta\varphi$, whereby it must be noted that this virtual work must be included in the work balance with a negative sign, since the moment M_0 and the virtual rotation $\delta\varphi$ have opposite directions of rotation. The virtual work is then

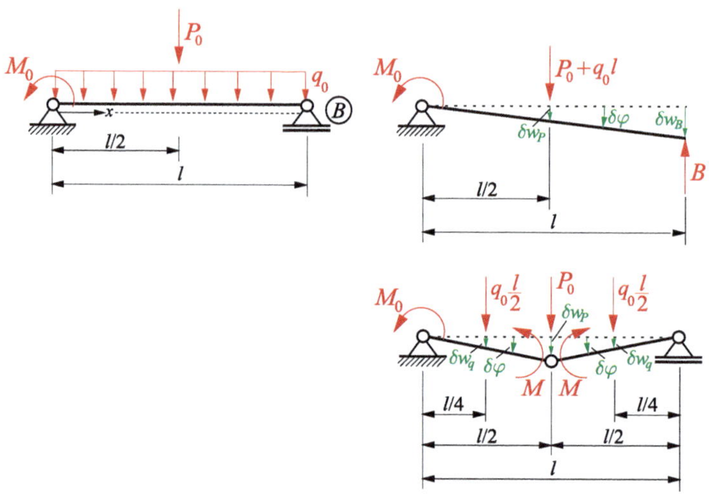

Fig. 7.6 Beam on two supports under line load q_0, force P_0 and moment M_0 (left), kinematics for the determination of the support force B (top right), kinematics for the determination of the internal moment M at the point $x = l/2$ (bottom right)

$$\delta W = -M_0\delta\varphi + (P_0 + q_0 l)\,\delta\varphi\frac{l}{2} - Bl\delta\varphi = 0, \tag{7.22}$$

or:

$$\left(-M_0 + (P_0 + q_0 l)\,\frac{l}{2} - Bl\right)\delta\varphi = 0. \tag{7.23}$$

The trivial solution $\delta\varphi = 0$ is not appropriate here, as it would assume that no virtual rotations have been applied to the system. We therefore set the bracketed expression to zero, which we can immediately solve for the support force B:

$$B = -\frac{M_0}{l} + \frac{P_0}{2} + \frac{q_0 l}{2}. \tag{7.24}$$

Apparently, this is the support force that we would also obtain from elementary equilibrium considerations. The principle of virtual displacements therefore obviously leads to a statement of equilibrium.

We proceed in the same way to determine the internal moment M in the centre of the beam at the point $x = \frac{l}{2}$. We release the required moment by inserting a hinge and consider the resulting virtual displacement figure, which results from the virtual rotation $\delta\varphi$ of the two beam segments of length $\frac{l}{2}$ around the respective support points. Both the released bending moment M and the applied moment M_0 then perform virtual work along these virtual rotations. In addition, the individual force P_0 undergoes the virtual displacement $\delta w_P = \frac{l}{2}\delta\varphi$, and the two resultants of the line load with the respective magnitude $q_0\frac{l}{2}$ on the left and right beam segment perform virtual work along the virtual displacement

$\delta w_q = \frac{l}{4}\delta\varphi$. The performed virtual work then results in

$$\delta W = \delta W_a = -2M\delta\varphi - M_0\delta\varphi + 2q_0 \cdot \frac{l}{2} \cdot \frac{l}{4} \cdot \delta\varphi + P_0 \cdot \frac{l}{2} \cdot \delta\varphi = 0. \tag{7.25}$$

This can be solved immediately for the bending moment M at the point $x = \frac{l}{2}$:

$$M = -\frac{M_0}{2} + \frac{q_0 l^2}{8} + \frac{P_0 l}{4}. \tag{7.26}$$

This result can also be determined by elementary equilibrium considerations.

When determining forces and moments according to the principle of virtual displacements, the system under consideration is made kinematically displaceable by releasing the desired quantity, whereby the resulting displacement figure can always be clearly defined. We refer to this virtual displacement figure as kinematic chain.

Example 7.1

Consider the beam shown in Fig. 7.7 under the force F_0, the moment M_0 and the uniform line load q_0. The support reactions and internal forces and moments C_V, B_V, A_V, M_A, M_B and M_D are to be determined using the principle of virtual displacements.

Solution:

All required virtual kinematic chains are shown in Fig. 7.8. The calculation is explained here using the support force C_V. The kinematic chain that results from cutting this support force free is shown in Fig. 7.8. The virtual work performed is then:

$$\delta W = \delta W_a = M_0\delta\varphi + q_0 l\delta w_1 - C_V\delta w_2 = 0. \tag{7.27}$$

From the kinematic relationships $\delta w_1 = \delta\varphi\frac{l}{2}$ and $\delta w_2 = \delta\varphi l$ then follows:

$$\left(M_0 + q_0 l\frac{l}{2} - C_V l\right)\delta\varphi = 0. \tag{7.28}$$

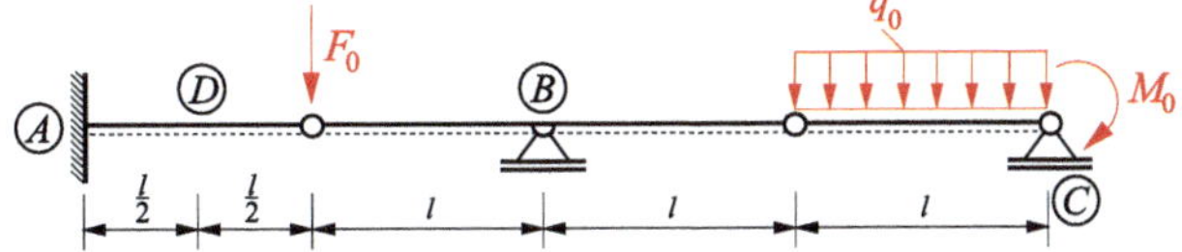

Fig. 7.7 Beam under force F_0, moment M_0 and uniform line load q_0

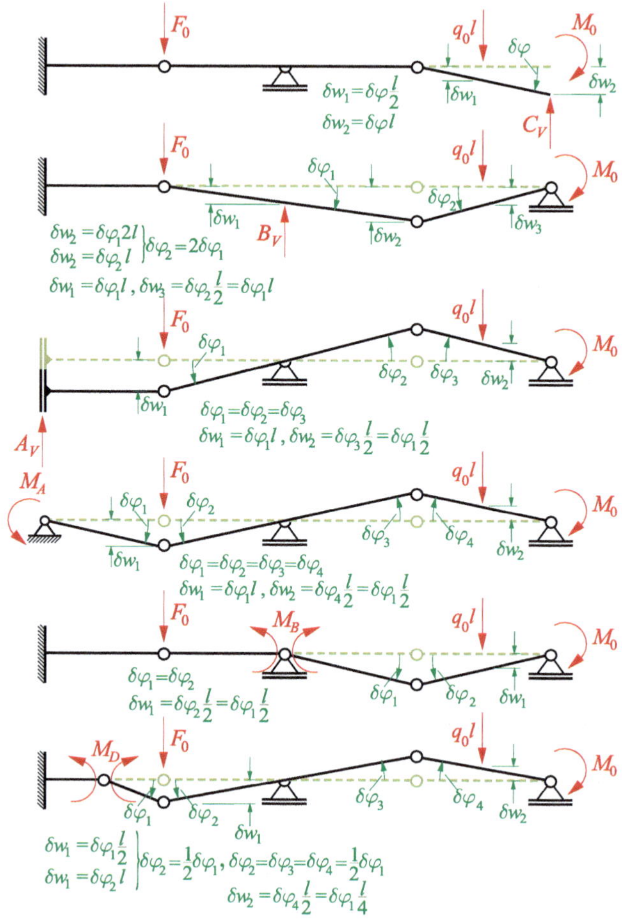

Fig. 7.8 Kinematic chains

Zeroing the content of the brackets gives the support force C_V as:

$$C_V = \frac{M_0}{l} + \frac{q_0 l}{2}. \tag{7.29}$$

The other forces and moments can be determined in the same way. We do not specify the analysis at this point and simply present the results:

$$B_V = q_0 l - 2\frac{M_0}{l},$$
$$A_V = F_0 - \frac{q_0 l}{2} + \frac{M_0}{l},$$
$$M_A = F_0 l - \frac{q_0 l^2}{2} + M_0,$$
$$M_B = M_0 - \frac{q_0 l^2}{2},$$
$$M_D = -\frac{F_0 l}{2} + \frac{q_0 l^2}{4} - \frac{M_0}{2}. \tag{7.30}$$

◀

Example 7.2

For the static system of Fig. 7.9, the support force B is to be calculated using the principle of virtual displacements. In addition, the internal moment M_D is to be determined.

Solution:

To calculate the support force B, we cut the support in question free and analyse the kinematic chain shown in Fig. 7.9, centre. The performed virtual work reads:

$$\delta W = 5ql\delta w_q - B\delta w_B - M_0\delta\varphi_2 - F_1\delta w_1 - F_2\delta w_2 = 0. \tag{7.31}$$

The virtual displacements δw_q and δw_B can be expressed by the virtual angle $\delta\varphi_1$ as follows:

$$\delta w_q = 2.5l\delta\varphi_1, \quad \delta w_B = 3l\delta\varphi_1. \tag{7.32}$$

A relationship between the two virtual angles $\delta\varphi_1$ and $\delta\varphi_2$ can be established from the virtual displacement of the hinge, i.e. the point of application of the moment M_0:

$$\delta w_G = 5l\delta\varphi_1 = 2l\delta\varphi_2, \tag{7.33}$$

which leads to:

$$\delta\varphi_2 = \frac{5}{2}\delta\varphi_1. \tag{7.34}$$

This means that the two virtual displacements δw_1 and δw_2 can also be expressed by the angle $\delta\varphi_1$:

$$\delta w_1 = l\delta\varphi_2 = \frac{5}{2}l\delta\varphi_1, \quad \delta w_2 = 2l\delta\varphi_2 = 5l\delta\varphi_1. \tag{7.35}$$

The work equation is then as follows:

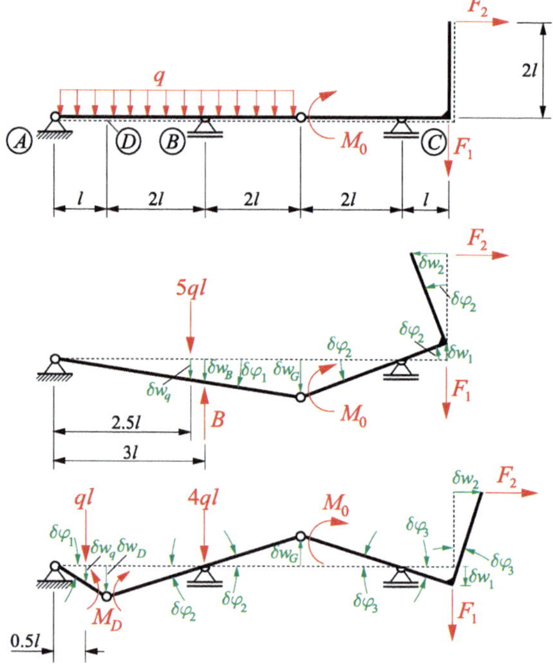

Fig. 7.9 Angled beam (top), kinematics for determining the support force B (centre), kinematics for determining the internal moment M_D (bottom)

$$\left(\frac{25}{2}ql^2 - 3Bl - \frac{5}{2}M_0 - \frac{5}{2}F_1 l - 5F_2 l\right)\delta\varphi_1 = 0. \tag{7.36}$$

Setting the bracket term to zero results in the required support force:

$$B = \frac{25}{6}ql - \frac{5M_0}{6l} - \frac{5}{6}F_1 - \frac{5}{3}F_2. \tag{7.37}$$

The kinematic chain, which is established by releasing the internal moment M_D by inserting a hinge at the relevant location, is shown in Fig. 7.9, bottom. The virtual work performed is then as follows:

$$\delta W = ql\delta w_q - M_D\delta\varphi_1 - M_D\delta\varphi_2 + M_0\delta\varphi_3 + F_1\delta w_1 + F_2\delta w_2 = 0. \tag{7.38}$$

The deflection of the moment point of application D can be used to establish a relationship between the two angles $\delta\varphi_1$ and $\delta\varphi_2$:

$$\delta w_D = l\delta\varphi_1 = 2l\delta\varphi_2. \tag{7.39}$$

From this we can deduce $\delta\varphi_2 = \frac{1}{2}\delta\varphi_1$. In exactly the same way, a relationship between the two virtual rotations $\delta\varphi_2$ and $\delta\varphi_3$ is obtained from the virtual displacement of the hinge G (point of application of the moment M_0):

$$\delta w_G = 2l\delta\varphi_2 = 2l\delta\varphi_3. \tag{7.40}$$

Obviously, $\delta\varphi_3 = \delta\varphi_2$. Then $\delta\varphi_3 = \frac{1}{2}\delta\varphi_1$ holds. The displacements δw_q, δw_1 and δw_2 can be expressed by the angle $\delta\varphi_1$ as:

$$\delta w_q = \frac{1}{2}l\delta\varphi_1, \quad \delta w_1 = l\delta\varphi_3 = \frac{1}{2}l\delta\varphi_1, \quad \delta w_2 = 2l\delta\varphi_3 = l\delta\varphi_1. \tag{7.41}$$

With the kinematic relationships found in this way, the internal moment M_D follows as $M_D = \frac{1}{3}ql^2 + \frac{M_0}{3} + \frac{1}{3}F_1 l + \frac{2}{3}F_2 l$ from the virtual work of the system.

◀

7.4 Pole Plans and Kinematic Chains

The non-rigid static systems dealt with so far were of a simple nature and could be treated by intuitively sketching the resulting kinematic chains. If, on the other hand, there is a system of greater complexity, this is not always as easy. In this section, we want to establish some rules for the determination of kinematic chains. This procedure is also often referred to as the so-called kinematic method, the aim of which is to create so-called pole plans that allow the kinematic chains to be created in a straightforward manner.

We consider a rigid disc (Fig. 7.10) which, for the sake of simplicity, we assume to be triangular with the vertices A, B and C. This disc is now rotated around the point P by the angle $\delta\varphi$. The δ−symbol again indicates that this is an infinitesimally small rotation which is only virtual, i.e. not real. Point P is fixed and therefore immovable, and hence we also refer to it as the instantaneous pole or main pole. The vertices of the disc have the distances or radii r_A, r_B, r_C to the main pole. The virtual displacements of the disc points are always perpendicular to the distance measured from the main pole. For the example in Fig. 7.10 we can therefore write:

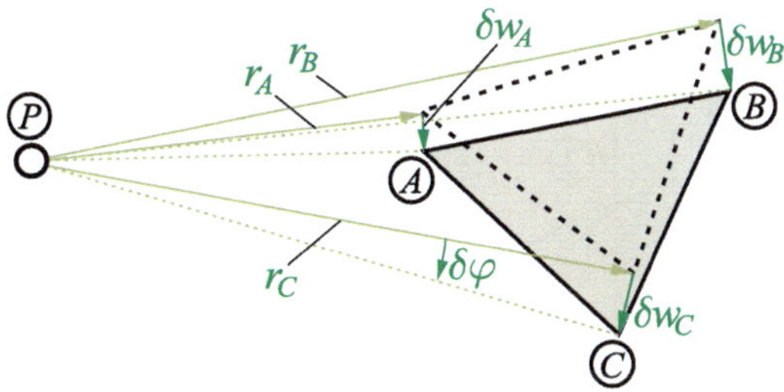

Fig. 7.10 Rigid disc (corner points A, B, C) under a virtual rotation $\delta\varphi$ around the point P

$$\delta w_A = r_A \delta\varphi, \quad \delta w_B = r_B \delta\varphi, \quad \delta w_C = r_C \delta\varphi. \tag{7.42}$$

The main pole is the absolute centre of rotation of a rigid disc, whereas the so-called common pole, secondary pole or relative pole of two bodies is the point where both bodies undergo the same displacement. For the disc in Fig. 7.10, the point P is a main pole because no displacement can occur at this point.

The major rules for determining a pole plan are illustrated in Fig. 7.11. We consider two rigid discs of arbitrary shape, which we will call discs 1 and 2. Let disc 1 have a two-valued hinged support at a given point. Disc 1 is connected to disc 2 by an ideal moment hinge. Let disc 2 also have a single-valued support at an arbitrary point. Obviously, this system is non-rigid and thus kinematically indeterminate.

Each non-displaceable support of a disc is always the main pole of this disc, and any movement of a point on this disc will then be perpendicular to the radius between the main pole and the point under consideration. The main pole does not move during a virtual displacement. For the example in Fig. 7.11, the main pole of disc 1 lies exactly in the two-valued support. The main pole is denoted by the number of the disc under consideration, which is enclosed in a circle.

A moment hinge between two discs is always their common secondary pole; the two discs connected in this way will always undergo the same displacement at this point due to the connection through the hinge. Secondary poles are given the numbers of the two connected discs (in this case 1, 2).

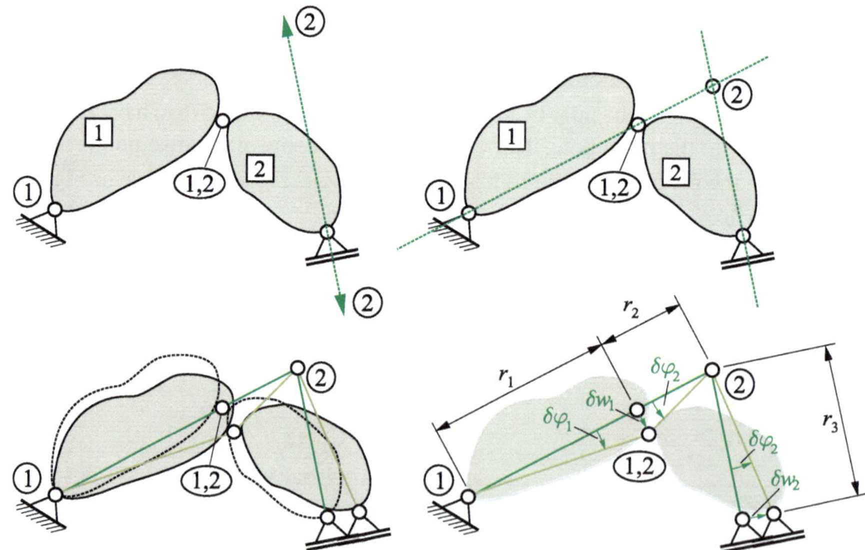

Fig. 7.11 Determination of a pole plan for a non-rigid system consisting of two rigid discs (top), resulting kinematic chain (bottom)

The main pole of a disc supported by a single-valued support is located on a line that runs through the support in the direction of the line of action of the support force. This is shown in Fig. 7.11 by the green line that runs through the support of disc 2.

The two main poles and the secondary pole of two discs always lie on a straight line. As the positions of main pole 1 and secondary pole of the two discs 1 and 2 have already been determined, the position of main pole 2 can be determined at the intersection of the two straight lines drawn. The resulting pole plan for this example is shown in Fig. 7.11, top right. The resulting kinematic chain is shown in Fig. 7.11, bottom left. The associated kinematics can be seen in Fig. 7.11, bottom right.

However, there are also situations in which poles are not always clearly defined, but can also lie at infinity. A corresponding example is shown in Fig. 7.12. Given is a system consisting of three straight beams. Let beams 1 and 3 be supported at their lower ends and connected to beam 2 at the upper end by moment hinges.

The main poles of beams 1 and 3 are obviously located in their lower support points. The secondary poles between beams 1 and 2 and beams 3 and 2 can also be easily determined in the upper moment hinges. Since the two main poles and the secondary pole of two beams must always be on a straight line (indicated by dashed lines in Fig. 7.12, left), but these two lines run parallel to each other in this example, the main pole of beam 2 must be at infinity. The resulting kinematic chain of Fig. 7.12, right, shows that beam 2 does not undergo any rotation, but suffers a pure rigid body displacement.

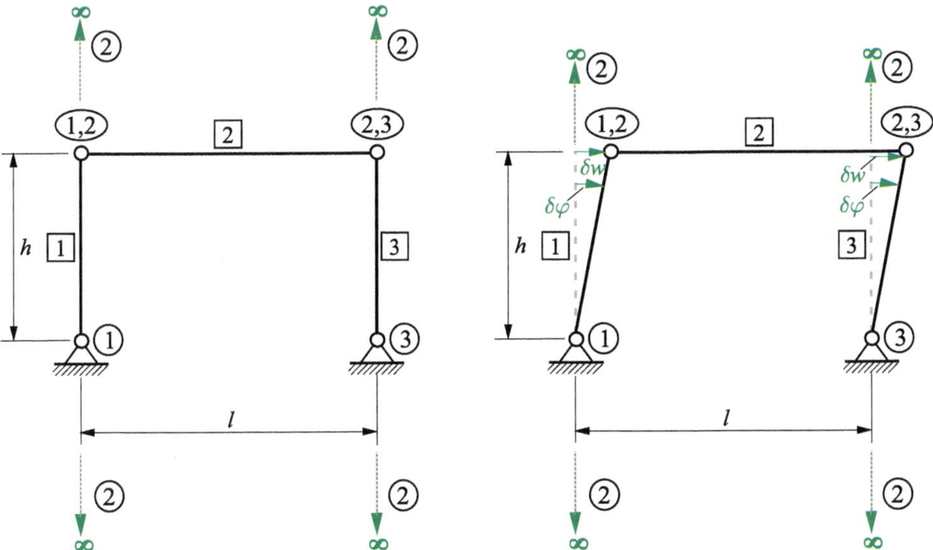

Fig. 7.12 Determination of a pole plan for a non-rigid system consisting of three connected beams (left), kinematic chain (right)

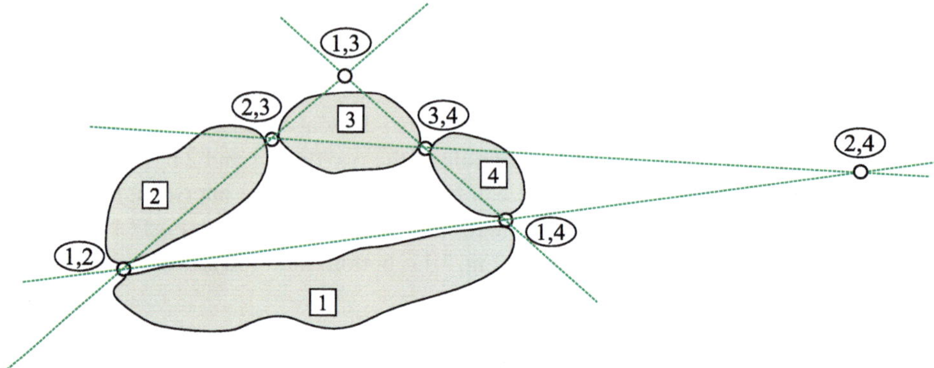

Fig. 7.13 Determination of secondary poles

The secondary poles of three discs always lie on a straight line, as shown in Fig. 7.13. Figure 7.13 shows four rigid discs, each of which is connected to each other by moment hinges. The secondary poles (1, 2), (2, 3), (3, 4) and (1, 4) are located in the joints between the discs. The remaining secondary poles can be determined by drawing straight lines through the relevant already known secondary poles and determining the missing secondary poles (1, 3) and (2, 4) at the corresponding intersection points.

If two beams are connected by a sliding joint, the secondary pole of these beams is perpendicular to the direction of movement of the joint at infinity (Fig. 7.14, left). The situation is similar with two beams connected by a normal force joint (Fig. 7.14, right). Two beams whose secondary pole is at infinity always rotate by the same angle.

Three discs that form a hinge triangle always behave like a rigid disc and are immovable (Fig. 7.15). This obviously also applies to three beams connected to each other by hinges.

All displacements and rotations that we consider in this chapter are assumed to be virtual and infinitesimally small. This is associated with the concept of projection constancy, which we explain using Fig. 7.16. Consider a beam on two supports, which is provided with a moment hinge at an arbitrary point. If finite, but not infinitesimally small displacements

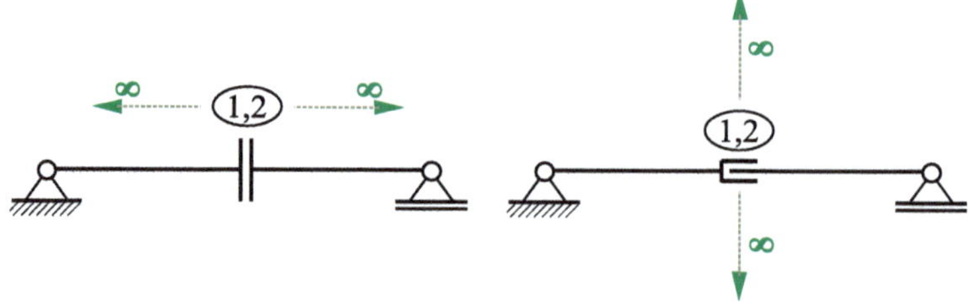

Fig. 7.14 Determination of secondary poles for shear force and normal force joints

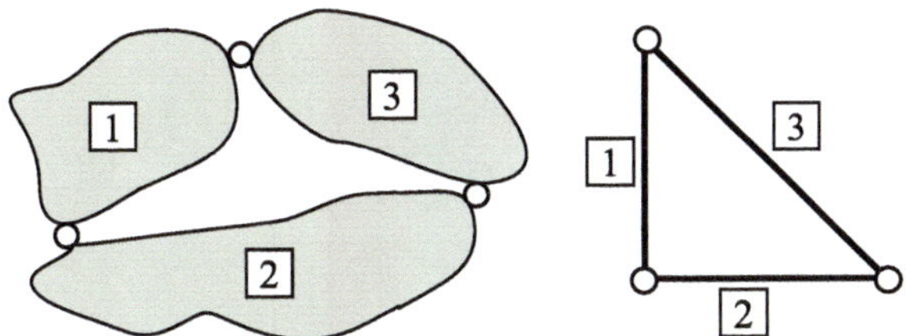

Fig. 7.15 Three discs connected by moment hinges (left), beam triangle (right)

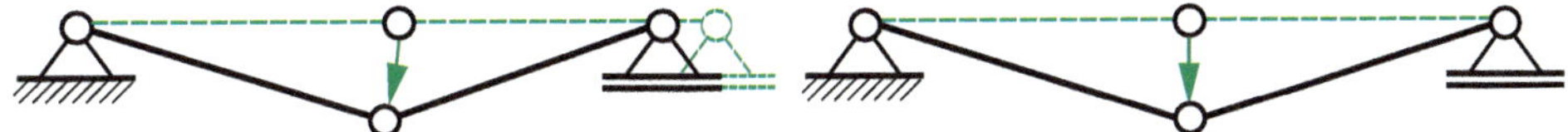

Fig. 7.16 Projection constancy

are considered on this displaceable beam, the deformation figure is as shown in Fig. 7.16, left. The hinge moves downwards on a circular path, and the movable support on the right side would move to the left due to this motion. However, since we assume virtual and infinitesimally small displacements and rotations, we can replace the real circular path with a straight downward line. This means that displacements are always replaced by their tangents, which are perpendicular to the radius to the main pole. Furthermore, in this specific case, the horizontal displacement of the right support is neglected.

With the explained rules, main and secondary poles can be found for any displaceable system and the corresponding pole plans and kinematic chains can be determined. In Fig. 7.17 some exemplary pole plans for selected systems and the corresponding kinematic chains are shown.

Example 7.3

For the system shown in Fig. 7.18, the following support reactions and stress resultants are to be determined using the principle of virtual displacements:

- Support moment M_A,
- Support force B,
- Internal moment M_D,
- Horizontal support force A_H,
- Internal moment M_C,
- Vertical hinge force G_V.

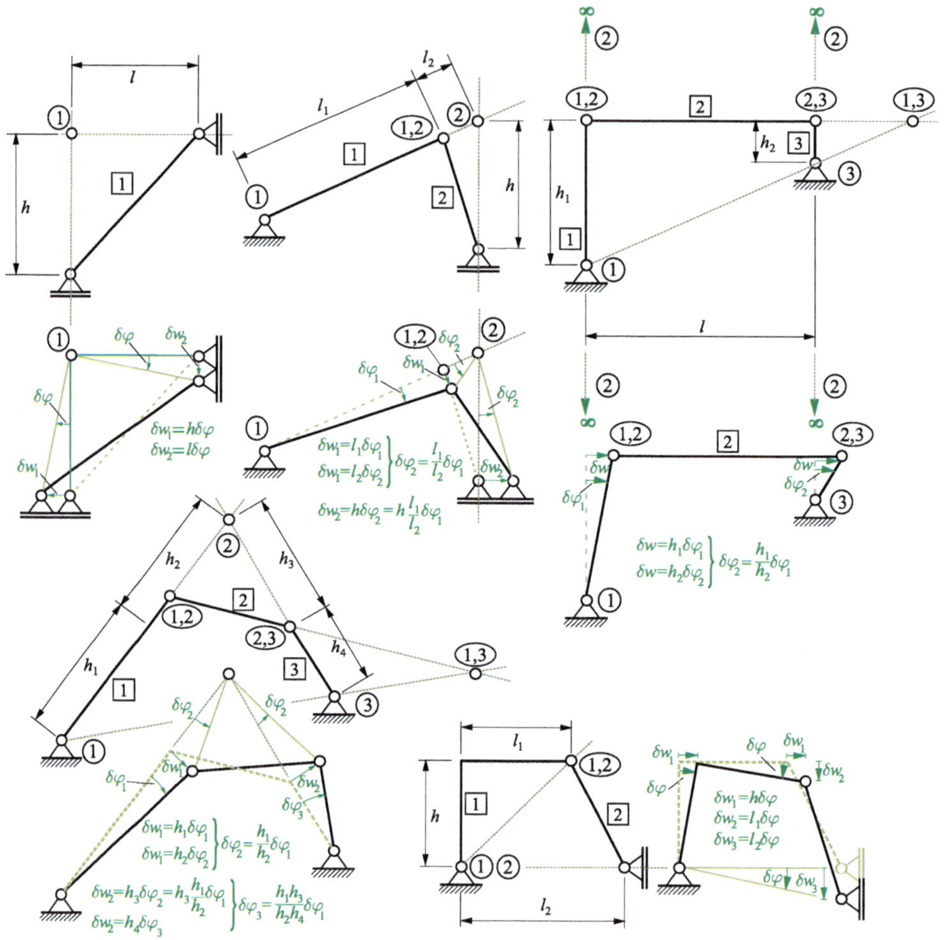

Fig. 7.17 Pole plans and associated kinematic chains

Fig. 7.18 Statically determinate frame under uniform line load q and point force F

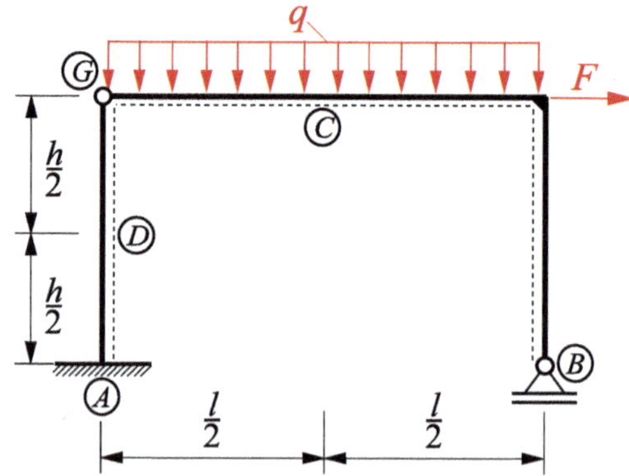

Solution:

The pole plans and kinematic chains associated with the quantities to be determined are shown in Fig. 7.19.

To determine the support moment M_A, the clamping is converted into a two-valued hinged support by releasing the moment M_A. The frame, which becomes displaceable as a result, consists of beams 1 and 2, and the main pole of beam 1 is located in the hinged support. The moment hinge located at the upper end of beam 1 is the secondary pole of the two beams 1 and 2. Main pole 2 must then be located on the line connecting main pole 1 and the secondary pole. At the same time, however, the main pole 2 must also be located on the straight line that runs through the displaceable support of beam 2 in the direction of the support force. These two straight lines run parallel to each other so that main pole 2 is

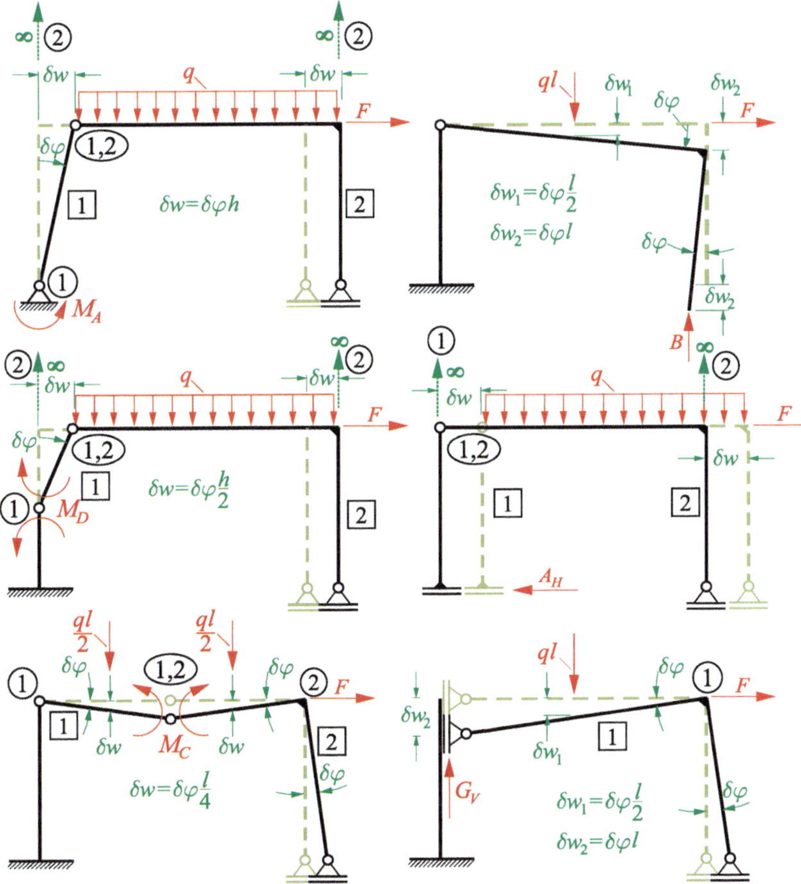

Fig. 7.19 Pole plans and kinematic chains

at infinity. The resulting kinematic chain is such that beam 1 performs a pure rotation $\delta\varphi$ around the lower two-valued support. Beam 2 shows a pure horizontal rigid body translation δw. The resultant of the line load q therefore does not perform any virtual work on this kinematic chain. The only virtual work performed results from the horizontal force F along the virtual displacement δw and from the support moment M_A along the virtual rotation $\delta\varphi$:

$$\delta W = F\delta w - M_A\delta\varphi = 0. \tag{7.43}$$

With $\delta w = \delta\varphi h$, the support moment M_A follows as $M_A = Fh$.

To determine the support force B, the support B is cut free and the support force B is released. Since the vertical beam is still firmly clamped at the lower end, it will not undergo any virtual displacements or rotations. The only movable part here is the angled beam on the right side, whose main pole is located in the moment hinge at the top left. Due to the virtual rotation $\delta\varphi$, the resultant of the line load q with the magnitude ql performs a virtual work along the virtual displacement $\delta w_1 = \delta\varphi\frac{l}{2}$. The force F does not perform any virtual work here, as can easily be seen from the resulting kinematic chain. The support force B undergoes the virtual displacement $\delta w_2 = \delta\varphi l$ and performs the negative virtual work $-B\delta\varphi l$ along it. The virtual work for this example is therefore

$$\delta W = ql\delta\varphi\frac{l}{2} - B\delta\varphi l = 0. \tag{7.44}$$

This can be solved for the required support force as $B = \frac{ql}{2}$.

The determination of the remaining forces and moments are not discussed here, and only the results are reported at this point:

$$M_D = -\frac{Fh}{2}, \quad A_H = F, \quad M_C = \frac{ql^2}{8}, \quad G_V = \frac{ql}{2}. \tag{7.45}$$

◀

Example 7.4

For the truss of Fig. 7.20 (see also Chapter 5, Fig. 5.1), the member forces of members 2, 3, 4 and 8 are to be determined using the principle of virtual displacements.

Solution:

The pole plans and kinematic chains required to solve the problem are shown in Fig. 7.21. To determine the bar force S_2, this bar is cut free. This makes the system displaceable, and the kinematic chain is shown in Fig. 7.21. Bars 6, 7, 9 as well as 5, 8, 9 and 3, 4, 8 each form triangles and are therefore immovable in themselves. We can combine these three triangles into a single inherently immovable disc (disc 2), which is connected to bar 1 by its

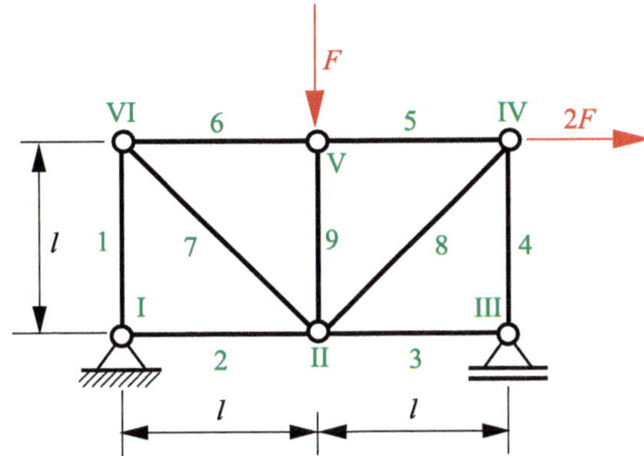

Fig. 7.20 Truss under two loads

secondary pole (1, 2), which is interpreted as disc 1. The main pole of disc 1 is located in the left non-displaceable support. Therefore, the main pole 2 must be located on the connecting line running vertically upwards. At the same time, main pole 2 must also be located on the line that runs vertically through the right movable support. Main pole 2 must therefore be at infinity, so that disc 1 undergoes a pure rotation $\delta\varphi$ around the main pole 1. Disc 2, on the other hand, shows a pure rigid body translation with the shift δw to the right. The principle of virtual displacements is as follows:

$$\delta W = 2F\delta w - S_2\delta w = 0. \tag{7.46}$$

The resulting work term can be solved for S_2 by avoiding the trivial solution $\delta w = 0$, and it follows:

$$S_2 = 2F. \tag{7.47}$$

We proceed in a very similar way for the member force S_3 which is released by cutting bar 3. Here, too, several joint triangles are formed, which we combine into a single rigid disc, which is referred to as disc 1. Bar 4, on the other hand, is interpreted as disc 2. Disc 1 has its main pole in the left-hand fixed support. The secondary pole with disc 2 is located in the moment joint between the two discs. Disc 2 also has a single-valued support, so its main pole must be located on the line running straight up through the support, where it crosses the connecting line between main pole 1 and secondary pole 1, 2. This means that main pole 2 is located in the moment joint that connects discs 1 and 2. Consequently, disc 1 will neither rotate nor translate and therefore does not change its position. Only disc 2 performs a pure rotation around its main pole, in such a way that the sliding support is shifted to the right. With the resulting kinematic chain, neither of the two acting concentrated loads performs any virtual work, so that the bar force S_3 is zero.

$$S_3 = 0. \tag{7.48}$$

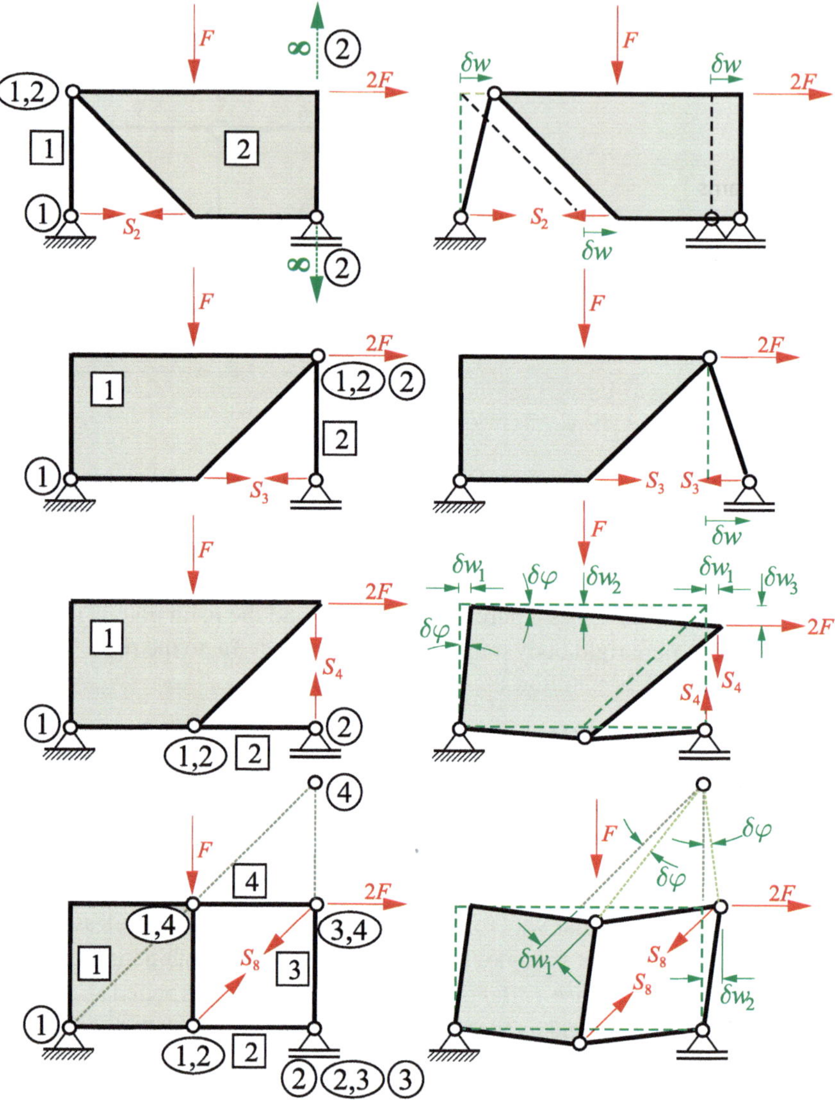

Fig. 7.21 Determination of member forces 2, 3, 4 and 8 using the kinematic method

The fact that bar 3 is actually a zero-value bar can also be easily understood by means of elementary equilibrium considerations (see Chap. 5).

If bar 4 is cut free, the result is again a rigid disc composed of several triangles, which is referred to as disc 1. Bar 3 is interpreted as disc 2. The main pole of disc 1 is located in the two-valued support on the left-hand side, and since the secondary pole between the two discs is located in the moment joint, the main pole of disc 2 is located in the support on the right-hand side. Disc 1 now undergoes a virtual rotation $\delta\varphi$, resulting in a horizontal virtual displacement δw_1. Along this displacement, the horizontal individual force $2F$ performs the virtual work $2F\delta w_1 = 2F\delta\varphi l$. In addition, the vertical individual force F is lowered by the amount δw_2, i.e. $\delta w_2 = \delta\varphi l$. Finally, the bar force S_4 at the upper end is lowered by the displacement δw_3, which can be expressed as $\delta w_3 = \delta\varphi 2l$. The bar force S_4 therefore performs the virtual work $S_4\delta\varphi 2l$. Finally, the principle of virtual displacements reads:

$$\delta W = 2F\delta\varphi l + F\delta\varphi l + S_4\delta\varphi 2l = 0. \tag{7.49}$$

This can be solved for the bar force S_4, and the result is

$$S_4 = -\frac{3}{2}F. \tag{7.50}$$

Finally, the bar force S_8 is determined. The bars 1, 2, 7 and 6, 7, 9 represent triangles, so that we can consider the left half of the static system as a rigid disc. We refer to this as disc 1 with the main pole in the left support. Disc 2 is rod 3, whose secondary pole is located in the hinge connecting it to disc 1. This means that the main pole of this disc is also fixed in the right-hand movable support. Disc 3 corresponds to rod 4. As the secondary pole of discs 2 and 3 is located in the movable support on the right, the main pole of disc 3 must also be located there. By looking at the main poles 1 and 3 as well as the secondary poles (1,4) and (3,4), the main pole of disc 4 is also found. To determine the required bar force S_8, a virtual rotation $\delta\varphi$ is applied to the system with respect to the main pole 4. This results in the virtual displacement $\delta w_1 = \delta\varphi\sqrt{2}l$ at the point of application of the vertically acting force F, so that this force is displaced by the amount $\frac{1}{\sqrt{2}}\delta w_1$. The virtual work performed by this force is therefore $F\delta\varphi l$. In exactly the same way, the horizontally acting force $2F$ is deflected by δw_2, which we can write as $\delta\varphi l$. Accordingly, this force performs the positive virtual work $2F\delta\varphi l$. The work performed by the bar force S_8 is considered in two parts. On the one hand, the horizontal component $S_8\frac{1}{\sqrt{2}}$ of the bar force shown above is shifted by the amount $\delta w_2 = \delta\varphi l$, so that this work component results as $-S_8\frac{1}{\sqrt{2}}\delta\varphi l$. On the other hand, the vertical component $S_8\frac{1}{\sqrt{2}}$ of the lower bar force S_8 is shifted downwards by $\delta\varphi l$, so that the virtual work performed here is $-S_8\frac{1}{\sqrt{2}}\delta\varphi l$. From the balance of all virtual work performed here follows

$$\delta W = F\delta\varphi l + 2F\delta\varphi l - S_8\frac{1}{\sqrt{2}}\delta\varphi l - S_8\frac{1}{\sqrt{2}}\delta\varphi l = 0. \tag{7.51}$$

This results in the required bar force S_8 as

$$S_8 = \frac{3}{\sqrt{2}} F. \tag{7.52}$$

◀

7.5 Equilibrium of Non-rigid Systems

In this section we discuss another application of the principle of virtual displacements, namely the analysis of equilibrium states in non-rigid systems. For illustration, we consider the example of Fig. 7.22. Consider a mass m suspended from a massless and inextensible rope of length l. Accordingly, the weight force mg acts, where g is the acceleration due to gravity. We now want to determine how large a horizontal force F must be so that the pendulum is in equilibrium at the angle φ to the vertical (Fig. 7.22, centre).

We apply the principle of virtual displacements and assume that an infinitesimal displacement from the equilibrium position by the infinitesimal angle $\delta\varphi$ has taken place. Due to the virtual angle $\delta\varphi$, the mass shifts by the virtual displacement $\delta w = l\delta\varphi$. The horizontal force F is therefore displaced by the virtual displacement $l\delta\varphi\cos\varphi$, whereas the weight force mg is displaced by the amount $l\delta\varphi\sin\varphi$. Accordingly, the force F performs positive virtual work, whereas the weight mg performs negative virtual work. The principle of virtual displacements is then as follows for the given example:

$$\delta W = Fl\delta\varphi\cos\varphi - mgl\delta\varphi\sin\varphi = (F\cos\varphi - mg\sin\varphi)\, l\delta\varphi = 0. \tag{7.53}$$

Zeroing the bracket term to avoid the trivial solution $\delta\varphi = 0$ finally results in the following expression for the force F:

$$F = mg\tan\varphi. \tag{7.54}$$

This result would also follow from the equilibrium of moments around the pendulum's suspension point.

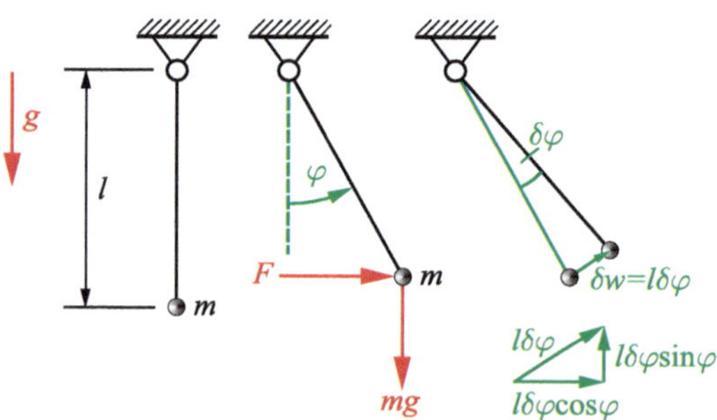

Fig. 7.22 Mathematical pendulum

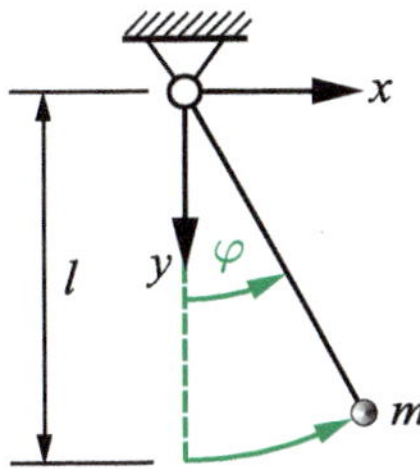

Fig. 7.23 Introduction of a coordinate system

The example shown above can also be solved using a more formal approach. To do this, we introduce a coordinate system (in Fig. 7.23 these are the $x-$ and the $y-$ axis) and describe the position vector $\underline{r}$ of the acting force F in this coordinate system, depending on the degree of freedom of the system (here the angle φ). For the location vector we then obtain:

$$\underline{r} = \begin{pmatrix} x \\ y \end{pmatrix} = \begin{pmatrix} l \sin\varphi \\ l \cos\varphi \end{pmatrix}. \tag{7.55}$$

The virtual displacement of the force application point can then be understood as an infinitesimal change in position, and the $\delta-$operator from the calculus of variations can be handled like a differential. The following then applies:

$$\delta\underline{r} = \frac{\mathrm{d}\underline{r}}{\mathrm{d}\varphi}\delta\varphi. \tag{7.56}$$

With this formal approach, the correct sign of the work does not have to be deduced beforehand, but is automatically correct. For the example we obtain:

$$\begin{aligned} \delta x &= \frac{\mathrm{d}x}{\mathrm{d}\varphi}\delta\varphi = l\cos\varphi\delta\varphi, \\ \delta y &= \frac{\mathrm{d}y}{\mathrm{d}\varphi}\delta\varphi = -l\sin\varphi\delta\varphi. \end{aligned} \tag{7.57}$$

The principle of virtual displacements then follows as (note that both forces point in positive axis directions):

$$\begin{aligned} \delta W &= F\delta x + mg\delta y \\ &= Fl\cos\varphi\delta\varphi - mgl\sin\varphi\delta\varphi = 0. \end{aligned} \tag{7.58}$$

Excluding $\delta\varphi$ and solving for the force F then leads to the already known result (7.54).

If there is a system with n degrees of freedom, then the position $\underline{r}\,(q_1, q_2, \ldots, q_n)$ of each force application point must be described as a function of the n generalised degrees of freedom, where $q_1, q_2, \ldots, q_n$ can be both displacements and rotations. The virtual displacements then follow analogously to the total differential of a function that depends on several variables:

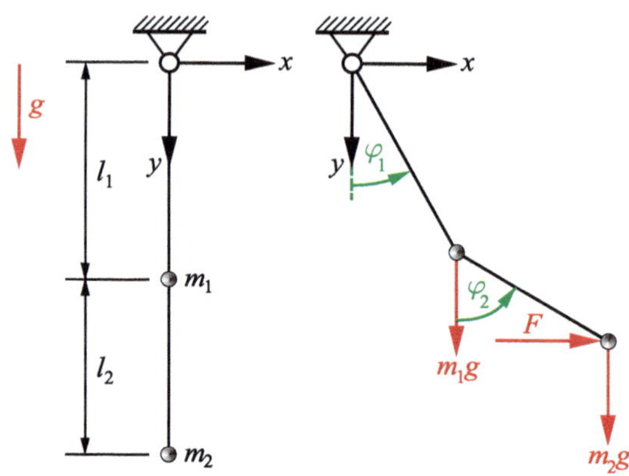

Fig. 7.24 Bar system

$$\delta \underline{r} = \frac{\mathrm{d}\underline{r}}{\mathrm{d}q_1}\delta q_1 + \frac{\mathrm{d}\underline{r}}{\mathrm{d}q_2}\delta q_2 + \cdots + \frac{\mathrm{d}\underline{r}}{\mathrm{d}q_n}\delta q_n. \tag{7.59}$$

As an example, consider the bar system in Fig. 7.24. Two point masses with the masses m_1 and m_2 are attached to two suspended massless and inextensible bars at their respective ends. The horizontal force F now acts on the lower mass. We are looking for the two angles φ_1 and φ_2 at which the system is in equilibrium. The two generalised degrees of freedom $q_1 = \varphi_1$ and $q_2 = \varphi_2$ are considered. The coordinates of the points of application of the masses are

$$\begin{aligned} x_1 &= l_1 \sin \varphi_1, \\ x_2 &= l_1 \sin \varphi_1 + l_2 \sin \varphi_2, \\ y_1 &= l_1 \cos \varphi_1, \\ y_2 &= l_1 \cos \varphi_1 + l_2 \cos \varphi_2. \end{aligned} \tag{7.60}$$

The position of the point of application of force follows as:

$$\begin{aligned} x_F &= x_2 = l_1 \sin \varphi_1 + l_2 \sin \varphi_2, \\ y_F &= y_2 = l_1 \cos \varphi_1 + l_2 \cos \varphi_2. \end{aligned} \tag{7.61}$$

The virtual displacements of the masses are then as follows:

$$\delta x_1 = \frac{dx_1}{d\varphi_1}\delta\varphi_1 + \frac{dx_1}{d\varphi_2}\delta\varphi_2 = l_1 \cos\varphi_1\delta\varphi_1,$$
$$\delta x_2 = \frac{dx_2}{d\varphi_1}\delta\varphi_1 + \frac{dx_2}{d\varphi_2}\delta\varphi_2 = l_1 \cos\varphi_1\delta\varphi_1 + l_2\cos\varphi_2\delta\varphi_2,$$
$$\delta y_1 = \frac{dy_1}{d\varphi_1}\delta\varphi_1 + \frac{dy_1}{d\varphi_2}\delta\varphi_2 = -l_1 \sin\varphi_1\delta\varphi_1,$$
$$\delta y_2 = \frac{dy_2}{d\varphi_1}\delta\varphi_1 + \frac{dy_2}{d\varphi_2}\delta\varphi_2 = -l_1 \sin\varphi_1\delta\varphi_1 - l_2\sin\varphi_2\delta\varphi_2, \quad (7.62)$$

and for the point of application of force:

$$\delta x_F = \delta x_2 = l_1\cos\varphi_1\delta\varphi_1 + l_2\cos\varphi_2\delta\varphi_2,$$
$$\delta y_F = \delta y_2 = -l_1\sin\varphi_1\delta\varphi_1 - l_2\sin\varphi_2\delta\varphi_2. \quad (7.63)$$

The principle of virtual displacements is then as follows:

$$\delta W = m_1 g\delta y_1 + m_2 g\delta y_2 + F\delta x_F = 0. \quad (7.64)$$

Inserting (7.62) and (7.63) then leads to:

$$(F\cos\varphi_1 - m_1 g\sin\varphi_1 - m_2 g\sin\varphi_1)\, l_1\delta\varphi_1$$
$$+\,(F\cos\varphi_2 - m_2 g\sin\varphi_2)\, l_2\delta\varphi_2 = 0. \quad (7.65)$$

Since the two virtual rotations $\delta\varphi_1$ and $\delta\varphi_2$ can be arbitrary and are also independent of each other, the requirement (7.65) can only be fulfilled if the two bracket terms disappear individually. The following therefore applies:

$$F\cos\varphi_1 - m_1 g\sin\varphi_1 - m_2 g\sin\varphi_1 = 0,$$
$$F\cos\varphi_2 - m_2 g\sin\varphi_2 = 0. \quad (7.66)$$

These two expressions can be solved for the two angles φ_1 and φ_2 as follows:

$$\tan\varphi_1 = \frac{F}{(m_1+m_2)\,g}, \qquad \tan\varphi_2 = \frac{F}{m_2 g}. \quad (7.67)$$

7.6 Potential

In this section we discuss the so-called potential forces. These are forces where the work performed depends only on the start and end points of the path, but not on the path travelled. Such forces are called potential forces or conservative forces. Examples of potential forces are weight and spring forces, which we will consider below.

We first consider a point mass m, which is initially located at the point A (position vector $\underline{r}_A$) (see Fig. 7.2). The point mass is now moved so that at the end of the motion it is at point

B, which has the position vector $\underline{r}_B$:

$$\underline{r}_A = \begin{pmatrix} x_A \\ y_A \\ z_A \end{pmatrix}, \quad \underline{r}_B = \begin{pmatrix} x_B \\ y_B \\ z_B \end{pmatrix}. \tag{7.68}$$

Now let the acceleration due to gravity be directed against the $z-$axis. We now consider the work differential $\mathrm{d}W$ as follows:

$$\mathrm{d}W = \underline{F}\mathrm{d}\underline{u} = \begin{pmatrix} 0 \\ 0 \\ -mg \end{pmatrix} \begin{pmatrix} \mathrm{d}x \\ \mathrm{d}y \\ \mathrm{d}z \end{pmatrix} = -mg\mathrm{d}z. \tag{7.69}$$

The work differential therefore only has a contribution with regard to the infinitesimal displacement in the $z-$ direction. The total work performed between the two points A and B is then

$$W = \int_A^B \mathrm{d}W = -mg \int_{z_A}^{z_B} \mathrm{d}z = -mg\left(z_B - z_A\right). \tag{7.70}$$

It can be seen that the total work performed between A and B corresponds to the negative weight force $-mg$ multiplied by the height difference $z_B - z_A$. Accordingly, horizontal displacements of the point mass have no influence on the performed work. The work performed here is negative if $z_B - z_A > 0$, i.e. if the point mass is displaced in the positive $z-$ direction against the effect of gravity. Similarly, the work is positive if $z_B - z_A < 0$, i.e. the mass is displaced in the opposite direction to the positive $z-$direction. The work performed here is independent of the path between A and B.

As a further example of a displacement-independent force, we consider a linear-elastic spring with the linear spring law $F_S = ku$ (Fig. 7.25), where k is the so-called spring constant, which is given in the unit of a force divided by a unit of length, e.g. in $\frac{\mathrm{N}}{\mathrm{m}}$. Let the spring be unstressed in its initial state ($u_A = 0$) and free of any spring force F_S, i.e. $F_S = 0$. We want to consider what work is performed by the spring force F_S when the spring is stretched from the unstressed state $u_A = 0$ to the length $u_B = u$. If the spring is now tensioned and a change in length u is applied, a spring force $F_S = ku$ occurs inside the spring, which is directed in the opposite direction to the change in length. The work increment is here:

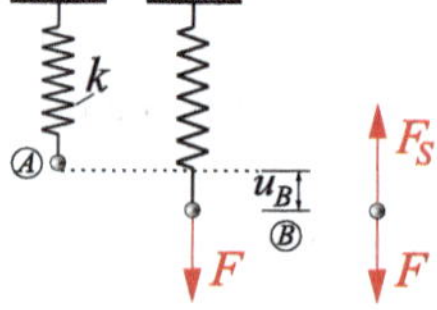

Fig. 7.25 Linear-elastic spring

$$dW = -F_S du = -kudu. \tag{7.71}$$

The total work done by the spring force is then W:

$$W = \int_A^B dW = -\int_0^u k\hat{u}d\hat{u} = -\frac{1}{2}ku^2. \tag{7.72}$$

Obviously, the work of the spring force is also a path-independent work, the knowledge of the initial and final state is sufficient for the determination of W.

The work of a linear-elastic rotational spring with the spring stiffness k_φ can be determined in the same way. Figure 7.26 shows an elastically clamped rigid beam under a force F, which deflects the beam from its horizontal position by the angle φ. Due to the rotation φ around the support point, a restoring moment $M_F = k_\varphi$ arises in the rotational spring, and the total work W of the spring moment M_S performed from the undeflected position $\varphi_A = 0$ to the deflected position $\varphi_B = \varphi$ reads:

$$W = \int_A^B dW = -\int_0^\varphi k_\varphi\hat{\varphi}d\varphi = -\frac{1}{2}k_\varphi\varphi^2. \tag{7.73}$$

This work is also only dependent on the start and end angles, but not on the path between these two positions.

At this point, we introduce the concept of potential or so-called potential energy Π. Conservative forces, i.e. forces for which the work performed does not depend on the trajectory and which are therefore path-independent, can be derived from a potential Π, which is defined as follows:

$$\Pi = -W = -\int_A^B \underline{F}d\underline{u}. \tag{7.74}$$

The concept of potential or potential energy has far-reaching significance for applied mechanics and will be of great use to us in the following Sect. 7.7 on stability problems, but also in Volumes 2 and 3.

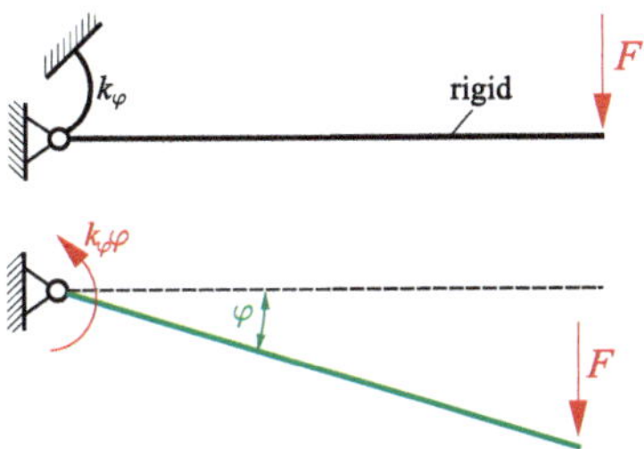

Fig. 7.26 Elastically clamped rigid bar under force F

As an example, we consider a weight that is shifted from the original position z_A to an arbitrary position z. The work performed according to (7.70) is then:

$$W = -mg\,(z - z_A)\,. \tag{7.75}$$

The potential is therefore:

$$\Pi = -W = mg\,(z - z_A)\,. \tag{7.76}$$

The weight force mg then follows from the negative potential if we derive Π with respect to z, i.e.:

$$-\frac{\mathrm{d}\Pi}{\mathrm{d}z} = -mg. \tag{7.77}$$

The negative sign is explained by the fact that mg acts in the opposite direction to the positive $z-$direction.

The potential of the spring force F_S of the linear-elastic displacement spring is calculated in the same way:

$$\Pi = -W = \frac{1}{2}ku^2. \tag{7.78}$$

Note that the potential is always positive, regardless of whether u is positive (the spring is lengthened) or negative (the spring is compressed). In both cases, energy is stored in the spring. The spring force F_S results from the negative derivative with respect to the displacement:

$$-\frac{\mathrm{d}\Pi}{\mathrm{d}u} = -ku. \tag{7.79}$$

The same applies to the linear-elastic rotational spring:

$$\Pi = \frac{1}{2}k_\varphi \varphi^2. \tag{7.80}$$

In general, it can be stated that the negative derivative of the potential with respect to the position parameter results in the potential force. Forces whose work is path-independent are potential forces. Frictional forces, on the other hand, are path-dependent and therefore not potential forces. Extending the principle of virtual displacements, it can be shown that a rigid body is in equilibrium under potential forces if the variation of the total potential vanishes with respect to the generalised coordinates $q_1, q_2,\ldots,q_n$:

$$\delta\Pi = \frac{\partial\Pi}{\partial q_1}\delta q_1 + \frac{\partial\Pi}{\partial q_2}\delta q_2 + \cdots \frac{\partial\Pi}{\partial q_n}\delta q_n = 0. \tag{7.81}$$

This then leads to n equilibrium conditions:

$$\frac{\partial\Pi}{\partial q_1} = 0, \quad \frac{\partial\Pi}{\partial q_2} = 0, \quad \ldots \quad \frac{\partial\Pi}{\partial q_n} = 0. \tag{7.82}$$

7.7 Stability of Equilibrium

The aim of the design of a structure is to ensure sufficient stiffness and strength while taking certain requirements into account. In practice, engineers are therefore often tasked with providing corresponding mathematical verifications, usually in the form of a strength verification, i.e. verification that the stresses in a structure, i.e. the internal forces, do not exceed permissible values. If we consider a bar under compression that is under the axially parallel compressive force F (Fig. 7.27), then a very important phenomenon arises. In many cases, there is no strength problem at all (i.e. the strength of the material is not reached), but several equilibrium states can occur depending on the level of the applied load, which describe different configurations of the bar. There is therefore not necessarily a clear relationship between the applied load and the resulting bar reaction. In addition to the straight configuration, configurations that are accompanied by a deflection of the bar are also conceivable. This phenomenon, i.e. the deflection of a compression member when a certain axis-parallel load is reached, is referred to as buckling. This deflection usually occurs abruptly when a certain load level $F = F_{\text{crit}}$ is reached, which we refer to as the critical load or buckling load. The overarching field of knowledge is the so-called stability theory, and the aim of a stability analysis is to determine the critical load or buckling load F_{crit} at which the previously straight bar transitions to a deflected position. The assessment of the stability of a structure and the classification of equilibrium positions is advantageously carried out using energy considerations. In Volume 2 we will go into more detail on very basic stability problems, but in the following we will consider the concept of stability in an elementary way, limiting ourselves here to systems with one degree of freedom.

We will take a closer look at the concept of stability and the associated stability of equilibrium positions below. To do this, we look at the so-called sphere analogy (see Fig. 7.28). A rigid sphere of mass m is considered, which is located on differently shaped surfaces. Gravity acts as shown. Firstly, we consider the case where the sphere lies on a concave surface (Fig. 7.28, top left). In this case, the sphere automatically comes to rest in its rest position at the lowest point of the surface and finds its natural equilibrium position there. If the position of the sphere is now briefly changed by an infinitesimal deflection and then left to its own devices (i.e. an infinitesimal displacement of the state of equilibrium is carried out), the sphere always returns to its original position under the influence of the earth's gravitational field and due to the influence of friction and resumes its state of rest at the

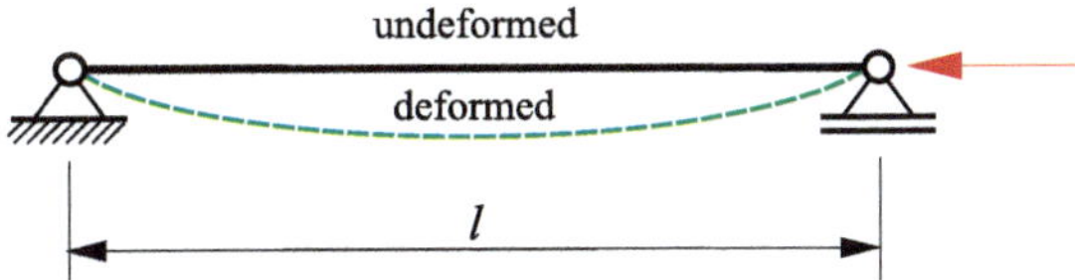

Fig. 7.27 Bar under compressive load

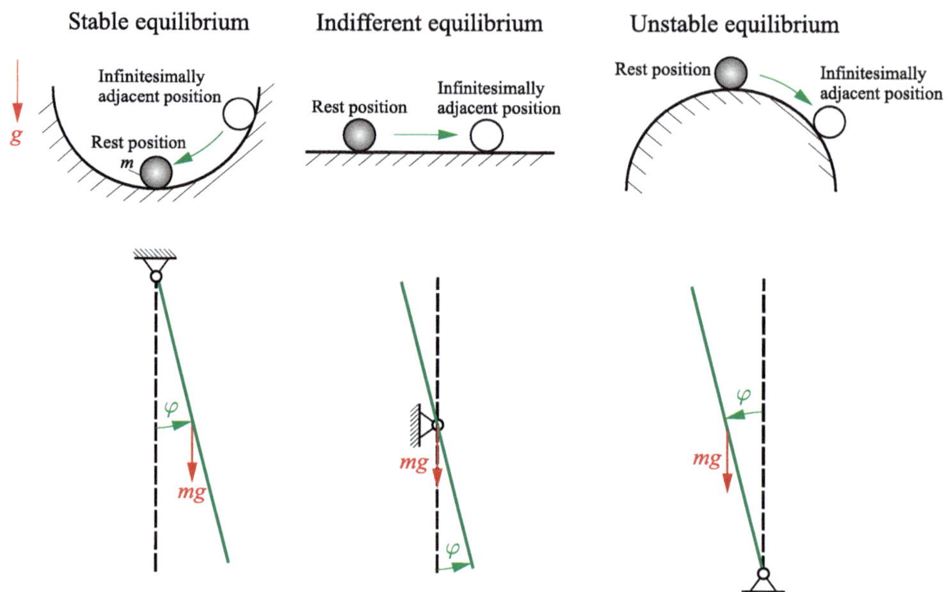

Fig. 7.28 Stability of equilibrium for a sphere on different surfaces and for the rigid bar with different support conditions

lowest point of the concave surface. Such an equilibrium position is referred to as a stable equilibrium. If the sphere is deflected upwards by δu from its equilibrium position at the lowest point of the trough, its potential changes by $\delta\Pi = mg\delta u$ and this change in potential is positive: potential energy has been gained through the upward movement.

This behaviour of the rigid sphere on a concave surface can be transferred to a bar that is hinged at its upper end and free at its lower end (see Fig. 7.28, bottom left). If an infinitesimal rotation is applied to this bar (i.e. the vertical equilibrium position is subjected to a displacement) and the bar is then left to itself, the bar will return to the vertical undeflected position by itself. This means that there is stable equilibrium in the vertical position.

We now consider the case where the sphere is in a state of equilibrium on a horizontal plane (Fig. 7.28, top centre). We now place the sphere at a different position on the plane and then leave it to its own devices. In this case, the sphere no longer returns to its old equilibrium position, but remains in its new rest position. In this case, both positions mentioned are therefore possible equilibrium positions and the equilibrium is no longer unique, but ambiguous - several equilibrium positions exist simultaneously. This type of equilibrium is referred to as indifferent. In this case, $\Delta\Pi = mgu = 0$ applies.

If this part of the sphere analogy is applied to the bar, which is now rotatably mounted exactly at its centre of gravity (Fig. 7.28, bottom centre), it can be seen that the resulting weight force mg always acts at the centre of gravity regardless of the rotation φ and the bar will therefore not return to its original vertical position by itself. In this case, an infinite number of equilibrium positions obviously exist simultaneously and there is indifferent equilibrium.

The last part of the sphere analogy concerns the case that the sphere rests on a convex surface (Fig. 7.28, top right). If the sphere is now subjected to an infinitesimal displacement and then left to its own devices, it no longer returns to its original resting position. The movement of the sphere follows the direction of the applied force. Such an equilibrium position is referred to as an unstable equilibrium. Unstable equilibrium is characterised by the fact that the ball leaves its resting position at the slightest disturbance and does not return by itself. For the change in potential, $\Delta\Pi = mg\delta u < 0$ applies here, potential energy is lost.

If we consider the bar shown in Fig. 7.28, bottom right, which is now supported at its base, this means that if this bar is rotated out of its vertical equilibrium position, it will move further and further away from its equilibrium position and will not return to its rest position by itself. The vertical equilibrium position of the bar supported at its foot point is therefore unstable.

The distinction as to whether an equilibrium position is stable, indifferent or unstable can be made by looking at the energy. Using the sphere analogy as an example, it can be seen that the potential energy of a sphere resting at the lowest point of a concave surface is a minimum - any other position is accompanied by an increase in potential energy. In concrete terms, this means that energy must be expended to move the ball from its resting position. A stable equilibrium is therefore associated with an energy minimum. Conversely, an unstable equilibrium means that energy is released when the ball is moved from its rest position. An unstable equilibrium is therefore associated with an energy maximum. This conclusion is consistent with the work principle, according to which the following must apply:

$$\delta\Pi = -\delta W = 0. \tag{7.83}$$

Since we are only considering systems with one degree of freedom x at this point, the following applies:

$$\delta\Pi = \frac{\mathrm{d}\Pi}{\mathrm{d}x}\delta x = 0. \tag{7.84}$$

Since δx is arbitrary and generally not zero, the following remains:

$$\frac{\mathrm{d}\Pi}{\mathrm{d}x} = 0. \tag{7.85}$$

If this condition is met, then an equilibrium situation exists. In the case of stable equilibrium, the potential has a minimum according to the above explanations. Similarly, there is a maximum in the case of unstable equilibrium. For the second derivative of the potential, it must therefore apply that a minimum exists if the second derivative is greater than zero:

$$\frac{\mathrm{d}^2\Pi}{\mathrm{d}x^2} > 0. \tag{7.86}$$

In this case, there is stable equilibrium. Unstable equilibrium follows analogously if the following applies:

$$\frac{d^2\Pi}{dx^2} < 0. \tag{7.87}$$

If the special case occurs that the second derivative of the potential becomes zero, i.e. if

$$\frac{d^2\Pi}{dx^2} = 0, \tag{7.88}$$

then further considerations must be carried out. Let $\Pi_0 = \Pi(x_0)$ be the potential of an equilibrium position. The potential of a neighbouring position $x_0 + \delta x$ can then be specified in the form of a Taylor series as follows:

$$\Pi(x_0 + \delta x) = \Pi_0 + \left.\frac{d\Pi}{dx}\right|_{x_0} \delta x + \frac{1}{2}\left.\frac{d^2\Pi}{dx^2}\right|_{x_0} \delta x^2 + \frac{1}{6}\left.\frac{d^3\Pi}{dx^3}\right|_{x_0} \delta x^3 + \cdots \tag{7.89}$$

The change of the potential $\Delta\Pi$ is then:

$$\Delta\Pi = \Pi(x_0 + \delta x) - \Pi_0 = \left.\frac{d\Pi}{dx}\right|_{x_0} \delta x + \frac{1}{2}\left.\frac{d^2\Pi}{dx^2}\right|_{x_0} \delta x^2 + \frac{1}{6}\left.\frac{d^3\Pi}{dx^3}\right|_{x_0} \delta x^3 + \cdots \tag{7.90}$$

Since the initial position x_0 is an equilibrium position, the first derivative of the total potential is identical to zero: $\left.\frac{d\Pi}{dx}\right|_{x_0} = 0$. The type of equilibrium then results from the type of second derivative $\frac{1}{2}\left.\frac{d^2\Pi}{dx^2}\right|_{x_0}$. However, if the second derivative is also zero, then the sign of the next higher derivative determines the type of equilibrium. However, if not only the second derivative, but also all higher derivatives are identical to zero, then the equilibrium is indifferent.

Example 7.5

Consider a massless rigid bar, which is hinged at its upper end and carries a mass m at its lower end (Fig. 7.29). For this pendulum, we want to determine possible equilibrium positions and investigate their nature.

Solution:

We consider the deflected position of the pendulum, as shown in Fig. 7.29, right. Due to the rotation φ around the support point, the weight force mg performs negative work:

$$W = -mgl\,(1 - \cos\varphi)\,. \tag{7.91}$$

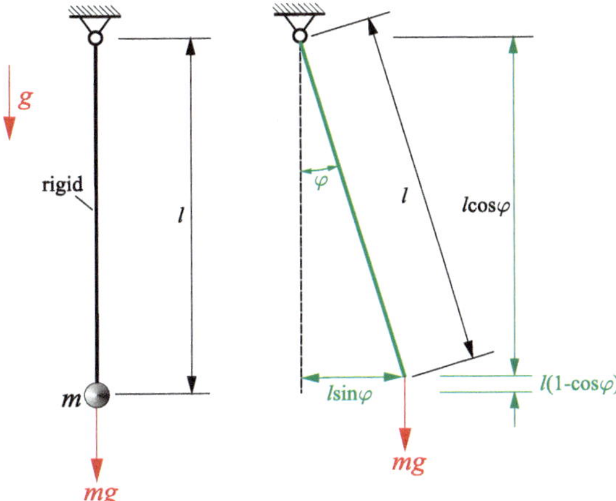

Fig. 7.29 Pendulum

Accordingly, the following potential exists:

$$\Pi = -W = mgl\,(1 - \cos\varphi)\,. \tag{7.92}$$

This corresponds to the potential energy of the mass if the reference level is set at the height of the undeflected position.

The equilibrium condition is obtained from the first derivative of the potential:

$$\frac{\mathrm{d}\Pi}{\mathrm{d}\varphi} = mgl\sin\varphi = 0. \tag{7.93}$$

This equation has two solutions, namely $\varphi_1 = 0$ (undeflected pendulum) and $\varphi_2 = \pi$ (pendulum is vertical above the support point). These positions represent the possible equilibrium positions here.

We obtain the type of equilibrium from the second derivative of the potential:

$$\frac{\mathrm{d}^2\Pi}{\mathrm{d}\varphi^2} = mgl\cos\varphi. \tag{7.94}$$

For $\varphi_1 = 0$, the second derivative of Π is positive, so that this position is in stable equilibrium. For $\varphi_2 = \pi$, on the other hand, there is a negative second derivative of Π, so that this position is an unstable equilibrium position. This result corresponds to our intuition: The pendulum would leave the position $\varphi_2 = \pi$ at the slightest disturbance, and energy would be released.

◀

Example 7.6

Let a rigid massless bar (Fig. 7.30) of length $l_1 + l_2$ be supported at a point in such a way that it can rotate freely around this point. The two masses m_1 and m_2 are attached to the ends of the bar at the distances l_1 and l_2 from the support point. Determine the equilibrium positions of this system and the types of equilibrium.

Solution:

If the reference level is measured from the horizontal position, the total potential of the given system reads:

$$\Pi = m_1 g l_1 \sin\varphi - m_2 g l_2 \sin\varphi. \tag{7.95}$$

The equilibrium condition follows from the first derivative of Π with respect to the angle φ:

$$\frac{d\Pi}{d\varphi} = (m_1 g l_1 - m_2 g l_2)\cos\varphi = 0. \tag{7.96}$$

This equation has two solutions. On the one hand, the content of the brackets can be set to zero, and we obtain:

$$m_1 l_1 = m_2 l_2. \tag{7.97}$$

This means that the given bar is always in equilibrium if $m_1 l_1 = m_2 l_2$, regardless of the angle φ. This result can be interpreted as meaning that the given system is in equilibrium for every angle φ as long as the condition (7.97) is fulfilled.

To solve (7.96), we can also demand that the cosine function becomes zero:

$$\cos\varphi = 0. \tag{7.98}$$

This is fulfilled for the angles

$$\varphi_1 = \frac{\pi}{2}, \quad \varphi_2 = \frac{3\pi}{2}. \tag{7.99}$$

Both angles describe the vertical position of the bar, whereby in the first case mass m_1 lies above the support point, whereas in the second case mass m_2 occupies the upper position.

We now analyse the type of equilibrium and form the second derivative of the potential Π:

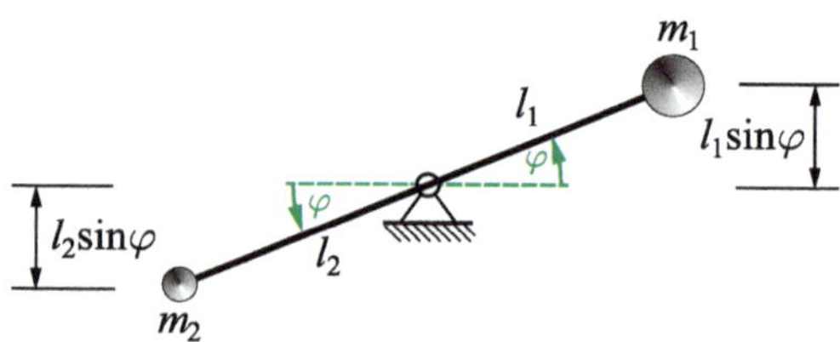

Fig. 7.30 Given system

$$\frac{d^2\Pi}{d\varphi^2} = -(m_1 g l_1 - m_2 g l_2)\sin\varphi = 0. \tag{7.100}$$

If we examine the case $m_1 l_1 = m_2 l_2$ further (see (7.97)), we find that the second derivative of the potential becomes zero. It is also easy to see that this also applies to all higher derivatives of Π as long as (7.97) is fulfilled. In this case, there is an indifferent equilibrium for any angle φ.

At this point, we assume that $m_1 l_1 > m_2 l_2$ applies. This shows that for the angle $\varphi_1 = \frac{\pi}{2}$ the second derivative of Π is negative. There is then an unstable equilibrium. For the angle $\varphi_2 = \frac{3\pi}{2}$, on the other hand, the second derivative of Π is positive, so that there is stable equilibrium. This statement is reversed if $m_1 l_1 < m_2 l_2$.

◀

Example 7.7

Let an ideally straight, ideally rigid, elastically supported massless bar of length l be loaded at its upper end by a compressive force F (Fig. 7.31). At its lower end, the bar is simply supported and also supported by a linear-elastic rotational spring (spring stiffness k_φ). The acting force F is directionally stable, i.e. it does not change its direction, regardless of the rotation of the bar. In this example, we want to investigate which equilibrium positions are possible depending on the magnitude of the force and what type they are.

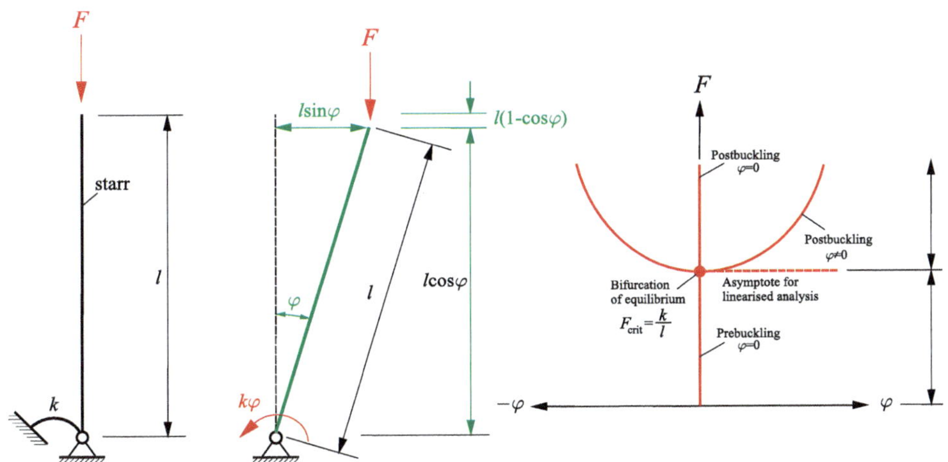

Fig. 7.31 Elastically supported, ideally rigid bar under compressive load F (left), deflected position (centre), force-rotation diagram (right)

Solution:

We consider the bar in the deflected state and determine the total potential Π. This is made up of a part due to the applied force F and a part due to the rotational spring. Assuming the zero level at height of the applied force, the result is

$$\Pi = \frac{1}{2}k\varphi^2 - Fl\,(1 - \cos\varphi)\,. \tag{7.101}$$

This means that the spring stores energy while the force F has lost potential energy, which explains the negative sign of the corresponding term in (7.101).

The given bar is a system with one degree of freedom φ, so that for equilibrium applies:

$$\frac{\partial\Pi}{\partial\varphi} = k\varphi - Fl\sin\varphi = 0. \tag{7.102}$$

This expression represents the equilibrium condition for the deflected bar. The type of equilibrium follows from the second derivative of the total potential Π:

$$\frac{\partial^2\Pi}{\partial\varphi^2} = k - Fl\cos\varphi. \tag{7.103}$$

Let us first consider the case of the undeflected bar, for which we set $\varphi = 0$, which is an obvious solution for (7.102):

$$\frac{\partial^2\Pi}{\partial\varphi^2} = k - Fl. \tag{7.104}$$

Three cases can be distinguished here:

- If $F < \frac{k}{l} = F_{\text{crit}}$, then $\frac{\partial^2\Pi}{\partial\varphi^2} > 0$. Then the total potential for $0 \le F < F_{\text{crit}}$ in the equilibrium position $\varphi = 0$ is a minimum. For the change to another possible position position, energy would have to be expended. Therefore, the equilibrium position $\varphi = 0$ for $0 \le F < F_{\text{crit}}$ is a stable equilibrium position. The force F_{crit} is referred to as the critical force for the bar under compressive axial load.
- For $F > \frac{k}{l} = F_{\text{crit}}$ $\frac{\partial^2\Pi}{\partial\varphi^2} < 0$ applies. The change to any other possible adjacent equilibrium position therefore means a loss of energy. Therefore, the equilibrium position $\varphi = 0$ for $F > \frac{k}{l} = F_{\text{crit}}$ is unstable. This means that for every small displacement from the straight position, the bar will immediately move to a neighbouring position and remain in this position.
- For the case $F = \frac{k}{l} = F_{\text{crit}}$ $\frac{\partial^2\Pi}{\partial\varphi^2} = 0$ a change to a possible neighbouring position is not associated with a loss of energy, nor is energy required for this. For $\varphi = 0$ at $F = \frac{k}{l} = F_{\text{crit}}$ there is indifferent equilibrium.

We now also investigate which equilibrium state follows for $\varphi \neq 0$. This state can only occur for forces $F > \frac{k}{l} = F_{\text{crit}}$. From $\frac{\partial\Pi}{\partial\varphi} = k\varphi - Fl\sin\varphi = 0$ follows $Fl = k\frac{\varphi}{\sin\varphi}$, which after insertion into $\frac{\partial^2\Pi}{\partial\varphi^2}$ results in the following expression:

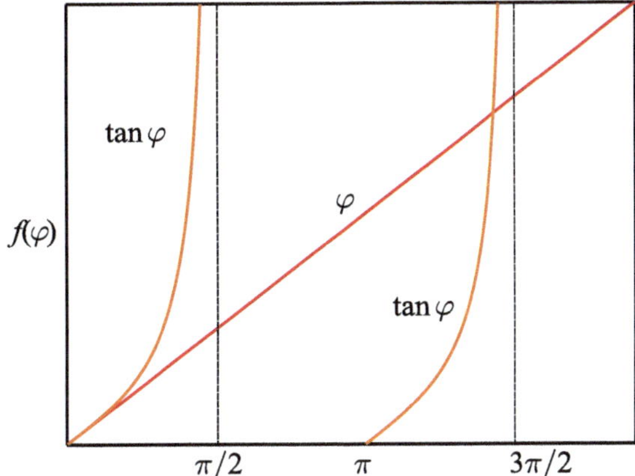

Fig. 7.32 Representation of the functions $f(\varphi) = \varphi$ and $f(\varphi) = \tan(\varphi)$

$$\frac{\partial^2 \Pi}{\partial \varphi^2} = k\left(1 - \frac{\varphi}{\tan\varphi}\right). \tag{7.105}$$

Here, $\frac{\varphi}{\tan\varphi} < 1$ (Fig. 7.32) applies if $|\varphi| < \frac{\pi}{2}$, and thus $\frac{\partial^2 \Pi}{\partial \varphi^2} > 0$ also applies for every angle $\varphi \neq 0$. The equilibrium position $\varphi \neq 0$ is therefore a stable equilibrium position.

The equilibrium types are shown in Fig. 7.33. The range $F < F_{\text{crit}}$ is also referred to as the prebuckling range. For $F > F_{\text{crit}}$ the bar is in the so-called postbuckling range. If F corresponds exactly to the critical load, this is referred to as a bifurcation of the equilibrium.

This example of a bar under compressive load belongs to the class of so-called buckling problems, which we will discuss in detail in Volume 2. The determination of critical loads of bars under compression must be given great attention in practical application and is an important task of engineers.

◀

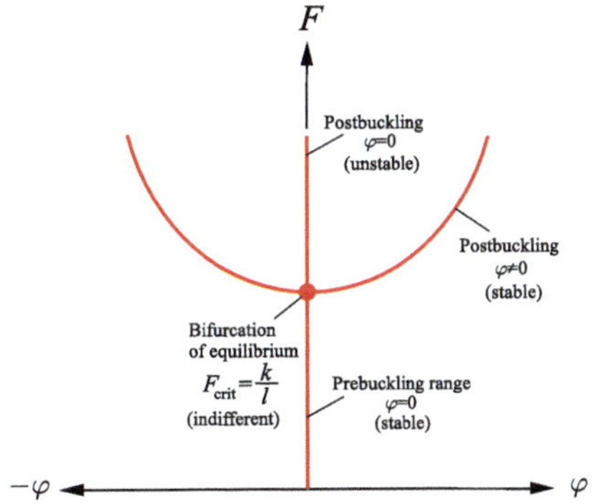

Fig. 7.33 Stable, unstable and indifferent equilibrium states

Example 7.8

We consider the bar system as shown in Fig. 7.34, top. We consider two ideally rigid and ideally straight bars of length l that are connected to each other by a hinge. A vertical force F acts in the joint. The left bar is supported at its lower end by a two-valued bearing, whereas the right bar is supported by a single-valued bearing. In addition, there is a linear elastic displacement spring with stiffness k on the right support, which supports the right bar laterally. In the unloaded state, both bars have an angle of inclination α to the horizontal, and in this state, the spring is unloaded. We want to determine the equilibrium positions and their type for the given situation.

Solution:

In the loaded state, the right support will shift to the right and the spring will be compressed. The rotation angle of the bar is denoted as φ (Fig. 7.34, bottom), which represents the only degree of freedom for this system. The angle φ is restricted by the requirement $-\frac{\pi}{2} \leq \varphi \leq \frac{\pi}{2}$. Due to the deformation, energy is stored in the spring, which can be calculated as follows

$$\Pi_{\text{spring}} = \frac{1}{2} k \, (2l)^2 \, (\cos\varphi - \cos\alpha)^2 = 2kl^2 \, (\cos\varphi - \cos\alpha)^2 \, . \tag{7.106}$$

The potential of the force F reads:s

$$\Pi_F = -Fl \, (\sin\alpha - \sin\varphi) \, . \tag{7.107}$$

In a state of equilibrium, the first derivative of the total potential disappears:

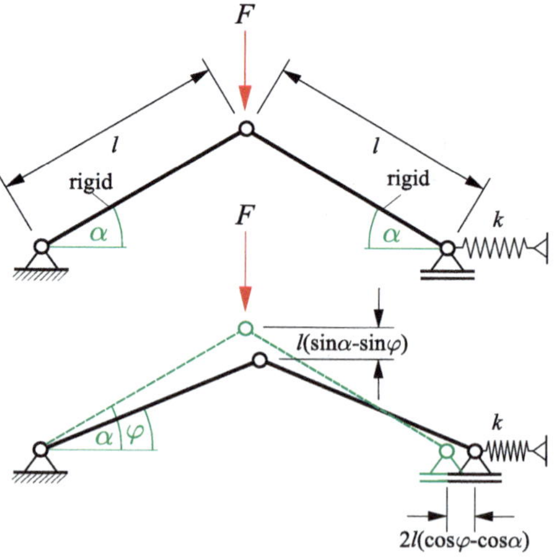

Fig. 7.34 Bar system under force F

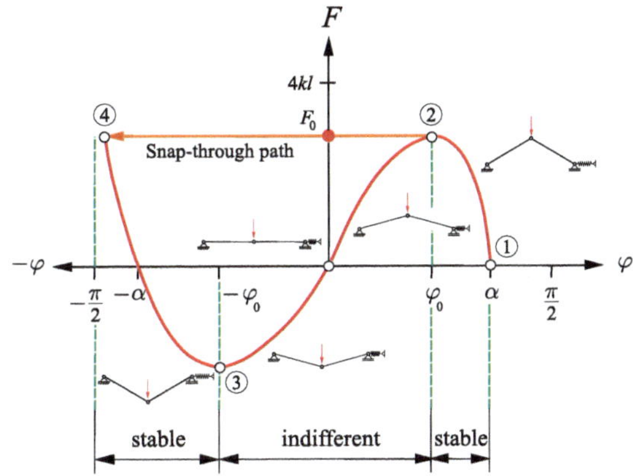

Fig. 7.35 Force-rotation diagram

$$\delta\Pi = \frac{\partial \Pi}{\partial \varphi} = -4kl^2 (\cos\varphi - \cos\alpha)\sin\varphi + Fl\cos\varphi = 0. \tag{7.108}$$

This results in the following expression for the force F:

$$F = 4kl(\cos\varphi - \cos\alpha)\tan\varphi. \tag{7.109}$$

The force F as a function of the angle φ (i.e. $F(\varphi)$) is shown in Fig. 7.35. In the initial state $\varphi = \alpha$, the force F is identical to zero (point 1). It increases as the angle φ decreases until it reaches a maximum at $\varphi = \varphi_0$ with $F = F_0$ (point 2). A further increase in load beyond this point is not possible, the system 'snaps through' until a new position is assumed at point 4. This is also referred to as a snap-through problem. All equilibrium positions between points 2 and 3 are unstable, and only above point 3 do stable equilibrium positions occur again.

Since the inclination of the curve $F(\varphi)$ is identical to zero at point 2, the angle φ_0 can be calculated as follows:

$$\frac{\partial F}{\partial \varphi} = -4kl\sin\varphi\tan\varphi + 4kl(\cos\varphi - \cos\alpha)\frac{1}{\cos^2\varphi} = 0. \tag{7.110}$$

The angle φ_0 follows from this as:

$$\cos^3\varphi_0 = \cos\alpha. \tag{7.111}$$

The associated force F_0 follows with (7.109):

$$\begin{aligned} F_0 &= 4kl(\cos\varphi_0 - \cos\alpha)\tan\varphi_0 \\ &= 4kl\left(\cos\varphi_0 - \cos^3\varphi_0\right)\tan\varphi_0 \\ &= 4kl\left(1 - \cos^2\varphi_0\right)\sin\varphi_0 = 4kl\sin^3\varphi_0. \end{aligned} \tag{7.112}$$

We determine the type of equilibrium from the second derivative of the total potential with respect to the angle φ:

$$\frac{\partial^2 \Pi}{\partial \varphi^2} = 4kl^2 \sin^2 \varphi - 4kl^2 (\cos \varphi - \cos \alpha) \cos \varphi - Fl \sin \varphi. \tag{7.113}$$

With (7.109) and taking into account $\sin^2 \varphi + \cos^2 \varphi = 1$ we obtain:

$$\frac{\partial^2 F}{\partial \varphi^2} = 4kl^2 \left(\frac{\cos \alpha}{\cos \varphi} - \cos^2 \varphi \right). \tag{7.114}$$

It is of particular interest to investigate the behaviour with respect to the angle φ_0, for which $\cos \varphi_0 = \sqrt[3]{\cos \alpha}$ and $\cos^2 \varphi_0 = \sqrt[3]{\cos^2 \alpha}$. Inserting into (7.114) shows that the equilibrium is stable for $\varphi_0 < \varphi \leq \alpha$ and $-\varphi_0 > \varphi > -\frac{\pi}{2}$. Unstable equilibrium occurs for $\varphi_0 > \varphi > -\varphi_0$. The bar would therefore switch to one of the two stable equilibrium positions for an equilibrium position in the interval $\varphi_0 > \varphi > -\varphi_0$ even with the smallest perturbation.

◀

7.8 Influence Lines for Forces and Moments in Statically Determinate Systems

An influence line is the graphical representation of a force or a moment in a static system in the case that the system is loaded by a moving load. The force or moment sought for every conceivable position of the moving load is then plotted over the entire considered system, and the resultant graph is the so-called influence line. As an introduction, we consider a beam on two supports (Fig. 7.36), which is under a moving load $F = 1$. We want to determine the influence line for the support force in the support point B (abbreviated as IL B). This influence line is already shown in Fig. 7.36, bottom.

The required influence line IL B can be determined elementarily from a simple observation of the given system. In this simple case, it is clear that the support force B assumes the value zero when the moving load is exactly above the left support. Similarly, it can be

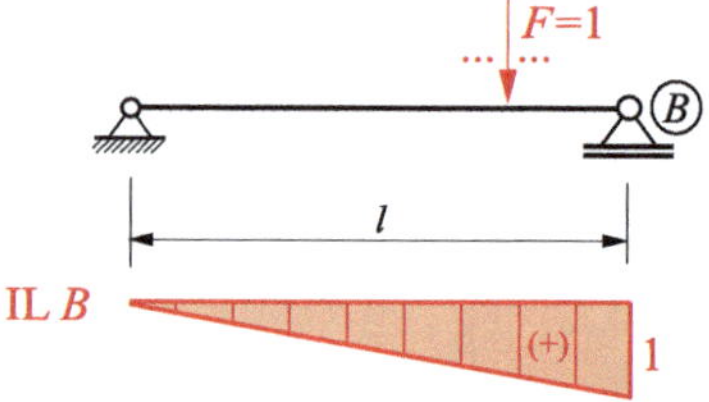

Fig. 7.36 Beam on two supports under moving load (top) and influence line for the support force B (bottom)

immediately concluded that the support force B assumes the value $B = 1$ if the moving load $F = 1$ is located exactly above the support point B. A linear line of influence IL B results between these two values.

For more complex systems than the one shown in Fig. 7.36 we need a general approach to determine such influence lines. Here we use the principle of virtual displacements and set the forces and moments for which we want to determine the influence line free, whereby the system becomes kinematically displaceable and a unique kinematic chain is created. The influence line we are looking for then corresponds exactly to the resulting kinematic chain, as we will show later. Since the kinematic chain is composed of the individual beam segments, all of which remain undeformed in themselves, the influence lines for forces and moments in statically determinate systems are always linear.

The procedure now consists of removing the bond that is energetically related to the quantitiy to be determined and applying the force or moment thus released to the system. Based on this, the kinematic chain of the system is determined, whereby we must ensure that the displacement variable assigned to the force variable being searched for assumes exactly the value -1 or that the released force variable performs negative work along the unit displacement variable. Then the influence line to be determined for the force quantity under consideration corresponds exactly to the resulting kinematic chain. This is referred to as Land's[2] theorem.

The proof of Land's theorem is shown in Fig. 7.37. Consider a beam on two supports, which is loaded by a moving load $F = 1$. We are looking for the bending moment M_S at an arbitrary point x. To do this, we introduce a moment hinge at point x that releases the required bending moment on both sides of the hinge. If we assume that the hinge rotates virtually by the angle $\delta\varphi$, then the moving load experiences the virtual displacement δw. The virtual work performed is then

$$\delta W_a = M_S \delta\varphi + F \delta w = 0. \tag{7.115}$$

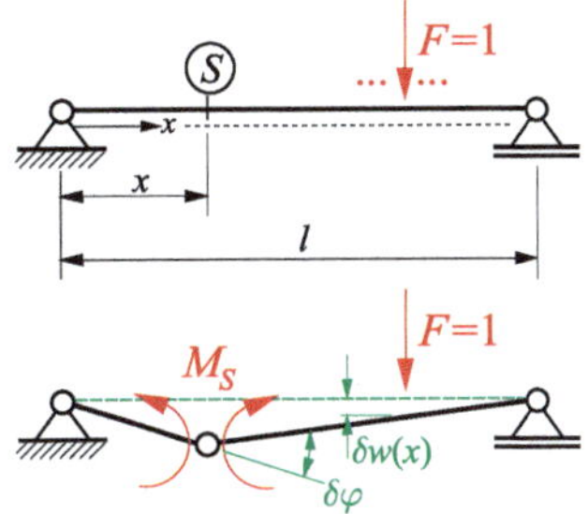

Fig. 7.37 On Land's theorem

[2] Robert Land, 1857–1899, German civil engineer.

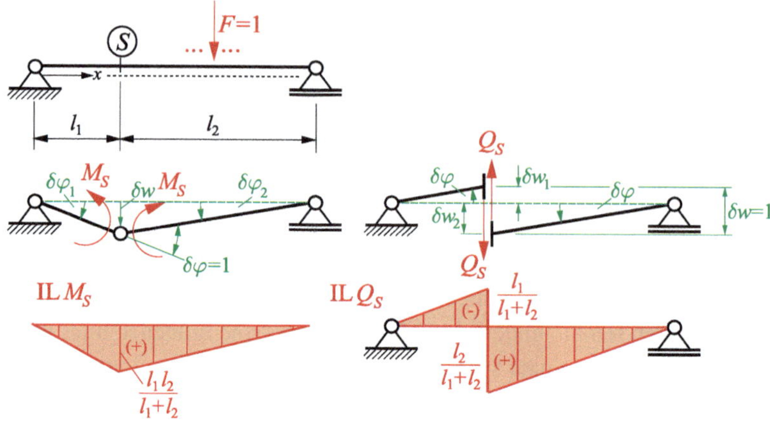

Fig. 7.38 Influence lines for a beam on two supports under moving load

Since we have assumed that the moving load is a unit load $F = 1$, we obtain

$$M_S \delta\varphi + \delta w = 0. \tag{7.116}$$

If we now apply the virtual rotation with $\delta\varphi = -1$ as described above, we obtain

$$-M_S + \delta w = 0, \tag{7.117}$$

or

$$M_S = \delta w. \tag{7.118}$$

This proves Land's theorem. It states that the influence line for a force or moment of a statically determinate system is identical to the kinematic chain resulting from the loosening of the corresponding bond when a correspondingly assigned unit deformation is applied.

As an elementary example, the beam in Fig. 7.38 under the moving load $F = 1$ is considered. The influence lines for the bending moment M_S and the shear force Q_S at point S are to be determined.

To determine the influence line IL M_S, the bending moment M_S is released and the virtual rotation angle is assumed to be $\delta\varphi = 1$, so that the released bending moment performs negative virtual work. The virtual hinge displacement δw can be used to determine a relationship between the angles $\delta\varphi_1$ and $\delta\varphi_2$:

$$\delta w = \delta\varphi_1 l_1 = \delta\varphi_2 l_2. \tag{7.119}$$

Hence:

$$\delta\varphi_2 = \delta\varphi_1 \frac{l_1}{l_2}. \tag{7.120}$$

The sum of these angles must be $\delta\varphi = 1$, i.e. $\delta\varphi_1 + \delta\varphi_2 = 1$. From this follows:

$$\delta\varphi_1\left(1+\frac{l_1}{l_2}\right)=1. \tag{7.121}$$

This can be solved for $\delta\varphi_1$:

$$\delta\varphi_1=\frac{l_2}{l_1+l_2}. \tag{7.122}$$

For $\delta\varphi_2$ thus follows:

$$\delta\varphi_2=\frac{l_1}{l_1+l_2}. \tag{7.123}$$

The virtual displacement δw of the hinge is then calculated as follows:

$$\delta w=\frac{l_1 l_2}{l_1+l_2}. \tag{7.124}$$

According to Land's theorem, the influence line IL M_S corresponds exactly to the kinematic chain that is created by the loosening of the corresponding bond. It is shown in Fig. 7.38, bottom left.

The influence line IL Q_S is determined in a very similar way, for the calculation of which a shear force joint is introduced at point S. This means that although virtual transverse displacements are possible at this point, virtual angular rotations and longitudinal displacements are prevented. The system now experiences the virtual total displacement $\delta w=1$, whereby the directions of the two partial displacements are chosen so that the shear force Q_S caused by the shear force joint performs negative virtual work. The two beam segments each experience identical virtual rotation angles $\delta\varphi$ around their respective support points. We denote the partial displacements as δw_1 and δw_2, their sum corresponds exactly to the applied displacement $\delta w=1$:

$$\delta w_1+\delta w_2=1. \tag{7.125}$$

In addition, the following relationships apply between δw_1, δw_2 and $\delta\varphi$:

$$\delta w_1=\delta\varphi l_1, \quad \delta w_2=\delta\varphi l_2. \tag{7.126}$$

This results in three equations for the three unknown virtual kinematic quantities. The result is:

$$\delta\varphi=\frac{1}{l_1+l_2}, \quad \delta w_1=\frac{l_1}{l_1+l_2}, \quad \delta w_2=\frac{l_2}{l_1+l_2}. \tag{7.127}$$

The influence line IL Q_S can be constructed from this, it is shown in Fig. 7.38, bottom right.

Example 7.9

Let us consider the beam discussed in Example 7.1, for which we now want to determine several influence lines under moving load (Fig. 7.39).

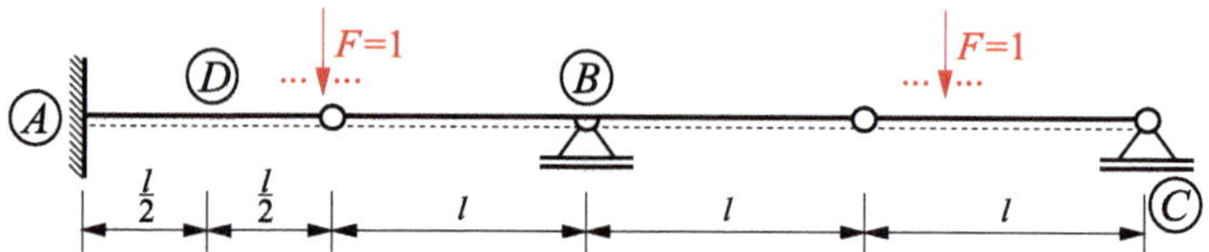

Fig. 7.39 Beam under moving load

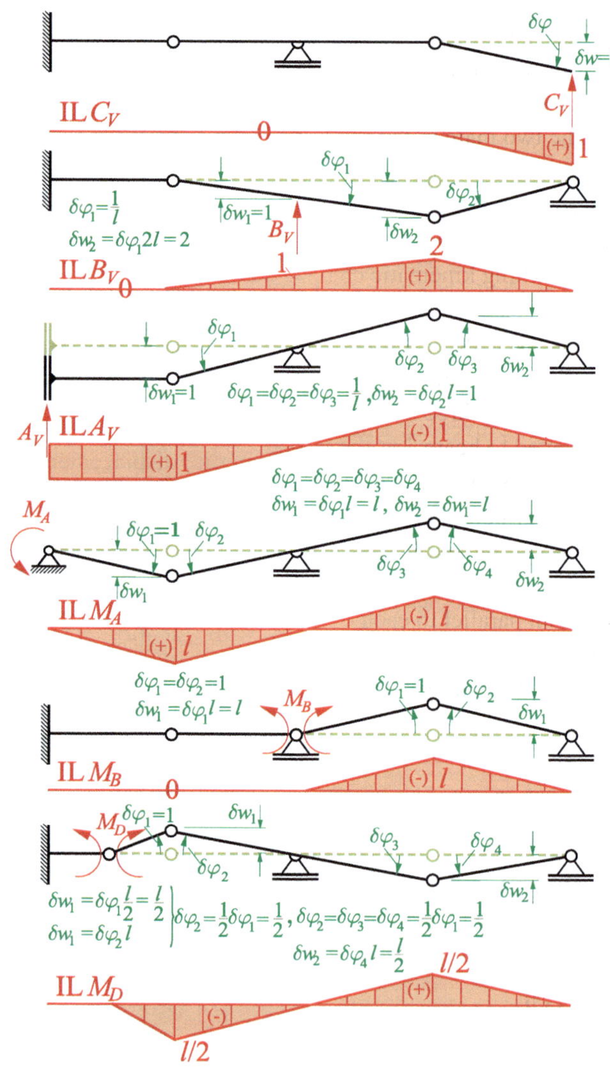

Fig. 7.40 Kinematic chains and influence lines

Solution:

The influence lines are shown in Fig. 7.40 next to the kinematic chains. A discussion of the solution is omitted at this point.

◀

Friction 8

In this chapter, we consider the topic of friction. After a basic introduction, we will deal with so-called Coulomb friction and consider both static friction and kinetic friction. The chapter concludes with the treatment of belt friction.

8.1 Fundamentals

Friction is an everyday phenomenon and is familiar to everyone from experience. For motivation, we consider a body with a rough surface lying on a rough base (Fig. 8.1). The body, which has the weight force mg (m =mass, g =acceleration due to gravity), is loaded by a force F that endeavours to pull the body to the right. It is clear that the state of the body depends on the magnitude of the horizontal force F acting on it. If the force F is sufficiently small, the body remains at rest and does not move. In fact, due to the surface roughness, a tangential force occurs in the contact area between the body and the surface, which we want to denote as H in the event that the body remains at rest. This state is called static friction: The body adheres to the surface. The force H, which can be visualised in a free-body diagram (Fig. 8.1, bottom left), is directed against the potential direction of movement x and is referred to as static friction force. The force balances can be formulated and obtained from the free-body diagram as follows:

$$\begin{aligned}\sum H = 0: &\quad H = F,\\ \sum V = 0: &\quad N = mg.\end{aligned} \tag{8.1}$$

C. Mittelstedt, *Engineering Mechanics 1: Statics*,
https://doi.org/10.1007/978-3-662-71852-0_8

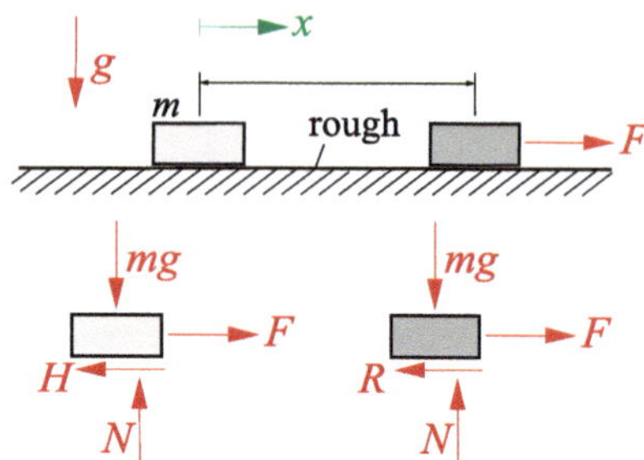

Fig. 8.1 Bodies on a rough surface (top), free-body diagrams (bottom)

The exact position of the vertical contact force N can be obtained from the moment balance, but this is not shown here.

However, experience also shows that the body starts to move when the force F exceeds a certain value. Here, too, a tangential force is generated in the contact surface between the body and the surface due to the surface roughness, which is referred to as the kinetic friction force R: The body slides on the surface. The kinetic friction force R is directed in the opposite direction to the movement, it is a result of the movement and must be overcome in order for the movement to be possible.

It is common to distinguish between the following friction states. The static friction state exists when the force F is below a limit value. The adhesive force H then occurs in the contact surface between the body and the surface. The body remains at rest and is in static equilibrium. If the force F is increased further and further, a clearly defined limit force H_0 is reached in the contact surface. In this state, the body is not yet moving. The adhesion condition $|H| \leq H_0$ therefore applies to static friction. However, if the force is increased slightly further to a maximum value F_{max}, the body is accelerated and starts to move. In this accelerated state, the force in the contact area corresponds to the kinetic friction force R_0 and there is a change from static to kinetic friction. At the end of the acceleration phase, the body enters a state of uniform motion (stationary kinetic friction) and the force in the contact surface is the kinetic friction force R, which is less than R_0 but also less than the static friction force H. In the state of uniform motion, F and R are in equilibrium with each other.

Both phenomena described here, i.e. static friction and kinetic friction, have in common that they are caused by the roughness of the bodies involved. Due to the roughness, adhesion and thus a balance between the force F and the static friction force H is only possible in the first place, whereby H represents a reaction force. The occurrence of friction would also not be possible without the roughness of the surfaces. It is clear that both the phenomenon of static friction and that of kinetic friction depend on the nature of the surfaces of the bodies involved. Friction can be desirable, as we can see from our own locomotion on foot: Without friction between the sole and the ground, walking would be impossible. However, friction is also undesirable in many technical systems and leads to a drop in performance and wear.

8.2 Coulomb's Theory of Friction

The formulation of friction laws goes back to Coulomb,[1] among others, and takes the thought experiment considered above into account. The body on a rough surface under the force F experiences a tangential force H in the area of contact with the surface. As long as the force F remains below a limit value, $H = F$, i.e. the body is in static equilibrium (see (8.1)) and remains at rest. The body remains at rest until a limit value H_0 is exceeded. It is known from experiments that the limit force H_0 is proportional to the contact force N, which is described by the so-called coefficient of static friction μ_0:

$$H_0 = \mu_0 N. \tag{8.2}$$

The coefficient of static friction depends on the nature of the surfaces involved and the material pairing as well as lubrication, but not on the size of the surface. The static friction condition follows directly from this

$$|H| \leq H_0 = \mu_0 N, \tag{8.3}$$

i.e. the body adheres as long as the magnitude of the tangential force H in the contact area does not exceed the value of the limiting adhesive force H_0. We use the magnitude of H in (8.3) because it is not always possible to predict in advance in which direction H will point in complex tasks.

Table 8.1 contains selected numerical values for the coefficient of static friction μ_0 for different material pairings.

We can transfer the previous findings on static friction to the case of kinetic friction, i.e. the case where the body is in uniform motion and friction occurs in the contact surface

Table 8.1 Coefficients of static and kinematic friction

Material pairing	Coefficient of static friction	Coefficient of kinetic friction
Steel on ice	0.03	0.015
Steel on teflon	0.04	0.04
Ski on snow	0.1...0.3	0.04...0.2
Steel on steel	0.15...0.5	0.1...0.4
Leather on metal	0.4	0.3
Wood on steel	0.5	0.15
Wood on wood	0.5	0.3
Car tire on street	0.7...0.9	0.5...0.8

[1] Charles Augustine de Coulomb, 1736–1806, French physicist.

between the body and the surface. The acting force F has therefore reached a level to cause this motion and overcome the static friction of the body. Accordingly, the tangential frictional force R occurs in the contact area (Fig. 8.1), which is directed in the opposite direction to the motion. Since the motion is uniform, i.e. unaccelerated, the following equilibrium conditions apply as in (8.1), whereby the static friction force H is now replaced by the kinetic friction force R, which is caused by the motion:

$$\sum H = 0: \quad R = F,$$
$$\sum V = 0: \quad N = mg. \tag{8.4}$$

In the case of kinetic friction, it can also be assumed in good approximation that there is a linear relationship between the kinetic friction force R and the contact force N as follows, which in turn is independent of the size of the contact surface and also of the velocity of movement:

$$R = \mu N. \tag{8.5}$$

We refer to the coefficient μ occurring here as the coefficient of kinetic friction. It is generally smaller than the static friction coefficient μ_0. Some numerical values are provided for different material pairings in Table 8.1.

Coulomb's law of static friction considered here is amenable to a particularly clear geometric interpretation. This is associated with the concept of the so-called static friction wedge or the so-called static friction cone. The static friction wedge or static friction cone indicates the maximum angle of inclination φ of a force F with respect to the vertical so that static friction is still possible and equilibrium is guaranteed. We consider Fig. 8.2, which shows a force F acting on a rough surface (coefficient of static friction μ_0) at an angle φ to the vertical. This force F causes an equivalent reaction force W, which acts at the same angle φ. This reaction force can be divided into the normal force N and the static friction force H. We now increase the static friction force until the limit state $H = H_0$ is reached. This is shown in Fig. 8.2, on the right. The force W_0, which is composed of H_0 and N_0, then acts at the angle φ_0, which can be determined as follows:

$$\tan\varphi_0 = \frac{H_0}{N_0}. \tag{8.6}$$

The following also applies:

$$\tan\varphi = \frac{H}{N}. \tag{8.7}$$

If this is inserted into the static friction law (8.2), we obtain:

$$\frac{H_0}{N_0} = \mu_0 = \tan\varphi_0 \geq \tan\varphi = \frac{H}{N}. \tag{8.8}$$

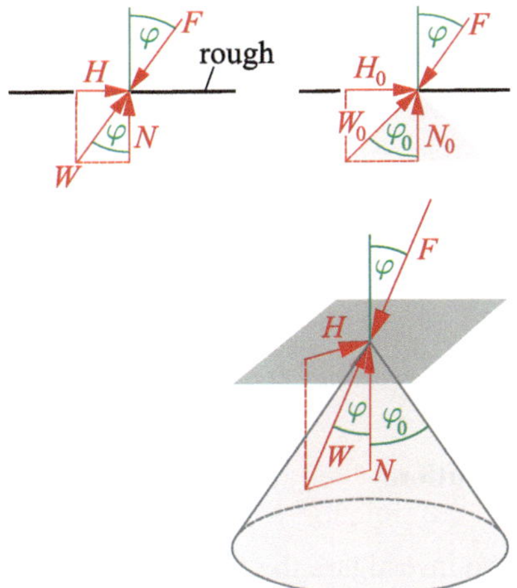

Fig. 8.2 Force F under inclination angle φ (left), static friction wedge (right)

Fig. 8.3 Static friction cone

Accordingly, $\varphi_0 \geq \varphi$ applies. The angle φ_0 is referred to as the angle of static friction. Accordingly, a system remains at rest, i.e. in a state of static friction, if the angle φ of a force is less than or equal to φ_0. If, on the other hand, φ_0 is exceeded, the body starts to move and kinetic friction occurs. The angle of static friction φ_0 results from (8.8) as:

$$\varphi_0 = \arctan \mu_0. \tag{8.9}$$

Interpreted graphically, this means that static friction occurs when the force W is within the area shown in Fig. 8.2, on the right, highlighted in grey. This area is referred to as the static friction wedge. However, if W is outside this area, kinetic friction occurs and the body starts to move.

The static friction law (8.2) can also be interpreted accordingly for the spatial case (Fig. 8.3). If the static friction wedge is rotated around the direction of action of the normal force N, the so-called static friction cone results in space. If the force W is located within this cone with the opening angle $2\varphi_0$, then static friction occurs.

Example 8.1

Consider a body with the weight force mg resting on a rough plane inclined at an angle α to the horizontal (Fig. 8.4) with coefficient of static friction μ_0. We want to determine the maximum angle α at which the plane may be inclined before the body starts to move and kinetic friction occurs.

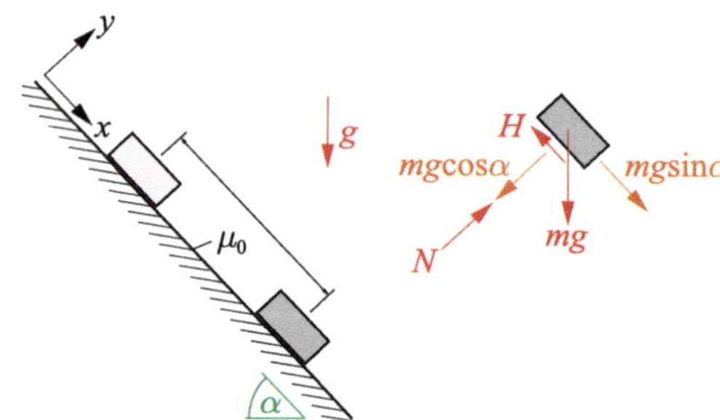

Fig. 8.4 Body on rough inclined plane

Solution:

We investigate the free-body diagram of Fig. 8.4, right, and form the sums of forces with respect to the $x-$ and $y-$directions:

$$\sum F_x = 0: \quad mg\sin\alpha - H = 0 \quad \rightarrow \quad H = mg\sin\alpha,$$
$$\sum F_y = 0: \quad N - mg\cos\alpha = 0 \quad \rightarrow \quad N = mg\cos\alpha. \tag{8.10}$$

We assume static friction so that the following condition applies:

$$H \leq \mu_0 N. \tag{8.11}$$

Inserting (8.10) gives the following result:

$$mg\sin\alpha \leq \mu_0 mg\cos\alpha \quad \rightarrow \quad \tan\alpha \leq \mu_0. \tag{8.12}$$

It can be seen that the critical angle depends exclusively on the coefficient of static friction μ_0, but not on the magnitude of the weight force mg. We obtain the following result:

$$\alpha = \arctan\mu_0. \tag{8.13}$$

Apparently, an increasing coefficient of static friction μ_0 leads to an increasing angle α which is a plausible result.

The above result can be interpreted as follows (Fig. 8.5). As long as the angle is smaller than the angle of static friction φ_0 (Fig. 8.5, left), the force W lies within the static friction wedge and the body remains at rest. If the angle α assumes exactly the value φ_0, the body remains at rest (Fig. 8.5, centre), and the force $H_0 = \mu_0 N_0$ acts. However, if the angle φ_0 is exceeded (Fig. 8.5, right), the body starts to move and the kinetic friction force R acts.

◀

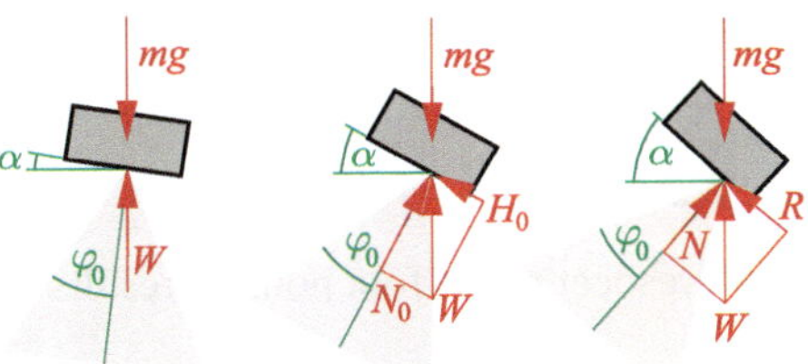

Fig. 8.5 Free-body diagram for different angles α

Example 8.2

A person with the weight force mg climbs up a ladder of length l (Fig. 8.6), which rests on a rough base at the base point A and leans against a rough wall at the top point B. The base and wall both have the static friction coefficient μ_0. Let the position of the person, measured from the base point A, be x. How large can the maximum distance $x_{\max}$ from the foot point A be before the ladder slips and a state of kinetic friction occurs?

Solution:

We consider the free-body diagram of Fig. 8.6, right, in which we have drawn the reaction forces N_A and N_B as well as the static friction forces H_A and H_B at the head and foot points. We determine the sum of the horizontal forces and obtain

$$\overset{\rightarrow}{\sum} = 0: \quad N_B - H_A = 0. \tag{8.14}$$

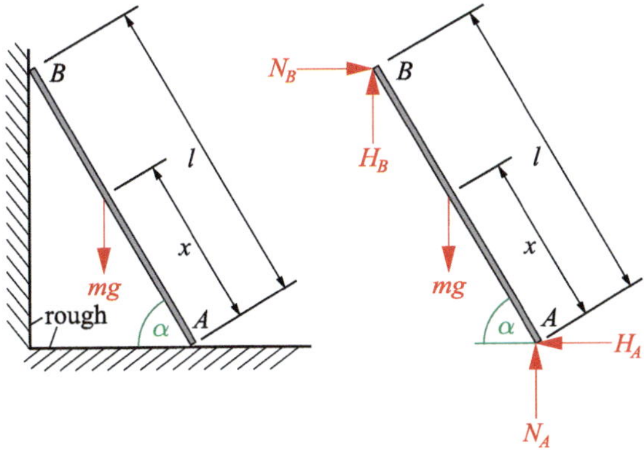

Fig. 8.6 Given situation

The sum of the vertical forces is as follows:

$$\sum^{\uparrow} = 0: \quad H_B - mg + N_A = 0. \tag{8.15}$$

The sum of moments with respect to the foot point A results in:

$$\sum^{\curvearrowleft} = 0: \quad mgx\cos\alpha - H_B l\cos\alpha - N_B l\sin\alpha = 0. \tag{8.16}$$

If we also consider the limit cases of static friction for the head and foot points, then the following applies:

$$H_A = \mu_0 N_A, \quad H_B = \mu_0 N_B. \tag{8.17}$$

This provides five equations for the five unknown variables N_A, N_B, H_A, H_B, x that can be solved as follows:

$$\begin{aligned} N_A &= \frac{mg}{1+\mu_0^2}, \\ N_B &= \frac{\mu_0 mg}{1+\mu_0^2}, \\ H_A &= \frac{\mu_0 mg}{1+\mu_0^2}, \\ H_B &= \frac{\mu_0^2 mg}{1+\mu_0^2}, \\ x = x_{\max} &= \frac{\mu_0 l\,(\mu_0 + \tan\alpha)}{1+\mu_0^2}. \end{aligned} \tag{8.18}$$

Obviously, the normal forces and static friction forces depend exclusively on the weight force mg and the coefficient of static friction μ_0. The maximum distance $x_{\max}$, on the other hand, shows no dependence on the weight force mg. If we want to ensure that the entire ladder can be climbed, we can determine the following relationship with $x_{\max} = l$:

$$x_{\max} = l = \frac{\mu_0 l\,(\mu_0 + \tan\alpha)}{1+\mu_0^2} \quad \rightarrow \quad \alpha = \arctan\left(\frac{1}{\mu_0}\right). \tag{8.19}$$

◀

8.3 Belt Friction

Many technical applications involve ropes being wrapped around bodies. This raises the question of the friction that exists between the rope and the wrapped body. This phenomenon is known as belt friction. If the rope does not move, then there is a state of static friction for rope, otherwise kinetic friction occurs. In the following, we will consider the case where a cylindrical body at rest is wrapped by a rope and kinetic friction does not occur. For this

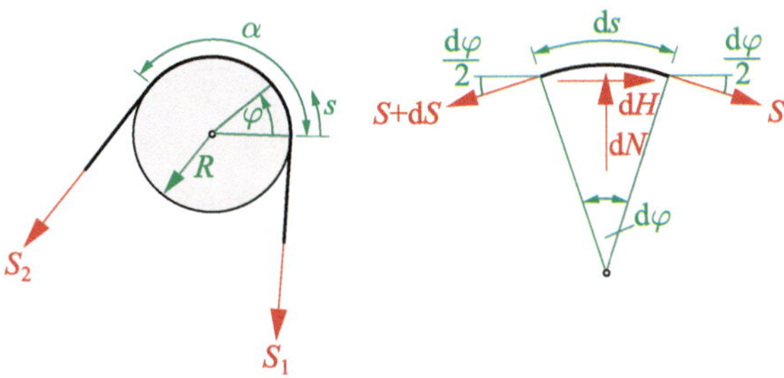

Fig. 8.7 Belt friction

purpose, we consider the situation shown in Fig. 8.7, left. Let a circular cylindrical body be given, which is wrapped by a rope, whereby the overlap is described by the angle α. Let the surface of the body be rough, so that there is friction between its surface and the rope. Since there is static rope friction, the rope forces on the left and right faces are different with the values S_2 and S_1, whereby we want to assume in the following that $S_2 > S_1$ applies. To describe the situation, we also use the angle φ and the circumferential coordinate s, where φ is measured in radians.

We consider the free-body diagram of Fig. 8.7, right, in which an infinitesimal rope section with the opening angle $\mathrm{d}\varphi$ is shown with the length $\mathrm{d}s$. The rope forces on the cut edges are given by S and $S + \mathrm{d}S$ respectively, and the contact force $\mathrm{d}N$ and the static friction force $\mathrm{d}H$ also act. We form the force equilibria with regard to the horizontal and vertical direction and obtain

$$\overset{\rightarrow}{\sum} H = 0: \quad S\cos\frac{\mathrm{d}\varphi}{2} - S\cos\frac{\mathrm{d}\varphi}{2} - \mathrm{d}S\cos\frac{\mathrm{d}\varphi}{2} + \mathrm{d}H = 0,$$

$$\overset{\uparrow}{\sum} V = 0: \quad \mathrm{d}N - S\sin\frac{\mathrm{d}\varphi}{2} - S\sin\frac{\mathrm{d}\varphi}{2} - \mathrm{d}S\sin\frac{\mathrm{d}\varphi}{2} = 0. \tag{8.20}$$

Assuming small angles leads to

$$\cos\frac{\mathrm{d}\varphi}{2} \simeq 1, \quad \sin\frac{\mathrm{d}\varphi}{2} \simeq \frac{\mathrm{d}\varphi}{2}, \tag{8.21}$$

so that after neglecting terms that are small compared to the other terms we obtain:

$$\mathrm{d}H = \mathrm{d}S, \quad \mathrm{d}N = S\mathrm{d}\varphi. \tag{8.22}$$

With $\mathrm{d}H = \mu_0 \mathrm{d}N$ this yields:

$$\mathrm{d}S = \mu_0 S\mathrm{d}\varphi \quad \rightarrow \quad \mu_0\mathrm{d}\varphi = \frac{\mathrm{d}S}{S}. \tag{8.23}$$

Integration of the left-hand term from 0 to the angle α and the right-hand term from S_1 to S_2 leads to:

$$\mu_0 \int_0^\alpha \mathrm{d}\varphi = \int_{S_1}^{S_2} \frac{\mathrm{d}S}{S} \quad \rightarrow \quad \mu_0\alpha = \ln\left(\frac{S_2}{S_1}\right). \tag{8.24}$$

This expression can be solved for the rope force S_2, and we obtain the so-called Euler[2]-Eytelwein[3] equation:

$$S_2 = S_1 e^{\mu_0 \alpha}. \tag{8.25}$$

In the case that $S_1 > S_2$ applies, then this relationship is as follows:

$$S_2 = S_1 e^{-\mu_0 \alpha}. \tag{8.26}$$

From these two relationships, we can derive the interval for S_2 for which the rope remains at rest for a given force S_1 and does not slip on the cylindrical surface:

$$S_1 e^{\mu_0 \alpha} \geq S_2 \geq S_1 e^{-\mu_0 \alpha}. \tag{8.27}$$

The angle α is inserted in [rad].

If the rope slips, the following applies for $S_2 > S_1$:

$$S_2 = S_1 e^{\mu \alpha}. \tag{8.28}$$

For $S_1 > S_2$ we have:

$$S_2 = S_1 e^{-\mu \alpha}. \tag{8.29}$$

[2] Leonhard Euler, 1707–1783, Swiss polymath.

[3] Johann Albert Eytelwein, 1764–1848, German engineer.

Index

C. Mittelstedt, *Engineering Mechanics 1: Statics*,
https://doi.org/10.1007/978-3-662-71852-0

<u>GPSR Compliance</u>

The European Union's (EU) General Product Safety Regulation (GPSR) is a set of rules that requires consumer products to be safe and our obligations to ensure this.

If you have any concerns about our products, you can contact us on ProductSafety@springernature.com

In case Publisher is established outside the EU, the EU authorized representative is:

Springer Nature Customer Service Center GmbH
Europaplatz 3
69115 Heidelberg, Germany

Batch number: 10371040

Printed by Printforce, the Netherlands